Flight Proven

E.B Cohen

First edition, 2026

Contents

Flight Proven

SpaceX and the New Race for Space

E.B Cohen

First edition, 2026.

This is a work of popular history and reportage. It rests on the public record of the modern space age — the launch webcasts and mission pages, the NASA and FAA documents, the corporate filings and financial disclosures around the 2026 public offering, the engineering talks and press conferences, and the reporting of journalists who have covered this industry closely, among them Eric Berger, Christian Davenport, Michael Sheetz, and the staffs of Ars Technica, Reuters, and *Scientific American*. It draws on book-length accounts by Berger (*Liftoff* and *Reentry*), Ashlee Vance, Walter Isaacson, and Lori Garver. It does not invent events or put words into mouths that the record does not support. Where a figure is disputed, where a familiar story turns out to be a myth, or where the evidence simply runs out, the text says so plainly.

For everyone who has ever looked up — and for the engineers who decided that looking up was not enough.

> "If something's important enough, you should try, even if the probable outcome is failure." — a founding creed of the reusable-rocket age
>
> "I would like to die on Mars. Just not on impact." — attributed, in the early years of the enterprise
>
> "Rockets are hard." — the industry's oldest and truest proverb, relearned by everyone

---

A Note on Numbers, Names, and Uncertainty

The people in this book did not know, while they were living it, whether any of it would work. That is worth holding onto, because hindsight flattens everything into inevitability. In 2008 it was entirely reasonable to expect that SpaceX would fail, that reusable rockets were a fantasy, and that the launch business would remain what it had been for half a century: a slow, costly, government-run enterprise in which every rocket was built at enormous expense and then thrown into the ocean. The story told here is dramatic precisely because, for a long time, the smart money was on the other side.

A word about numbers. Much of this story is unusually well documented — launches are filmed, telemetry is public, and NASA contracts are matters of record — but some of the most interesting quantities are not. SpaceX was a privately held company for almost the whole of the period this book covers, and its revenues, its internal valuations, and the true economics of reuse and of Starlink were closely guarded.

Where this book gives such figures, they are the best public estimates, and are named as estimates. The one great exception is the company's June 2026 initial public offering, which for the first time forced a flood of real financial data into the open; the book ends there in part for that reason.

A word about the fast-moving present. This book was completed in mid-2026, and several of its threads were still unspooling as it went to press. Starship's twelfth flight had just ended with its upper stage splashing down intact in the Indian Ocean and its booster breaking apart over the Gulf, and the vehicle was grounded pending an FAA investigation. Blue Origin's New Glenn had just been badly set back by an explosion on its only launch pad. NASA's Artemis timeline had slipped again, and China's crewed lunar program was pressing toward its own deadline. The reader who comes to these pages later will know how some of these contests turned out; the author, writing now, does not. Where the book reaches the edge of the known, it says so, and stops.

Finally, a caution about the founder. It is impossible to tell the story of SpaceX without Elon Musk, whose drive, judgment, and money made the company possible, and whose politics and public conduct in the 2020s became entangled with its reputation and its relationships with government. This book neither lionizes nor prosecutes him. It tries to do the harder thing: to describe what he and the thousands of people around him actually built, to give credit where the record supports it, and to be honest about the costs and the unfinished business — the crowded skies, the concentrated power, and the promise of Mars that remains, for now, a promise.

The truth of this story is strange enough without embellishment. A company that could not land a rocket learned to land them routinely. A nation that had given up on its own astronauts started launching them again from a private capsule. A sky that held a few thousand satellites came to hold more than ten thousand belonging to one firm. And a race that everyone thought had ended in 1972, when the last men left the Moon, quietly began again. This is how it happened.

Chapter 1: A Rocket in the Desert

On September 28, 2008, a small white rocket stood on a steel platform on a speck of coral in the middle of the Pacific Ocean, and a company's entire future stood with it. The rocket was called Falcon 1. It was slim and unremarkable to look at, a little over twenty-one meters tall, about the size of a modest office building laid on its side, and it had already failed three times. The island beneath it, Omelek, was part of the Kwajalein Atoll in the Marshall Islands, some seven thousand kilometers southwest of Los Angeles, a green dot in a blue immensity, reachable only by boat or helicopter and hot enough that the engineers who worked there peeled and burned and drank warm water from plastic bottles. They had shipped the rocket across the ocean in pieces. They had bolted it back together in a tin shed. Now they were going to try, for the fourth time, to put it into orbit, and everyone on the island understood that if it failed again there might not be a fifth.

The company that owned the rocket was called Space Exploration Technologies Corporation, SpaceX for short. It had been founded six years earlier by a thirty-seven-year-old software entrepreneur named Elon Musk, who had made a fortune selling internet companies and had decided, against a great deal of expert advice, to spend it on rockets. By the standards of the aerospace industry SpaceX was a rounding error: a few hundred employees, a single unproven vehicle, and a founder who was, by his own later admission, nearly out of money. The established giants of the business — Boeing, Lockheed Martin, the great national space agencies — measured their programs in decades and their budgets in the tens of billions. SpaceX measured its remaining life in launches. It had money, Musk would say afterward, for about three or four attempts. This was the fourth.

To understand why grown engineers stood on that island with their careers and their savings riding on a single ignition, you have to understand what the launch business was in 2008, and what it had been for the half-century before. It was, in a word, expendable. Every rocket ever flown to orbit had been built at staggering cost and then destroyed in the course of a single mission — the first stage tumbling into the sea, the upper stage burning up or drifting off as junk. It is often said, by way of illustration, that this was like an airline building a new airplane for every flight, flying it once from New York to London, and then throwing the airplane away; the comparison is not an exaggeration but a literal description of the practice. No one would fly under such terms. No one could afford to. Yet that was precisely how humanity had reached space since the first satellites of the 1950s, and the price reflected it: to put a kilogram of anything into low Earth orbit cost something on the order of ten thousand dollars or more, and that number had barely moved in forty years. Space was a place you visited at ruinous expense, on government business, a few times a year.

Musk's wager — the wager that had drained his fortune and brought his engineers to this island — was that this did not have to be true. He believed the cost of reaching space could be cut not by tenths but by factors, and that the key was to stop throwing the rockets away. If a rocket could be recovered and flown again, like an airplane, the economics of everything above the atmosphere would change. Satellites, space stations, the Moon, eventually Mars — all of it was gated behind the price of launch, and the price of launch was gated behind the mad habit of building a machine and using it once. Falcon 1 was not yet a reusable rocket; its first stage was not designed to survive the journey home. It was the small, cheap, deliberately humble first step: prove you can build a rocket at all, prove you can reach orbit, earn the right to attempt the harder thing. And it kept blowing up.

The first flight, in March 2006, had ended in fire seconds after liftoff, a fuel leak igniting at the base of the engine, the rocket arcing over and falling back to smash into the reef near the launch site. The cause, when the investigators finally ran it to ground, was almost insultingly small: a single aluminum nut on the engine, no larger than a thumbnail, that the salt air of the atoll had corroded until it cracked. The second flight, a year later, had climbed much higher, into space itself, before the second stage began to slosh and wobble and finally shut down short of orbital velocity. The third, in August 2008, had been the cruelest. It had flown beautifully; the first stage had done its job; and then, in the instant of separation, a newly upgraded engine had produced a small residual thrust that nudged the spent booster forward into the base of the second stage, a soft, fatal kiss that sent both tumbling. Three rockets, three ways to lose. Each one had carried real payloads — satellites belonging to customers and to the U.S. military, and on the third flight even the cremated ashes of two hundred and eight people, among them a Mercury astronaut and the actor who had played Scotty on *Star Trek* — and each had carried something less tangible and more important: the diminishing patience of a company running on fumes.

The people on Omelek in September 2008 were tired in a way that is hard to convey. Many of them had spent weeks or months on the island, away from families, working in humidity that rusted tools overnight and corroded the rocket's aluminum skin. They had learned to improvise: to fix a cracked component with parts flown in commercial luggage, to dry out water-damaged electronics, to work through the night because the launch window and the weather and the money would not wait. They were, most of them, young — engineers in their twenties and thirties who had chosen the risky startup over a safe career at an aerospace prime, and who now had to reckon with the possibility that they had chosen wrong. Above them, in California and around the world, the wider economy was collapsing. Two weeks before the launch, on September 15, the investment bank Lehman Brothers had filed for bankruptcy, and the great crash of 2008 was underway. It was, by any measure, a terrible time to be asking anyone for money.

Musk himself was not on the island. He was in Los Angeles, and he was, by his own accounts and those of the people around him, close to breaking. His other company, the electric-car maker Tesla, was also nearly out of cash and lurching toward bankruptcy in the same weeks. His marriage was ending. He was, he later said, waking in the night in a physical state close to panic. He had put roughly a hundred million dollars of his own money into SpaceX — the great majority of what he had made from selling his stake in PayPal — and he had watched three rockets fail. He had told his team that he would fund a fourth flight, and, if he could scrape the money together, perhaps one beyond that. But everyone understood the arithmetic. A company cannot lose forever. The fourth flight was not a test. It was a verdict.

The rocket that stood on Omelek that day was, in one sense, the most carefully prepared of the four. The engineers had traced the third failure to its precise cause — the residual thrust at staging — and had fixed it with a change so simple it was almost embarrassing: they would wait a few seconds longer, letting the leftover thrust die away before the stages parted. They had checked and rechecked the vehicle. They had, in the ruthless logic of a company with one shot left, stripped away everything that could go wrong and could be removed. And they had loaded it not with a customer's precious satellite but with a hundred-and-sixty-five-kilogram mass simulator, a dummy payload of aluminum, so that this flight at least would be about one thing only: reaching orbit. There was nothing left to do but light the engine.

When the countdown reached zero, the single Merlin engine at the base of the rocket ignited, and Falcon 1 lifted off its platform on a column of pale flame, climbing away from the island. In the control room, a converted shipping container crowded with young engineers, and in the SpaceX headquarters in California where more of them watched the telemetry stream in, there was near-total silence. The first stage burned for a little under three minutes, pushing the rocket up and out over the Pacific and toward the speed it would need. Then came the moment that had killed the previous flight — separation — and this time the spent booster fell cleanly away, the coast timed correctly, and the second stage's Kestrel engine lit against the black. It burned, and burned, and did not slosh, and did not shut down. The numbers on the screens climbed toward orbital velocity and held. Nine minutes and thirty-one seconds after liftoff, the engine cut off exactly as planned, and Falcon 1 was in orbit — circling the Earth at an altitude of roughly three hundred and thirty kilometers, within a fraction of a degree of its intended path.

For a moment no one quite believed it. Then the control room erupted — cheering, embracing, weeping, the pure release of people who have been holding their breath for years. In California, Musk, who had steeled himself for another failure, allowed himself to feel that the thing he had bet everything on might actually work. Addressing the exhausted company a short time later, he made a promise that has followed him ever since: "For my part, I will never give up, and I mean never." Falcon 1 had become the first privately developed liquid-fueled rocket to reach Earth orbit, a feat that until that day had belonged exclusively to national governments — to the United States and the Soviet Union, to a handful of great powers with the resources of nations behind them. A company of a few hundred people, run out of a warehouse in a Los Angeles suburb, had done it on its fourth try and its last dollar.

It is tempting to end the scene there, on the cheering, because that is how such stories are usually told. But the truth of that day is more precarious and more interesting. Reaching orbit did not save SpaceX. It is often assumed that the historic flight secured the company's future in a single stroke; in fact it did nothing of the kind. The company was still nearly broke. A single successful launch of a small rocket, however historic, did not generate the cash to keep hundreds of engineers employed and to develop the far larger vehicle that Musk already knew he needed to build. Falcon 1 had proven a point, but a proven point does not pay a payroll. In the weeks after the flight, Musk was still scraping together money from every source he could find, splitting his last reserves between SpaceX and Tesla, borrowing from friends, staring into the same abyss he had been staring into before the launch. The victory on Omelek bought the company something priceless — credibility, and the right to be taken seriously — but it did not, by itself, buy survival.

Survival came a few weeks later, and it came from an unexpected quarter: the very institution the industry assumed SpaceX was trying to disrupt. Three days before Christmas, on December 23, 2008, NASA awarded the company a Commercial Resupply Services contract to fly cargo to the International Space Station — a deal worth about a billion and a half dollars over a series of a dozen flights. Musk would describe the phone call bringing the news as one of the best moments of his life, and would later recount that he blurted out to the NASA official on the line, more or less involuntarily, that he loved the agency. The contract did not merely rescue SpaceX from the immediate crisis. It gave the company a customer with deep pockets and a genuine need, a reason to build the larger rocket, and a runway of years rather than weeks. The same final week of 2008, Tesla closed the financing round that saved it from its own bankruptcy, the two deals landing within days of each other at the close of the worst year of the financial crisis. Musk has said that had either gone the other way, he might have lost both companies. The rocket in the desert had reached orbit; the contract from Washington let the company that built it live long enough to build the next one.

This book is about what happened next — how a firm that could barely afford a fourth launch went on, over the following two decades, to land rockets on their tails and fly them again, to carry astronauts to orbit from American soil when the nation's own space agency could not, to fill the sky with more satellites than any government had ever launched, to build a stainless-steel colossus taller than the Saturn V, and, in the process, to reopen a race for space that most people believed had ended in 1972. You meet the descendants of that small white rocket almost every night now, if you know where to look: the trains of Starlink satellites gliding across the dusk, the reused Falcon 9 boosters returning to their pads on pillars of fire, are the grandchildren of the machine that kept falling into the Pacific. But to understand how the company on Omelek came to exist at all — why a software millionaire was standing at the edge of ruin to put a small rocket into orbit — we have to go back further, to a young man in South Africa who read too much science fiction, and to the peculiar chain of events that turned a fortune made on the internet into a bet on the oldest dream there is. That is where the story begins: not with a rocket, but with a boy who could not stop looking up.

Chapter 2: The Man Who Sold PayPal

The founder of the company that would remake spaceflight was born in Pretoria, South Africa, on June 28, 1971, into a country in the grip of apartheid and a childhood that he has described, in the sparing way he describes such things, as difficult. Elon Musk was a bookish, socially awkward boy, the eldest of three children of an engineer father and a model-turned-dietician mother whose marriage dissolved into acrimony when he was young. He was bullied at school badly enough that on one occasion a group of boys beat him and threw him down a flight of concrete stairs, leaving him hospitalized with injuries that required corrective surgery decades later. He retreated, as such children often do, into books, and he read at a pace and a range that astonished the adults around him. By his own account he would work through the local library's shelves and then the encyclopedia, absorbing whole reference volumes for the pleasure of it, and there is a much-repeated family story that as a small child, when asked a factual question no one could answer, he would simply supply it, having read it somewhere and not forgotten.

The retreat into books was matched by an early plunge into machines. When he was around ten, Musk taught himself to program on a primitive home computer, working through the manual — meant to be studied over months — in a matter of days. At twelve he wrote a simple space-shooter video game he called Blastar, a few hundred lines of code in which the player blasted alien freighters carrying hydrogen bombs, and sold the listing to a South African computing magazine for something like five hundred dollars. It is a small episode, but a telling one: the boy who would one day try to make spaceflight cheap had, as a child, already written a game about firing rockets at spacecraft and been paid for it. The pattern was there early — a mind that turned enthusiasm into something built, and built into something sold.

He would later say that much of his moral and imaginative furniture came from the novels of Isaac Asimov and Robert Heinlein, from the *Foundation* series with its vision of civilization preserved as knowledge across the collapse of a galactic empire, and from Douglas Adams's *The Hitchhiker's Guide to the Galaxy*, whose great joke — that the hard part is not the answer but the question — became, for the young Musk, something close to a philosophy. If you could frame the right question about the future of humanity, the reasoning went, the answers might follow. It is worth pausing on the *Foundation* influence, because it is not decorative. Asimov's premise is that a civilization can foresee its own dark age and act deliberately to shorten it, to preserve the seed of what has been built so that the recovery takes centuries rather than millennia. That idea — of a small, driven group working to make a catastrophe survivable — is a recognizable ancestor of the argument Musk would later make for going to Mars.

The question he settled on, in adolescence and then more firmly in his twenties, was a grand one: what were the things that would most affect the future of human civilization? The list he assembled — he would recite versions of it for the rest of his life — included the internet, sustainable energy, and the extension of human life beyond a single planet. It is easy, from the far side of his success, to read that list as retroactive myth-making, the tidy origin story of a man who went on to found an internet company, an electric-car company, and a rocket company. But the list appears, in one form or another, in accounts from people who knew him long before he was famous, and it is consistent with the choices he actually made. Whether or not one finds the man congenial — and by the 2020s a great many people did not — it is a matter of record that he identified spaceflight as a civilizational priority years before he had the means to do anything about it.

He left South Africa at seventeen, partly to avoid compulsory military service in the apartheid regime's army, a prospect he found morally intolerable. He travelled first to Canada, where he could claim citizenship through his Canadian-born mother, and spent a lean stretch working odd jobs — including, at one point, shoveling grain and cleaning boilers in a lumber mill — before enrolling at Queen's University in Ontario and then transferring to the University of Pennsylvania. There he studied physics and economics, the latter at the Wharton School, and paid part of his way by throwing large ticketed parties out of a rented fraternity-style house that he and a housemate ran, for a time, as a makeshift nightclub. Physics mattered to him in a particular way: not as a body of facts but as a method, a way of reasoning about the world from first principles rather than by analogy. Reasoning by analogy, he would later argue, is how most people and most companies think — we did it this way before, so we will do it this way again — and it is efficient but conservative, forever bounded by what already exists. Reasoning from first principles means stripping a problem down to the physical facts and building your conclusions back up from there, which is slower and harder but occasionally lets you see that something everyone believes to be impossible is merely expensive, or that something everyone believes to be necessary is merely habitual. This habit of mind would matter enormously when he turned to rockets.

In 1995 Musk arrived in California intending to begin a doctorate at Stanford in materials science, aiming at the physics of energy storage — the batteries that might one day make electric cars practical. He lasted, by most accounts, a matter of days, deferring the program almost as soon as he had entered it. The internet was erupting, and he could not bear to watch the greatest technological wave of his lifetime from a laboratory. He dropped out and, with his younger brother Kimbal, founded a company called Zip2, which built online city guides and business directories for newspapers at a moment when newspapers were only beginning to grasp what the web would do to them. It was a hard, lean beginning; the brothers could not afford an apartment, so they slept on a futon in the small office, worked in shifts on a single computer, and showered at a nearby YMCA. But the timing was right, and in February 1999 the computer maker Compaq bought Zip2 for around three hundred and seven million dollars in cash. Musk's roughly seven-percent share came to some twenty-two million dollars. He was twenty-seven, and, by the standards of ordinary life, rich beyond any need to work again.

He did the opposite of retiring. He took the great majority of the Zip2 money — a reported twelve million dollars of it — and poured it into a new venture, an online bank he called X.com. He had a lifelong attachment to the letter X, and would return to it again and again across his career, and he was betting that financial services were ripe for reinvention by the internet. X.com merged in 2000 with a rival called Confinity — a startup founded by Peter Thiel and Max Levchin that had built a slicker product for sending money by email — and the combined company eventually took the name of that product: PayPal. The history of PayPal is a saga in itself, full of boardroom coups and near-death technical crises, and it contains an episode that is often glossed over in the triumphant retellings. In the autumn of 2000, while Musk was aboard a flight bound for Australia, part vacation and part delayed honeymoon, the company's board moved against him and replaced him as chief executive with Thiel. It is often said that Musk built and ran

PayPal; in fact he was pushed out of its top job before it reached its greatest success, a bruising lesson about control and about boards that he would not forget, and that would shape how he structured every company he founded afterward. But when the online marketplace eBay bought PayPal in 2002 for a billion and a half dollars in stock, Musk was still its largest single shareholder. His payout, after taxes, came to something on the order of a hundred and seventy-six million dollars. He was thirty years old, and for the second time he had turned a stake in a young company into a fortune.

What he chose to do with the PayPal money is, in retrospect, the hinge of his whole life, and it is worth registering how narrowly it was reasoned. He did not diversify into safe investments, as prudence and every financial adviser would have counseled. He did not buy the trappings of the newly rich and settle into the comfortable role of an angel investor. Instead he treated the fortune as venture capital for the projects on his adolescent list, and he did the arithmetic coldly: he would commit the bulk of it and keep a modest reserve, knowing that both companies might fail and take the money with them. By 2008 that reserve was nearly gone, and he was borrowing to keep the lights on. The point is that the recklessness the world would later marvel at was not the impulsiveness of a man who did not understand risk; it was the deliberate wager of a man who understood the risk precisely and accepted it.

Most people in that position, twice over, would have concluded that they had a gift for internet businesses and should keep making them. Musk concluded something stranger. He went back to his adolescent list — internet, energy, space — and decided that, having done the internet, he would spend the fortune it had given him on the other two. He would build an electric-car company, which became Tesla, to attack the energy problem. And he would build a rocket company to attack the space problem. It was, by any conventional measure of prudence, an act of financial recklessness bordering on the absurd. The automobile and aerospace industries were the two most capital-intensive, most regulated, most entrenched manufacturing businesses in the world, dominated by century-old giants and littered with the corpses of well-funded challengers. To attack even one of them with a couple of hundred million dollars was optimistic. To attack both at once was, in the view of nearly everyone who knew him, a plan to lose all his money — and, in the crisis years of 2008, it very nearly was exactly that.

The space idea, in particular, seemed to his friends like a mid-life folly dressed up as a mission. When Musk first began talking about it around the turn of the millennium, he did not, at first, intend to build rockets at all. His initial idea was philanthropic and symbolic: a project he called "Mars Oasis," which would send a small robotic greenhouse to the surface of Mars, grow the first plants on another world in a chamber of Earth soil and nutrient gel, and beam back the image — a green shoot against the red desert — in the hope of reigniting public enthusiasm for space and pressuring governments to fund human exploration. He had noticed, with something between disappointment and alarm, that NASA had no serious plan to send people to Mars, that the timeline for it kept receding into a vague future, and that public interest in space, so incandescent during the Apollo years, had cooled into indifference. He went so far as to search NASA's own website for its Mars schedule, expecting to find one, and was startled to find nothing. He thought a striking image and a bold private gesture might change that. It was a rich man's idea for spending a few million dollars to move the culture.

The trouble, as he quickly discovered, was that even his modest stunt ran headlong into the same wall that blocked everything else in space: the cost of launch. To send even a small greenhouse to Mars, he would need a rocket, and rockets were absurdly, prohibitively expensive — the launch alone threatening to cost many times more than the payload it carried. As he investigated further — reading textbooks, talking to engineers, calculating — he became convinced that the high cost of launch was not a law of nature but a failure of the industry, a failure rooted in the expendable habit and in a business culture that had never been forced to compete on price. And he began to see that the symbolic gesture he had planned was addressing a symptom while the disease went untreated. The problem was not that the public lacked inspiration. The problem was that space was too expensive, and it was too expensive because no one had seriously tried to make it cheap. That reframing — from a public-relations problem to an engineering and business problem — is the pivot on which SpaceX was born.

It is worth registering, before we follow him east, just how improbable the whole undertaking looked to the people who wished him well. Musk was not a rocket scientist. He had no aerospace training, no engineering degree, no experience of the industry beyond a few months of frantic reading. He was a two-time internet entrepreneur, celebrated in the technology press but unknown in the world of propulsion and orbital mechanics, and he was proposing to spend the bulk of a hard-won fortune on a business that had humbled far better-resourced men. His friends staged something close to an intervention; his own family worried; and the aerospace establishment, when it noticed him at all, regarded him as a curiosity — a dot-com millionaire with a hobby that would soon relieve him of his money. That he took the plunge anyway, with clear eyes about the odds, is the thread that runs through everything that follows. To see how it happened, we have to follow Musk to Moscow, where he went to buy a rocket and came home determined to build one instead.

Chapter 3: Mars on the Mind

In the autumn of 2001, Elon Musk flew to Moscow to buy a rocket. He went with two companions: Jim Cantrell, a space consultant with a long résumé in the industry, and Adeo Ressi, a college friend and fellow entrepreneur who thought the whole venture faintly mad and had come partly to keep an eye on him. Their plan was straightforward, if audacious. Musk had concluded that the cheapest way to get his Mars Oasis greenhouse off the ground was not to build a new rocket but to buy an old one — specifically, a refurbished Russian intercontinental ballistic missile, of which the collapse of the Soviet Union had left a great surplus. These missiles, the giant SS-18s that NATO had once nicknamed "Satan," were being converted into satellite launchers under the commercial name Dnepr, and in the lean years after the Cold War the Russians were willing to sell such things to more or less anyone with hard currency. Musk had the currency. He needed the missiles. Characteristically, though he had tens of millions of dollars in the bank, he flew to Moscow in economy class, unwilling to spend on a first-class seat what he could spend on a rocket.

The meetings, by all accounts, did not go well. The Americans were received with a mixture of suspicion and condescension. Musk was young, unknown, and manifestly not a member of the aerospace establishment; to the Russian officials and engineers across the table he appeared to be exactly what he was, an internet millionaire with more money than sense, dabbling in a business he did not understand. There were long lunches heavy with vodka and, on a second trip in February 2002, a scene that has passed into the company's founding lore. This time Musk brought along Mike Griffin — a formidable aerospace engineer who had worked for the CIA's venture-capital arm and who would, three years later, become the Administrator of NASA and the architect of the very program that would one day save SpaceX. Over the toasts — "To space!" "To America!" — Musk asked point-blank what a missile would cost. Eight million dollars each, the Russians said. Musk countered by offering eight million for two. The proposal was met with something between a laugh and a dismissal; by one account a negotiator asked, in effect, whether the young American was serious. On a later reckoning the asking price for the rockets he wanted had climbed to a level that made the whole scheme uneconomic. Musk flew home from Russia empty-handed, and by the accounts of those with him, quietly furious — not so much at the Russians as at the situation. He had come to buy a way to space and had been told, in effect, that the price of a ticket was whatever the sellers felt like charging.

It was on the flight back, the story goes, that the decisive shift occurred. Musk had brought a laptop, and instead of brooding he began to build a spreadsheet. He had spent months by then reading everything he could find about rocketry — famously working his way through textbooks and picking the brains of every engineer who would talk to him — and he now had a rough but genuine grasp of the physics and the materials. In the spreadsheet he laid out the raw cost of the ingredients of a rocket: the aluminum and the alloys, the propellant, the electronics, the engines. Rockets are not, at bottom, made of exotic or precious substances. They are mostly metal tanks holding fuel and oxygen, with an engine at the bottom and a guidance system at the top. And when Musk added up the cost of the materials that actually go into a launch vehicle, he arrived at a number that stopped him cold. The materials came to something like two or three percent of the price at which rockets were sold. The other ninety-seven percent or so was, in a sense, overhead: the cost of a way of doing business that had never been forced to become efficient, the cost of building each vehicle by hand in small numbers, of armies of contractors and subcontractors, of the expendable habit that meant every rocket's development had to be paid back over a single flight.

This was the first-principles habit of mind applied to the oldest and most intimidating of industries, and it produced a conclusion that was either delusional or revolutionary, depending on who was doing the judging. If the physical stuff of a rocket cost only two or three percent of the selling price, then the selling price was not a law of physics. It was a business failure — or, more precisely, a business opportunity. It is worth being careful here, because the "materials are only two percent" argument has since been repeated so often that it is sometimes taken to mean that rockets are secretly easy, a matter of simply buying aluminum and welding it together. That is not what the figure means, and Musk did not believe it. Turning raw metal into a machine that survives the most violent few minutes in engineering is monstrously hard; the two-percent figure measures not the difficulty but the inefficiency, the vast gap between what the inputs cost and what the industry charged. A company that could manufacture its own rockets, in-house, with modern methods, without the layers of contractors and the cost-plus incentives of the traditional aerospace primes, might be able to close that gap — to sell launches for a fraction of the going rate and still make money. And if it could then learn to reuse the hardware rather than throwing it away, the savings could be greater still. Musk did not, on that flight, have a design for a reusable rocket. What he had was the conviction, grounded in arithmetic, that the wall everyone accepted as permanent was in fact made of habit and could be pushed through by anyone willing to build rockets the way Silicon Valley built software companies: vertically integrated, iterating fast, controlling their own destiny.

The decision to build rather than buy was not, it should be said, obviously correct, and the people around Musk mostly thought it was a catastrophe in the making. It is often assumed that Musk set out from the first to found a rocket company; in truth he tried hard to avoid it, and turned to building his own vehicles only after the market for buying one had rebuffed him. Cantrell, the consultant, has recounted that he privately gave the venture almost no chance and worried that his friend was about to immolate his fortune. Ressi went further; by his own telling he was so alarmed that he assembled a montage of rockets exploding — a reel of the industry's most spectacular failures — and sat Musk down to watch it, a last attempt to talk him out of the whole thing. Everyone knew the grim history. The list of well-financed attempts to build cheaper rockets was long and almost uniformly tragic. Governments with the resources of nations had struggled for decades to bring launch costs down and had largely failed. The idea that a software entrepreneur with a couple of hundred million dollars — a large fortune by ordinary standards, but a trivial one by the standards of aerospace — could succeed where they had not seemed to demand a special kind of arrogance. Musk himself did not disagree about the odds. He would later say, in various forms, that he thought SpaceX was more likely to fail than to succeed, that he had assumed he would probably lose the money, and that he had done it anyway because the alternative — a future in which humanity stayed forever on one planet, one asteroid strike or one civilizational stumble away from extinction — was worse than the loss of a fortune.

That last point is worth pausing on, because it is the emotional and philosophical engine of everything that followed, and it is easy for a skeptical reader to dismiss as grandiosity or salesmanship. Musk framed SpaceX, from the beginning and consistently thereafter, not as a business opportunity but as a hedge against the extinction of the species. His argument, which he repeated in speeches and interviews for two decades, ran roughly like this. For all of history, human beings and all their works have existed on a single planet. A single planet is a single point of failure. Whether through a natural catastrophe like the asteroid that ended the dinosaurs, or a self-inflicted one — nuclear war, engineered plague, some technology gone wrong — there is a non-trivial chance that Earth-bound civilization could be destroyed or set back to a level from which it might never recover. The only way to make the story of consciousness robust against such an ending is to spread it to more than one world, to make humanity, in his phrase, a multi-planetary species. He liked to put it in terms of probability rather than prophecy: not that Earth was doomed, but that a backup was cheap insurance against a small chance of an infinite loss. Mars was the nearest candidate: a world with a thin atmosphere, with water ice, with a day close to Earth's, harsh but not impossible. And the gating obstacle to Mars was, once again, the cost of launch. Bring that down far enough, and a self-sustaining city on Mars moved from science fiction toward engineering. That was the mission. The rockets, the satellites, the money — all of it was, in Musk's telling, in service of that single end.

There is a self-consciously modest version of the argument that Musk liked to use, and it disarms the charge of grandiosity better than the cosmic one. He would say that going to Mars was not a matter of certainty that Earth was doomed but of buying insurance — that a civilization capable of becoming multi-planetary and choosing not to would be like a person wealthy enough to insure a house against fire who declined the policy to save a little money. The premium, in this telling, was the cost of the rockets; the payout, in the unlikely event of catastrophe, was the survival of everything human beings had ever made or thought. Framed that way, the project was not the fever dream of a man who wanted to leave Earth behind but a hedge taken out on its behalf. Whether the premium was worth paying, and whether Mars was the right policy to buy, are fair questions; but the logic was coherent, and it was the logic that carried Musk through years in which the rockets did nothing but explode.

One does not have to accept this framing whole to recognize that it functioned. Whatever one thinks of the Mars goal as a use of human effort — and there are serious people who regard it as a distraction from the problems of the planet we already have — the mission gave SpaceX something its competitors lacked: a reason to take risks that a normal company would never take, and a way to attract a particular kind of person. Over the following years SpaceX would draw thousands of the most talented young engineers in the world, many of whom could have earned more and worked less at established firms, and it would ask them to work brutal hours under enormous pressure. The thing that kept many of them there, by their own testimony, was not the pay or the stock but the sense that they were doing something that mattered on a historic scale — that they were, in the most literal sense, building the future. You can still hear the residue of that pitch in the way SpaceX engineers describe their work today, and in the way recruits speak of turning down safer, better-paid jobs to join it. Missions of that kind are rare, and companies that have them can demand things that companies without them cannot. Musk understood this, and he built it into the company's DNA from the first day.

It is often said, in the compressed retellings, that Musk simply decided one day to build rockets, as if the resolve arrived whole. The reality was a months-long apprenticeship in a subject he had no business mastering. He acquired and worked through classic texts of propulsion and astronautics — among them the standard references that engineers cut their teeth on — and he pressed the consultants and moonlighting professionals around him with questions until, by several of their accounts, he had absorbed enough that they could no longer easily tell they were talking to an amateur. Cantrell has described being startled by how quickly

Musk went from ignorance to genuine fluency, quizzing him relentlessly and returning each time having plainly done the reading. This was the first-principles method in its rawest form: not trusting that rockets must be expensive because experts said so, but going to the physics and the textbooks and the cost of aluminum, and reasoning up from the ground. The decision to build rather than buy was the conclusion of that self-education, not a whim that preceded it.

So Musk came home from Russia not with a rocket but with a decision. He would start a company, hire the best propulsion and structures engineers he could find, and build rockets from scratch — cheaper, faster, and under his own control. He would begin small, with a modest vehicle to prove the concept and earn a foothold, and he would use whatever he learned to build larger ones. He set aside a substantial slice of his PayPal fortune, an initial commitment reported at around a hundred million dollars, understanding that it might all be gone in a few years. And in the early months of 2002, in a rented warehouse in the Los Angeles suburb of El Segundo, in the shadow of the aerospace giants whose business he intended to steal, he began to build the company that would, against nearly everyone's expectations, change what it costs to leave the Earth. He needed engineers, and he needed them fast. The first person he had to find was someone who could build a rocket engine — the hardest single component, the heart of the machine. He found him working nights and weekends in a garage, building the world's largest amateur liquid-fuel engine for the sheer love of it.

Chapter 4: Founding a Rocket Company

The man Elon Musk needed most in the early months of 2002 was building rocket engines in a garage on the weekends. His name was Tom Mueller, and by day he worked at TRW, one of the established aerospace contractors, on propulsion systems of the kind that took armies of engineers and years of paperwork to produce. By night and on weekends, Mueller belonged to a subculture almost invisible to the outside world: the amateur high-power rocketry community, and in particular the Reaction Research Society, hobbyists who designed and fired their own engines in the California desert for the pure joy of it. Mueller was among the best of them. He had built, in his spare time, what was then the largest amateur liquid-fuel rocket engine in the world — a machine that weighed only about eighty pounds yet produced some thirteen thousand pounds of thrust, which he had hauled out to a friend's warehouse and the open desert to test. When a mutual acquaintance introduced him to Musk at a gathering of space enthusiasts, and Musk — who had come specifically hoping to meet the man behind that engine — began asking pointed questions about how cheaply a real orbital engine could be built, Mueller found himself talking to someone who was not merely curious but serious, someone with money and an intention to spend it. Within a short time Mueller had left the security of TRW to become the founding employee of SpaceX and, in effect, the architect of its engines.

Mueller's defection captures something essential about how SpaceX was built. Musk did not recruit primarily from the ranks of aerospace management, the program directors and executives who knew how the industry worked. He recruited builders — hands-on engineers, often young, often frustrated by the slowness and timidity of the traditional firms, people who wanted to design a thing on Monday and test it on Friday rather than spend three years writing requirements documents. He was looking, he often said, for evidence that a person had actually built something difficult, and he was skeptical of credentials divorced from creation; a favorite interview tactic was to ask a candidate to describe in detail a hard problem they had personally solved, on the theory that the people who had truly done the work could always go a level deeper than those who had merely supervised it. The early SpaceX was full of people who had won engineering competitions, built experimental aircraft, machined their own parts, or, like Mueller, fired their own engines in the desert. It was a culture organized around making, and it was deliberately hostile to the layers of process that Musk believed had strangled the incumbents.

The company that took shape in the El Segundo warehouse, and soon after in a much larger factory in nearby Hawthorne — a low, sprawling building that had once been used to assemble fuselage sections for Boeing's 747 — was built on a principle that ran against the grain of the entire industry: vertical integration. The traditional aerospace primes were, in large part, assemblers and managers of vast supply chains. A rocket from Boeing or Lockheed was stitched together from components made by hundreds of subcontractors, each adding its own margin, its own schedule, its own paperwork, and each insulated by contract from the pressure to make its piece cheaper. Musk's insight — the same insight that had produced the two-percent figure on the flight home from Moscow — was that this structure was where the cost lived. So SpaceX would make as much as it possibly could itself. It would build its own engines, its own airframes, its own avionics, its own flight software, its own guidance systems. Where the incumbents bought a radiation-hardened flight computer from a specialist supplier for a fortune, SpaceX engineers would design one from ordinary commercial electronics for a small fraction of the price and make it reliable through clever redundancy. Where the incumbents treated each part as a fixed given, SpaceX treated each part as a question: does this need to exist, can it be simpler, can we make it ourselves for less? The approach was not entirely without precedent, and it is an overstatement — one the company's own mythology sometimes encourages — to say that SpaceX built literally everything in-house; it still bought raw materials, some avionics components, and specialized parts from outside. But it made far more of its own vehicle than any competitor, on the order of seventy to eighty percent, and that gave it control over cost, schedule, and design in a way the industry had never seen.

Mueller was the first, but he was quickly joined by a small band of engineers who would form the technical spine of the early company. Chris Thompson, a former Marine who had built structures at Boeing and TRW, took charge of the airframe. Hans Koenigsmann, a soft-spoken German-born engineer with a background in small satellites, arrived to lead avionics and flight software and would go on to become the company's guardian of mission assurance, the man whose job in later years was to worry, professionally, about everything that could go wrong. Musk involved himself in the hiring to a degree almost unheard of in a chief executive; for years he personally interviewed nearly every engineer the company took on, well into the hundreds and by some accounts the thousands, probing not for pedigree but for the unmistakable sound of someone who had actually solved hard problems with their own hands. The result was a workforce selected less for where it had trained than for what it had made, and a factory floor where a fresh graduate might find themselves entrusted, within weeks, with responsibility that a traditional prime would have withheld for a decade.

The names the company chose for its hardware told you where its heart was. The rocket was called Falcon — not, as is sometimes assumed, after the bird of prey, but after the Millennium Falcon of *Star Wars*, an affectionate nod to the science fiction that had shaped Musk's imagination. The engines, by contrast, were named for birds: Merlin and Kestrel, both small, fast falcons. It was a small thing, but it captured the sensibility of the place — an aerospace company that wore its boyhood enthusiasms openly, that thought reaching space should be an adventure and not merely a procurement.

Alongside Mueller and the engineers, Musk made one hire in 2002 that would prove as consequential as any piece of hardware. Gwynne Shotwell was an engineer and businesswoman who had worked at the Aerospace Corporation and then at a small firm called Microcosm, and who had a rare combination of technical fluency and commercial instinct. As the story is told at SpaceX, she came to drop off a colleague who was interviewing, was given a tour, remarked bluntly that the company's business-development effort needed work — and was hired, as roughly the seventh employee, to fix it. She turned out to be extraordinary at it. Over the following years Shotwell would rise to become the company's President and Chief Operating Officer, the person who ran SpaceX day to day while Musk set the vision and made the largest bets. If Musk was the company's accelerator — pushing always for more, faster, harder, sometimes past the point of prudence — Shotwell was its steering and, at times, its brakes. She managed the customers, negotiated the contracts, dealt with the government, and held the organization together through crises that might have shattered it. In an industry and a company culture famous for volatility, she became a figure of unusual steadiness, and many of SpaceX's most important deals, including the ones that saved it, bear her fingerprints. The partnership between Musk and Shotwell — his relentless ambition disciplined by her operational command — is one of the underappreciated engines of the whole story.

The first product was deliberately modest. Musk did not begin by trying to build a giant rocket; he began with a small one, Falcon 1, designed to carry a few hundred kilograms to orbit. The logic was that of a startup building a minimum viable product. A small rocket would be cheaper to develop, would let the company learn to build and launch on a survivable budget, and would prove to customers and to the government that SpaceX could do the thing it claimed. The engine at its heart, designed by Mueller's team, was the Merlin, and it burned a refined form of kerosene called RP-1 with liquid oxygen — an old, well-understood, relatively cheap propellant combination, chosen for exactly those reasons rather than for the higher performance of the exotic cryogenic hydrogen that some rockets used. The upper stage used the smaller Kestrel. Everything about the design reflected the company's obsessions: simplicity, cost, and the ability to build in-house. Where the incumbents might have reached for an expensive, high-performance solution, SpaceX reached for the cheapest thing that would work, and iterated.

The physical setting reinforced the ethos. There were no walls of separate offices insulating management from the work; Musk's own desk sat out on or beside the factory floor, within sight and earshot of the machines and the people building them, so that a design decision and its consequences were never far apart. Engineers did not hand a drawing to a distant shop and wait; they walked a few steps to the machinists who would make the part, argued it out, and had a piece of hardware in hand in days. This compression of the distance between thinking and making — between the person who designed a component and the person who cut the metal and the person who would watch it fly — was itself a competitive weapon. It meant that when something failed on the test stand, the fix could be conceived, fabricated, and tried again before a traditional program would have finished scheduling the meeting to discuss it.

The culture that produced these machines was, by every account including those of the people who loved it, extreme. The hours were punishing; twelve-hour days and six- or seven-day weeks were normal during development pushes, and the pressure from the top was unrelenting. Musk was a demanding and sometimes brutal boss, capable of great inspiration and of withering, personal criticism, given to impossible deadlines that he seemed to believe in even when everyone around him knew they could not be met. The phrase that came to be associated with SpaceX's schedules, "Elon time," captured the gap between his announced timelines and reality — a gap that could stretch from months to years. And yet the deadlines, however unmet, functioned. By demanding the impossible, Musk pushed his engineers to accomplish things they would have deemed out of reach under a reasonable schedule. There is a well-worn observation about this method: that a team told it has three years will find a way to need three years, while a team told it has one will discover, in the scramble, how much of the received wisdom about how long things "must" take was never true to begin with. People burned out and left; people also did the [illegible] and most consequential work of their lives. Both [illegible] true, and any honest account of SpaceX has to [illegible]ther.

[illegible]ailure deserves particular emphasis, be[illegible]it that most sharply distinguished SpaceX [illegible] around it and the one that would matter [illegible] just ahead. In the traditional aerospace [illegible]t blew up was a catastrophe and often a [illegible]grams were structured above all to avoid [illegible]de them slow, conservative, and expen[illegible]ding failure means testing endlessly on paper and in simulation before committing hardware. SpaceX inverted the logic. It would build cheaply, fly early, and expect some of what it flew to break, treating each explosion not as a disgrace to be hidden but as a lesson bought at the price of a rocket — often the fastest and truest lesson available, since the real world has a way of revealing what no simulation anticipated. This was only affordable because the hardware was cheap, and the hardware was cheap because the company built it itself, so the philosophy and the vertical integration reinforced one another. It was a coherent system, and it was about to be put to the harshest possible test.

There is a temptation, in telling the story of a company that succeeded so spectacularly, to romanticize the early hardship, to present the brutal hours and the founder's tantrums as the necessary price of greatness. That temptation should be resisted. The costs were real and fell on real people; some of the practices that the company's admirers celebrate as boldness were, from the inside, closer to recklessness or cruelty, and more than a few talented engineers left with their health and marriages the worse for it. But it is equally dishonest to pretend that the culture did not produce results. SpaceX did things that better-funded, better-credentialed, more humane organizations had failed to do, and it did them in part because it was willing to move fast, break things, and absorb failures that would have ended careers elsewhere. The early years in El Segundo and Hawthorne established the pattern that would define the comp y: a small, brilliant, exhausted team; a founder whose visi volatility were inseparable; a willingness to build in- what others bought; and a tolerance for failure as the of speed. That last trait — the embrace of failure as inf tion rather than disgrace — would be tested to its li the years to come, on a small island in the Pacific, whe first fruit of all this labor would be trucked to a launc and, three times running, would fail to reach the sky. E the triumph on Omelek that opened this book, there the three strikes that nearly ended everything. To tho now turn.

Chapter 5: Falcon 1

A rocket is, in its simplest description, a machine for throwing a small thing very fast in one direction by throwing a large amount of hot gas very fast in the other. The physics was worked out three centuries ago by Newton — every action has an equal and opposite reaction — and refined a hundred years ago by the Russian schoolteacher Konstantin Tsiolkovsky, whose famous equation lays out the brutal arithmetic that governs everything SpaceX would ever do. To reach orbit, a vehicle must accelerate not merely upward but sideways, to a speed of roughly twenty-eight thousand kilometers an hour, fast enough that as it falls toward the Earth it keeps missing, circling instead of crashing. It is a point worth dwelling on, because the popular imagination gets it backward: people picture spaceflight as the act of going up, when in truth the going up is the easy part and the going sideways is nearly the whole battle. A powerful cannon could loft a shell a hundred kilometers straight into space, high enough to cross the notional boundary of the atmosphere, and the shell would simply fall back down. To stay in space you must go fast enough to fall around the planet forever, and reaching that speed is what makes rockets enormous.

Achieving it requires burning an enormous quantity of propellant, and the propellant itself has weight that must also be accelerated, which requires yet more propellant, in a vicious spiral that Tsiolkovsky's equation makes precise. The upshot is that a rocket is mostly fuel tank. Consider the proportions: on the launch pad, something like ninety to ninety-five percent of a well-designed rocket's mass is propellant, only a few percent is the structure — the tanks, the engines, the skin — and the payload, the satellite or capsule or the reason for the whole enterprise, is a sliver, often just one or two percent of the total sitting on the pad. Picture a soda can, thin-walled and nearly weightless when empty, filled to bursting and then flung; that is closer to the truth of a rocket than the solid, missile-like object of the imagination. This is why rockets are hard. There is almost no margin. Every kilogram of unnecessary structure is a kilogram of payload you cannot carry, and the difference between a rocket that reaches orbit and one that does not can be a matter of small errors compounding.

Tsiolkovsky's equation is worth stating plainly, because its cruelty explains the shape of every rocket ever built. It says that the change in velocity a rocket can achieve depends on two things: how fast its engine hurls exhaust out the back, and the ratio between the vehicle's fully fueled mass and its empty mass — and that this second quantity enters not linearly but through a logarithm, which is a mathematician's way of saying that the demands grow savagely as the target speed rises. To reach low orbit a vehicle must gain something like nine and a half kilometers per second of velocity once the drag of the atmosphere and the pull of gravity are counted in, and no chemical engine throws its exhaust fast enough to manage that with a comfortable margin. The equation punishes you: to go a little faster you must carry a lot more fuel, and the fuel has weight, and so you must carry still more. The escape from this trap is staging — throwing away the empty first stage, tanks and engine and all, once its fuel is spent, so that the second stage does not have to drag the dead weight up to orbital speed. Every expendable rocket is, in this sense, a confession that the arithmetic left no other choice, and the whole later dream of reusability is an attempt to keep the benefit of staging without paying its price in discarded hardware.

Falcon 1 was SpaceX's attempt to build such a machine as cheaply and simply as it could be built while still reaching orbit. It stood a little over twenty-one meters tall and weighed some twenty-seven tonnes fully fueled, and it was designed to place a few hundred kilograms into low orbit — a figure that grew, as the design matured, toward roughly six or seven hundred kilograms. It was a two-stage rocket. The first stage, the larger lower section, held the propellant and the single Merlin engine that would do the initial work of lifting the vehicle off the pad and pushing it up through the thick lower atmosphere. The second stage, sitting atop the first, held its own smaller tanks and the Kestrel engine, and its job was to take over after the first stage had exhausted itself, carrying the payload the rest of the way to orbital velocity. Between them and around the payload was the usual apparatus of plumbing, valves, avionics, and guidance. By the standards of the industry it was a small vehicle, capable of lifting only a modest satellite, not the big communications spacecraft that were the industry's bread and butter. But it was never meant to be a commercial workhorse. It was meant to be a school. And it undercut the going rates dramatically: SpaceX advertised a Falcon 1 launch for a few million dollars at a time when even a small orbital launch commonly ran to tens of millions, a price that seemed impossible to competitors and that the company could contemplate only because it built almost everything in-house. Whether it could actually deliver on that price was, in 2005 and 2006, an open question — a promise printed on a brochure and not yet proven by a rocket that reached orbit. But the number itself was a statement of intent, a public declaration that the old cost structure was a choice rather than a law, and it put SpaceX on a collision course with an industry that had never had to compete on price and did not, at first, take the challenge seriously.

The two-stage architecture reflected the discipline Tsiolkovsky's equation imposed. The first stage did the brute work of getting the vehicle off the pad and up through the dense lower air, then fell away spent; the second, lighter and more efficient in the near-vacuum where it lit, carried the payload the rest of the way to orbital speed. Around the payload sat a protective nose cone, or fairing, that shielded the delicate satellite from the battering of the atmosphere and was jettisoned once the rocket was high enough that the air could no longer harm it. Every one of these elements — stage, engine, fairing, the avionics that steered it all — was a place where a small error could end the flight, and Falcon 1 would, over its first three attempts, find a new one each time.

The Merlin engine was the heart of the design and the focus of Mueller's team. Building a rocket engine is one of the most demanding tasks in all of engineering. Inside the combustion chamber, propellant and oxidizer mix and burn at temperatures hot enough to melt the very metal that contains them, at pressures that would burst an ordinary vessel, and the whole violent process must be controlled precisely enough to produce steady, directed thrust rather than an explosion. Engines fail in a hundred ways: combustion instabilities that shake them apart, cooling systems that cannot keep the chamber walls from melting, turbopumps — the machinery that forces propellant into the chamber at tremendous rates — that shatter under the strain. Mueller's team made a deliberate choice to reduce that risk by borrowing a proven idea. At the center of the Merlin was a device called a pintle injector, a single central element that sprayed the propellants together in a controlled cone rather than through the hundreds of tiny orifices used in most engines. The design was not new; it traced directly to the descent engine of the Apollo Lunar Module, which had lowered astronauts onto the surface of the Moon, and which had been built by TRW — the very company where Mueller had spent his days before SpaceX. The pintle was famously stable, resistant to the combustion oscillations that had wrecked so many engines, and comparatively forgiving to develop. It was a characteristic SpaceX move: reach for the robust, well-understood solution, then make it cheap and make it yours.

The early Merlin was cooled ablatively, its chamber lined with a material designed to char and erode away in a controlled fashion, carrying heat off as it went — simple, but good for only a single firing. Later the team would replace it with a regeneratively cooled version, the Merlin 1C, which circulated cold propellant through channels in the chamber wall to keep the metal from melting, a more sophisticated and reusable approach. The Merlin was tested to destruction, over and over, at a sprawling site the company leased outside the small town of McGregor, Texas — a place whose residents grew accustomed to the ground-shaking roar of engines being run and, not infrequently, blown apart in the course of learning. The site itself carried a warning from the recent past: it had previously belonged to Beal Aerospace, a well-funded private venture that had set out, only a few years earlier, to build a large commercial rocket and had folded in 2000 without ever reaching orbit, defeated by cost and by the difficulty of competing against government-subsidized incumbents. SpaceX took over the test stands of a company that had tried something similar and failed, a reminder, quite literally poured in concrete, of how many had walked this road before and not made it to the end. The Merlin they eventually produced was not the most sophisticated engine in the world, but it was robust, relatively cheap, and, crucially, something SpaceX could build itself and improve rapidly. Over the following two decades the Merlin would be upgraded again and again, growing several times more powerful than its original form, and it would become one of the most flown and most reliable rocket engines in history. But in the Falcon 1 years it was still young, still being wrestled into submission.

The choice of where to launch reflected the same mix of ambition and constraint that shaped everything else. Musk had initially hoped to launch from Vandenberg Air Force Base in California, but the base was preparing to fly a large classified national-security payload from a neighboring pad, and the military would not permit an unproven rocket from a startup to sit nearby and risk the schedule of a mission worth many times more. Faced with delays and restrictions the young company could not tolerate, SpaceX looked for somewhere it could control, and settled on Omelek, a tiny island in the Kwajalein Atoll in the Marshall Islands, part of a U.S. Army missile-testing range in the middle of the Pacific. Kwajalein — "Kwaj" to the Americans who worked there — had genuine advantages. It sat near the equator, where the Earth's rotation is fastest and gives a rocket a helpful running start; it offered open ocean downrange in every useful direction; and it was under American control without the congestion of a major spaceport. But its disadvantages were equally real. It was staggeringly remote, thousands of kilometers from the Hawthorne factory, so that every part, every tool, every engineer had to be flown or shipped across the ocean. It was hot and humid and salty, an environment actively hostile to the delicate machinery of a rocket, and the SpaceX teams fought a constant war against corrosion and moisture. And it was primitive; the launch infrastructure had to be built more or less from scratch, in improvised sheds, by the same small crew that was building the rocket. Photographs from the era show young engineers in shorts and sunglasses hauling rocket components across coral, sleeping in cramped quarters, working through tropical nights. It was, depending on one's temperament, either a nightmare or the adventure of a lifetime, and for many of the people who lived it, it was both.

The schedule, as always at SpaceX, was wildly optimistic. Musk had announced early on that Falcon 1 would fly within a year or so, a timeline that provoked open skepticism from industry veterans who knew how long such things really took. He was, of course, wrong — the first launch would not come until March 2006, years behind his original projections — but the wrongness was, in a sense, productive. By committing publicly to schedules that were impossible, Musk kept his team sprinting when a realistic timeline would have let them walk. This pattern, established with Falcon 1, would repeat throughout the company's history: announce an impossible date, miss it, but arrive far sooner than a sane plan would have allowed. Critics saw dishonesty or delusion; defenders saw a management technique for extracting the maximum from talented people. Both readings have merit. What is undeniable is that the technique, whatever its human cost, produced hardware.

By early 2006, after years of design, building, testing, shipping, and rebuilding, Falcon 1 stood on its pad on Omelek, ready for its first flight. The company had spent a large share of its founding capital getting to this point. It had a handful of customers who had signed up to fly small satellites on the new rocket, including the U.S. military, which had an interest in cheap, responsive launch. It had the attention, mostly skeptical, of an industry that regarded the whole enterprise as a rich man's hobby. And it had, for the first time, a real rocket on a real pad, fueled and ready. The engineers who had built it gathered to watch, exhausted and hopeful, on March 24, 2006, as the countdown proceeded and the Merlin engine roared to life and Falcon 1 rose off the platform for the first time. For a little over half a minute it flew. Then, at the base of the engine, a fuel leak — later traced with almost cruel precision to a single small aluminum fitting, a "B-nut," that the salt air had corroded and cracked — caught fire. The flame spread, the engine lost power, and the rocket, which had climbed only a short way into the sky, arced over and fell back toward the island, crashing into the reef and scattering debris across the lagoon. Some of it landed, in a final indignity, near the very shed the team had worked from. The first Falcon 1 was gone in under a minute, undone not by the exotic physics of combustion or guidance but by a corroded nut the size of a fingertip. It was the beginning of the hardest stretch in the company's history, three failures in a row that would bring SpaceX to the edge of extinction and test whether Musk's tolerance for failure was wisdom or folly.

Chapter 6: Kwajalein

To build the first SpaceX rocket was one thing; to launch it was another, and the story of where and how Falcon 1 was flown deserves a chapter of its own, because it captured the improvised, seat-of-the-pants character of the early company more vividly than any engineering diagram could. The place was Omelek, a coral island of a few acres in the Kwajalein Atoll, part of a scattered ring of some ninety islets enclosing the largest coral lagoon in the world, in the middle of the Pacific Ocean, thousands of kilometers from anywhere. It belonged to the U.S. Army's Ronald Reagan Ballistic Missile Defense Test Site, a range built to track and intercept the dummy warheads that the military fired across the ocean from California — a place, in other words, whose entire reason for existing was to watch things streak in from space and fall into the sea. SpaceX had secured permission to launch from it after the delays and restrictions of the California site had proven intolerable. It is sometimes imagined that the company chose a remote tropical island out of some romance of isolation, or a desire to work unobserved; the truth is more prosaic. Omelek was not chosen for its beauty but taken for its availability, the one place SpaceX could get to a pad quickly and control on its own terms. On a map it is a speck, a comma of green in a vast blue field. In the lives of the young engineers who worked there, it became a whole world — hot, remote, beautiful, and brutally hard. Years later, many of them would speak of it the way people speak of a war they survived: with a mixture of horror at the conditions and a fierce, almost protective affection for the time and the people, unwilling to have it romanticized by outsiders yet unable to describe it without a trace of longing.

The logistics alone were punishing. Everything had to come from the Hawthorne factory in California: the rocket in pieces, the tools, the propellant, the ground-support equipment, the people. The main Army base on the island of Kwajalein was itself more than four thousand miles from Los Angeles, and Omelek lay a further boat or helicopter ride across the lagoon. Components travelled by air and sea across the width of the Pacific, and when something was missing or broke — as things constantly did — the replacement might be days or weeks away, or might have to be improvised on the spot. The engineers learned to carry critical parts in their personal luggage on the long flights out, to fabricate what they could not ship, and to nurse the rocket through the assaults of an environment that seemed designed to destroy it. For Omelek was not just remote; it was corrosive. The salt air ate at the rocket's aluminum skin and its fittings, rusting bolts overnight, seeping into electronics, so that the team fought a constant, grinding battle against the elements just to keep the vehicle sound. It was corrosion, in the end — a fuel-line fitting eaten by the salt air — that helped doom the first launch. The island itself was trying to kill the rocket, and the engineers stood between them.

The Marshall Islands themselves carried a heavy history that hung, largely unspoken, over the SpaceX camp. This was the region where the United States had conducted its Pacific nuclear tests in the postwar years — Bikini and Enewetak lie in the same island nation — and Kwajalein Atoll had for decades been the terminus of the country's long-range missile trials, the target zone into which intercontinental warheads, unarmed, were fired from thousands of miles away to test the accuracy of the arsenal and, later, the defenses meant to stop it. The main island of Kwajalein was a small piece of transplanted America, with its bowling alley and its baseball diamond and its Army-run commissary, home to the technicians and contractors who ran the range. Omelek, by contrast, was raw. It had been little more than a scrap of coral and scrub before SpaceX arrived to build a launch pad on it, and the company's presence there was an incongruity within an incongruity: a Silicon Valley startup renting a corner of a Cold War missile range to fly a commercial rocket. The engineers who worked there were, whether they thought about it or not, standing at the crossing point of two ages of spaceflight — the old one of warheads and superpowers, and the new one they were themselves trying to bring into being.

The single most vivid illustration of what the distance cost came not on the island but on the way to it, in the frantic weeks before the fourth flight. To save time, SpaceX chartered a giant Boeing C-17 transport to fly a Falcon 1 first stage across the Pacific rather than send it by ship. The rocket's tanks were sealed against moisture, vented only through a narrow quarter-inch line, and the engineers had calculated that they could equalize against the cabin's changing pressure as the aircraft climbed and descended. But the C-17 came down toward Kwajalein far faster and steeper than the figures in the aircraft's manual had led them to expect, and the outside pressure rose faster than the thin vent line could bleed into the tank. In the cargo hold the crew heard loud, terrible popping and groaning sounds, and turned to see the last rocket the company could afford beginning to crumple — the thin metal of the tank buckling inward under a pressure it had not been built to resist, like a plastic bottle in a fist. It was, for a few minutes, the sound of the company dying. Engineers scrambled, tearing into the packaging to open the tank and let the pressures equalize, and the stage survived, dented and frightening but repairable. They nursed it back to health on the island and flew it, three weeks later, to orbit. That a rocket could very nearly be destroyed by the descent of the airplane carrying it tells you everything about the razor margins the company lived on.

Life on Omelek was a strange mixture of hardship and idyll. The engineers, many of them in their twenties and early thirties, lived in cramped, improvised quarters, worked impossibly long hours in tropical heat and humidity, and were cut off from the ordinary comforts and connections of home for weeks at a stretch. They ate what could be brought in, drank warm water, burned and peeled under the equatorial sun, and pushed themselves to the edge of exhaustion under relentless schedule pressure. And yet, by many of their own later accounts, it was also the adventure of their lives — a band of young idealists on a tropical island, building a rocket with their own hands, doing something no private company had ever done, bound together by shared hardship and shared purpose in a way that the comfortable careers they had forsaken could never have provided. They fished and swam in the lagoon, snorkeled over the reef in the rare hours they could steal, watched spectacular equatorial sunsets, and formed friendships forged in the crucible of a shared, improbable mission. The camaraderie of the Omelek team became part of SpaceX's founding legend, the origin story of a culture that prized dedication, improvisation, and belief above comfort or convention.

The physical apparatus of a launch, which at a government spaceport fills whole buildings, was on Omelek reduced to almost comic austerity. The launch was controlled from a converted shipping container, hot and cramped and crowded with young engineers hunched over screens, that sat a safe distance from the pad. Fuel and liquid oxygen came from tanks the team had plumbed themselves. When the rocket needed to be raised upright or laid down for work, it was done with a hydraulic cradle the crew operated by hand. The team slept in trailers and a low bunkhouse, and in the dead hours between tasks they did what young people marooned on a beautiful island do: they swam the warm lagoon, speared fish off the reef, and watched the debris of their own failures — after the first flight, pieces of the rocket lay scattered across the coral and even near the buildings — become, over weeks, just another feature of the landscape. The strangeness of it was not lost on them. They were living like castaways and working like a NASA control room, and the two conditions coexisted hour by hour.

The improvisation was not a charming quirk but a necessity, and it revealed something essential about how SpaceX worked. A traditional aerospace program, launching from an established spaceport, would have had the infrastructure, the supply chains, the standing armies of technicians and specialists that decades of government spending had built up. SpaceX had none of that. It had a small team, a remote island, and a mandate to make it work with whatever was at hand. So the engineers did everything themselves — assembled the rocket, prepared the pad, fueled the vehicle, ran the countdown, fixed what broke — wearing many hats, learning by doing, taking on responsibilities that at a larger organization would have been divided among dozens of specialists. A propulsion engineer might spend the morning tracing a wiring fault and the afternoon driving a forklift; a guidance specialist might find himself on his back under the rocket with a wrench. This breadth of hands-on experience made them extraordinary engineers, people who understood the whole vehicle and the whole operation rather than a single narrow slice, and it built the deep, practical, integrated competence that would serve SpaceX for decades. The very lack of resources that made Omelek so hard also made it a school of a kind that no well-funded program could offer. Many of the people who lived through it would go on to lead the programs that landed rockets and flew astronauts, carrying the lessons of the island into everything that came after.

There was a darker side to all of it, and honesty requires acknowledging it. The hours were inhuman, the pressure relentless, the toll on personal lives and health real. The three failures that preceded the fourth flight's success ground the team down, and the constant nearness of the company's death — the knowledge that each failure brought bankruptcy closer — created a stress that no amount of camaraderie could entirely offset. Some people burned out and left. Marriages and relationships strained under the long absences and the all-consuming intensity. The romance of the tropical rocket island was real, but so was the exhaustion and the fear, and the two were inseparable. The culture that Omelek forged — of total dedication, of subordinating everything to the mission, of accepting punishing sacrifice as the price of doing something great — would remain a defining and double-edged feature of SpaceX ever after, celebrated by those who thrived in it and criticized by those who saw in it a pattern of overwork and burnout that exacted a genuine human cost.

It is worth drawing out why the island mattered beyond the romance, because there is a durable lesson in it that outlived the rocket. The engineers who came off Omelek had been forced, by scarcity, to become generalists — to understand propulsion and avionics and structures and ground operations all at once, because there were not enough people to specialize. That breadth turned out to be exactly the kind of competence a fast-moving rocket company needed, and many of the individuals who suffered through the tropical years rose, over the following decade, into the senior ranks that would build Falcon 9, land boosters, and fly astronauts. The atoll functioned, in effect, as a brutal graduate school in end-to-end rocketry, and its graduates carried its habits — the improvisation, the ownership, the refusal to wait for permission — into every program that came after. When people later marveled that SpaceX could do things no one else could, part of the answer traced back to a handful of acres of coral where a small group had been forced to learn everything.

When Falcon 1 finally reached orbit on that fourth flight in September 2008, the triumph belonged, more than to anyone, to the small band who had lived through the

Omelek years — who had shipped the rockets across the ocean, fought the corrosion and the heat, saved a crumpling booster in the belly of a transport plane, improvised their way through failure after failure, and refused to give up when the company's death seemed near. They had proven, on a coral speck in the Pacific, that a private company could build a rocket and reach orbit, and they had done it not with the vast resources of a government program but with their own hands, their own ingenuity, and their own willingness to sacrifice. The Omelek chapter was brief — Falcon 1 flew only five times, and the company soon moved on to Falcon 9 and to established launch sites with real infrastructure at Cape Canaveral in Florida and Vandenberg in California, and eventually to its own spaceport in South Texas. But the culture it forged, and the lessons it taught, stayed with SpaceX as it grew into a giant. The scrappy, improvising, all-in spirit of the island became the DNA of the company, for better and for worse, and it is worth remembering, amid the later triumphs of landed rockets and trillion-dollar valuations, that it all began with a handful of exhausted young engineers on a tropical island, betting their careers and their savings and their youth on a small white rocket that kept, at first, falling out of the sky.

Chapter 7: Three Strikes

The failure of the first Falcon 1, on March 24, 2006, was painful but not, in itself, alarming. First flights of new rockets fail routinely; the history of spaceflight is a graveyard of maiden launches that ended in fire, and even the mightiest programs — the American and the Soviet alike — littered their early years with wreckage. The SpaceX engineers dissected the telemetry and the debris, traced the fuel leak to its corroded source, and set about the long process of building the next vehicle and shipping it across the ocean to Omelek. There was, at this stage, a certain grim optimism. The rocket had at least left the pad and flown; the failure had an identifiable cause; the fix was straightforward. There was even a strange grace note to the disaster: the small satellite the rocket had carried, a student-built spacecraft from the Air Force Academy called FalconSAT-2, survived its plunge back to the island largely intact, coming to rest not far from the very shed the team worked out of — a reminder, however bittersweet, that the payload had been robust even if the rocket had not. The company had money for more attempts, and everyone assumed that each flight would teach them what they needed to know to succeed on the next. That assumption was correct in the end, but the road proved longer and steeper than anyone hoped, and it would come nearer to swallowing the company whole than even the pessimists on the team feared that spring. What none of them yet knew was that the three failures ahead would each teach a different and more expensive lesson than the last, and that the cushion of money and patience they were counting on would be all but gone by the time the fourth rocket stood on the pad.

The second flight came a year later, on March 21, 2007, and it went far better — right up until it didn't. This time Falcon 1 climbed cleanly off the pad, the first stage burned through its work, separation occurred, and the second stage's Kestrel engine ignited and carried the vehicle up into space, higher and faster than any SpaceX rocket had gone — past the edge of the atmosphere, to an altitude approaching three hundred kilometers, closing on the speed it needed. For several minutes it looked like success. Onboard cameras sent back footage of the blue Earth curving away beneath the rocket, and in the control room hope curdled into something like belief. Then, in the final phase of the second-stage burn, the vehicle began to wobble. The problem was propellant slosh: the liquid fuel in the second stage's tanks, no longer held firmly by the acceleration of a full burn, began to slosh back and forth, and the sloshing set up an oscillation that the guidance system could not damp; a control issue compounded it, letting the rocket begin to roll. The engine's thrust no longer pointed cleanly along the flight path, and control was lost. The second stage shut down short of the speed it needed. Falcon 1 had reached space but not orbit, falling agonizingly short of the goal. It was a more sophisticated failure than the first — the rocket had done almost everything right — but a failure nonetheless, and the second in a row.

There was, in the second failure, a foretaste of a habit that would later define the company's public character: its willingness to show the world its own footage. Falcon 1's second flight carried cameras that streamed the ascent, and SpaceX released the video, so that anyone could watch the blue curve of the Earth fall away and then, minutes later, the horizon begin to spin as the doomed second stage lost its composure. Traditional aerospace concerns guarded such images jealously, releasing polished success and burying failure. SpaceX, by temperament and by the frankness of its founder, tended to let people see the thing go wrong. The payload on that flight was a demonstration mass for the Pentagon's advanced-research agency rather than a paying customer's satellite, which softened the commercial blow, but the footage of a rocket tumbling in space was seen by a wide audience, and it did the company's reputation no favors in the eyes of an industry already inclined to doubt it.

Even now, the mood at SpaceX was more determined than despairing. Slosh was a known problem with known solutions; baffles could be added to the tanks to break up the motion of the liquid, and the control laws refined. The company pressed on. But the failures were beginning to cost more than money. They were costing credibility and time, and time was the one resource Musk could not manufacture. Each flight took the better part of a year to prepare, and each consumed a chunk of the founding capital. The wider context, too, was darkening. Musk's fortune was not infinite, and it was now stretched across two hungry companies, SpaceX and Tesla, both of which devoured cash and neither of which yet made a profit. The financial cushion that had seemed generous in 2002 was thinning alarmingly by 2007. And the industry's skeptics, who had predicted from the start that the internet millionaire would fail, were beginning to look prescient.

The third flight, on August 3, 2008, was the one that came closest to breaking the company's spirit — and it did so in the cruelest possible way, by failing at the very last moment through a mistake that came directly from success. Between the second and third flights, SpaceX had upgraded the Merlin engine to a more powerful and more efficient version, the regeneratively cooled Merlin 1C, which used the propellant itself to keep the chamber from melting. This new engine, the fruit of the team's relentless improvement, was better in almost every way. But it had one subtle new behavior: after the first stage's engine shut down and the stages separated, the new engine continued to produce a tiny residual thrust for a moment, a faint cough of leftover power as the last propellant in its lines burned off. On the old engine, this had not happened; the separation sequence had been designed around an engine that stopped cleanly. On the new engine, that small residual thrust was enough to push the spent first stage forward, up into the base of the second stage that was just pulling away above it, about two minutes and twenty seconds into the flight. The two stages, meant to part cleanly, instead collided — a gentle, catastrophic touch. The second stage was knocked off course, its engine unable to save the flight, and Falcon 1's third attempt ended, like the first two, in the ocean.

It is often assumed that the three failures were simply bad luck, the random misfortunes that dog any hard endeavor. They were not. Each had a distinct, traceable, and different cause — a corroded nut, then fuel slosh, then a staging collision — and the third was not bad luck at all but the direct and unforeseen consequence of the team's own progress. They had built a better engine, and the better engine had killed the flight. It was the kind of failure that makes engineers question whether they will ever get everything right at once, whether the margins in this brutal business will ever line up in their favor. And this time the failure carried real cargo to the bottom of the Pacific. The third Falcon 1 had been flying not a dummy mass but actual payloads: Trailblazer, an experimental satellite for the U.S. Air Force's operationally responsive space program; two small spacecraft for NASA, called NanoSail-D and PRESat; and a memorial payload from a company called Celestis carrying the cremated remains of two hundred and eight people. Among those ashes were those of Gordon Cooper, one of the original Mercury Seven astronauts, and of James Doohan, the actor who had played Scotty, the miracle-working engineer of *Star Trek*. All of it was lost. That a rocket meant to open space to ordinary people should carry, on the flight that failed, the ashes of a real astronaut and a fictional engineer, is the kind of detail a novelist would be told to cut for being too neat.

For SpaceX, the situation after the third failure was close to terminal. The company had now failed three times. It had, by Musk's own reckoning, money for perhaps one more attempt, maybe two if every economy was made. The global financial crisis was breaking over the world that autumn, drying up capital and making it nearly impossible to raise more. Tesla was simultaneously teetering, and Musk was pouring his last personal reserves into keeping both companies alive, splitting his remaining money between them, borrowing where he could, mortgaging his future. He would later describe this period as the worst of his life, a stretch of sleepless nights and physical distress — he has spoken of waking with what felt like a panic attack, and of friends fearing for his health — in which he genuinely believed he might lose everything he had built and everything he had. The employees felt it too. Morale, which had survived the first two failures on a diet of determination, began to fray. People wondered, reasonably, whether they were witnessing the slow death of the company.

Musk's public posture in those days was, by any measure, remarkable. Rather than retreat into silence or spin, he issued a statement that acknowledged the failure plainly and then pivoted immediately to resolve, declaring in effect that SpaceX was in the launch business for the long haul and would not falter — that the company had the determination and the resources to see the problem through. Whether this was genuine conviction, a leader steadying a frightened crew, or a founder talking himself off a ledge, it was almost certainly all three at once. Privately he was closer to breaking than the public words let on; the two things, the outward defiance and the inward dread, were not in contradiction but were the daily condition of running the company in 2008.

And yet — and this is the part of the story that reveals something real about how SpaceX worked — the response to the third failure was not paralysis but acceleration. Musk gathered the team and, rather than announcing retreat, told them they would fly again, and soon. The cause of the third failure was understood within days: the residual thrust, the too-quick separation. The fix was almost embarrassingly simple — wait a little longer before separating the stages, letting the residual thrust die away before the second stage pulled clear, a change of a few seconds in the flight computer's timeline. And crucially, the company had, in inventory, most of the parts for a fourth rocket, built in parallel with the third in a bet that they would need it. Rather than take months to prepare the next flight, the team resolved to do it in weeks — to build up the fourth Falcon 1, ship it to the pad on Omelek, and launch before the money ran out entirely. It was an insane schedule by any normal standard, a matter of six or seven weeks to prepare an orbital launch that would ordinarily take many months, and it was during this frantic sprint that the fourth booster very nearly crumpled in the belly of the transport plane carrying it across the Pacific. The engineers, exhausted and frightened and somehow still believing, threw themselves at it anyway.

There is a detail in the recovery that captures the company's whole method. The team was able to move in weeks rather than months only because it had, months earlier, decided to build a fourth rocket in parallel with the third — an expensive hedge that a cash-strapped startup might reasonably have declined to make, but that meant the parts were largely on hand when disaster struck. Manufacturing capacity, the ability to have another vehicle nearly ready before you know whether you need it, turned out to be as decisive as any clever piece of engineering. It is a theme that would recur throughout SpaceX's history: the company won not only by designing well but by building fast and building many, so that a failure was a setback of weeks rather than years. The fourth Falcon 1 existed to be flown quickly precisely because someone had gambled, before the third even launched, that it would be needed.

That decision — to respond to the company's most demoralizing failure with its most aggressive schedule — is the hinge of the whole early history. A more cautious organization would have paused, regrouped, spent months on analysis, and, in all likelihood, run out of money before flying again. SpaceX did the opposite, betting that speed was survival, that the fastest path through the crisis was straight ahead. It was a bet born partly of desperation and partly of the culture Musk had built, a culture that treated failure as data and speed as the supreme virtue. The fourth flight, prepared in a frantic sprint and launched in late September 2008, would either vindicate that culture or end it. As the countdown ticked toward zero on Omelek that night, with the company's life in the balance, no one watching could have told you which. What happened next — the clean separation, the steady second-stage burn, the numbers climbing to orbital velocity and holding — we have already seen. Falcon 1 reached orbit on its fourth attempt, and SpaceX, against all the odds and most of the expert opinion, survived. But reaching orbit was not the same as being safe, and the weeks after the fourth flight would bring the rescue that truly secured the company's future.

Chapter 8: The Fourth Flight

The success of Falcon 1's fourth flight, on September 28, 2008, was a genuine milestone in the history of spaceflight, and it deserves to be understood clearly for what it was and was not. For the first time, a rocket developed and built by a private company, funded with private money rather than a national treasury, had reached Earth orbit. This had never happened before. Every previous vehicle to reach orbit — American, Soviet, European, Chinese, Japanese — had been the product of a government program, backed by the resources and patience of a state. Falcon 1 was different in kind. It had been built by a few hundred people in a Los Angeles warehouse, on the fortune of a single entrepreneur, in six years, for a sum that was a small fraction of what a comparable government program would have spent — the total development cost of the vehicle running to roughly a hundred million dollars, against the billions such an effort would traditionally consume. That was the historic achievement, and it was real.

The scale of what the small team had done becomes clearer by comparison. A conventional government rocket program of the era might absorb billions of dollars and the better part of a decade before its first flight, sustained by the patience of a national budget; SpaceX had reached orbit on roughly a hundred million dollars of private money in six years, having designed and built its own engine, its own airframe, its own avionics, and its own launch operation more or less from nothing. Cheapness of this kind was not merely a matter of thrift; it was the whole thesis of the company, the proof that the wall of cost Musk had run into on the flight home from Moscow really could be pushed through. But it was not, by itself, a viable business, and everyone at SpaceX knew it. Falcon 1 was a small rocket with a small market. The satellites that made money were mostly large — heavy communications spacecraft bound for distant orbits — and Falcon 1 could not lift them. The company had always understood the small rocket as a stepping stone, a way to learn and to earn credibility, not as the product that would sustain it. The real plan, from early on, had been to use what SpaceX learned from Falcon 1 to build a far larger rocket, Falcon 9, capable of carrying serious payloads and, eventually, of being reused. But building Falcon 9 would cost far more than building Falcon 1 had, and after four flights and six years, the founding capital was nearly gone. Reaching orbit had proven the concept; it had not filled the bank account. In the autumn of 2008, SpaceX was a company that had just done something extraordinary and was still, simultaneously, close to death.

The rescue, when it came, arrived from the institution that the whole enterprise had implicitly set out to challenge: NASA. To understand why the agency threw SpaceX a lifeline, one has to understand a quiet revolution that had been building inside NASA for several years — a shift in how the agency thought about getting cargo and, eventually, people to orbit. For decades, NASA had operated in the traditional aerospace mode: it designed spacecraft to its own detailed specifications and paid contractors on a "cost-plus" basis, reimbursing their expenses plus a guaranteed profit. It is worth being concrete about what that arrangement did, because the perversity of it is the whole point. If a contractor was reimbursed for its costs and then paid a percentage on top, then every dollar of overrun, every delay, every inefficiency, increased the contractor's profit rather than reducing it. The incentive ran exactly backward: the model rewarded spending, not thrift. It produced magnificent machines — the Apollo hardware, the Space Shuttle — but it left NASA owning and operating everything at enormous expense, and it gave no one any reason to make spaceflight cheap. With the Shuttle approaching retirement and no replacement ready, a group of reformers inside and around the agency began to argue for a different approach. Instead of designing and owning the vehicles, NASA would become a customer. It would tell private companies what it needed — cargo delivered to the International Space Station, say — and pay them a fixed price for the service, results only, leaving the companies to design, build, own, and operate the rockets and capsules themselves. This was the philosophy behind a program called Commercial Orbital Transportation Services, or COTS, launched in 2006, and its successor, the Commercial Resupply Services program.

There is an irony buried in this story that is too good to leave unremarked. The NASA Administrator who championed the commercial approach, who threw the agency's weight behind the idea of buying rides from startups rather than building its own, was Michael Griffin — the same Mike Griffin who, a few years earlier, had flown to Moscow at Elon Musk's side as part of the doomed effort to buy a Russian missile. The man who had watched Musk get laughed out of the room by rocket sellers was now, from the Administrator's office, helping to steer the contract that would keep Musk's company alive. It is a small reminder of how tightly interwoven this world was, and of how much the outcome turned on a handful of people who kept crossing one another's paths.

COTS was, in a sense, tailor-made for a company like SpaceX, though it had not been designed with SpaceX specifically in mind. It rewarded exactly the things SpaceX was built to do: develop hardware quickly and cheaply, own the whole stack, and sell a service rather than run up a cost-plus bill. In August 2006, on the strength of its promise rather than any proven track record — this was two years before Falcon 1 would even reach orbit — SpaceX had won a COTS award to develop the capability to carry cargo to the space station, a milestone-based partnership worth a few hundred million dollars that helped fund the development of Falcon 9 and a cargo capsule called Dragon. That award had been a vote of confidence at a crucial moment, and it kept the lights on through the darkest stretch. But the truly decisive event came on December 23, 2008, in the weeks after the fourth Falcon 1 flight, when NASA awarded SpaceX a full Commercial Resupply Services contract: a commitment to buy a series of a dozen cargo flights to the space station, worth on the order of a billion and a half dollars.

For SpaceX, this contract was salvation. It provided not just cash but the thing a young company needs most: a committed customer with deep pockets and a genuine, recurring need. It gave the company a concrete reason to build Falcon 9 and Dragon, and the financial runway to do it. Musk has described learning of the award as one of the happiest moments of his life; by his own account he was so overcome, so unguarded in the rush of relief, that he blurted out to the NASA official on the phone that he loved the agency — an outburst from a man who was, in that instant, being pulled back from the edge of ruin. The timing could hardly have been tighter. That same December, in the final hours of the year, Tesla was closing a financing round that saved it from its own bankruptcy, a deal that came together so late that Musk has said the money arrived on Christmas Eve, when he had days of cash left. If either the SpaceX contract or the Tesla financing had failed, he has said, he might have lost both companies. Both came through, within days of each other, at the close of the worst year of the financial crisis. It is one of the narrowest escapes in the history of modern business, and it is worth sitting with, because the entire subsequent story — the landed rockets, the astronauts, the satellite constellation, the new space race — hangs on the fact that in December 2008 a small number of decisions went the right way instead of the wrong one.

It is worth dwelling for a moment on what SpaceX proposed to build with the reprieve, because the design choices carried the lessons of the Falcon 1 years inside them. Falcon 9 took its name from the nine Merlin engines clustered in its first stage, and that number was not arbitrary. Nine engines meant that the vehicle could, in principle, lose one to failure and still complete its mission — an "engine-out" capability reminiscent of the Saturn V that carried the Apollo astronauts, and a direct answer to a company whose small rocket had been undone, more than once, by a single point of failure. Building nine copies of an engine SpaceX already knew how to make was also cheaper and faster than developing one large new engine, and it let the same Merlin that flew on Falcon 1 mature into the workhorse of a far bigger machine. The Dragon capsule that would ride atop it was designed from the outset to be more than a cargo box: it was built with the eventual goal of carrying people, its pressure vessel and heat shield and systems sketched with astronauts in mind years before anyone would trust it with a life. The company was, in other words, using a cargo contract to build the bones of a vehicle that could one day fly crew — thinking, as it always did, several moves ahead of the game it was ostensibly playing.

There is a larger point buried in the story of the CRS contract, and it complicates the popular image of SpaceX as a purely private, government-defying enterprise. The company that would come to symbolize the triumph of private industry over government bureaucracy was, at the moment of its greatest vulnerability, saved by the government. NASA's willingness to become a customer rather than an owner — to bet on a small, unproven company and to structure the deal so that SpaceX could survive and grow — was as essential to the outcome as anything Musk or Mueller did. The relationship was genuinely symbiotic. NASA needed a cheaper way to reach the space station, especially with the Shuttle retiring and the alternative being total dependence on Russian rockets to supply an American-led outpost. SpaceX needed a customer. Each got what it needed from the other. In the years to come, this partnership would deepen and would prove, again and again, decisive — through the development of Falcon 9, through the cargo flights, and eventually through the far bigger prize of carrying astronauts. Those who tell the SpaceX story as a simple tale of heroic private enterprise, and those who tell it as a tale of a company living off government contracts, are both partly right and both miss the truth, which is that the achievement was a collaboration, a new kind of relationship between the state and private industry that neither could have managed alone. You can see the model everywhere now — it became the template by which the agency would later buy rides for its astronauts and its Moon-bound landers — but in 2008 it was an experiment, and SpaceX was its first great proof.

The COTS program deserves to be recognized as one of the more consequential policy experiments in the modern history of spaceflight, and not only for what it did for SpaceX. It was, at bottom, a bet by a handful of officials that the way to lower the cost of reaching space was to stop micromanaging how it was done and instead pay for results — a bet that ran against decades of institutional habit and against the interests of the established contractors, who understandably preferred the old arrangement in which their expenses were guaranteed. The program was structured so that the government paid only as companies hit agreed milestones, which capped the taxpayer's exposure and shifted the risk of failure onto the firms themselves. For SpaceX this was demanding but survivable; the company was built to hit milestones cheaply. For NASA it was a way to seed a capability it badly needed without writing the kind of open-ended check that the cost-plus world required. That the model worked — that it produced, within a few years, a functioning private cargo service to the space station — is the reason it would later be extended to the far more sensitive business of carrying human beings, and the reason it is now remembered as a turning point rather than a footnote.

With the CRS contract in hand, SpaceX turned decisively away from the small rocket that had nearly killed it. Falcon 1 would fly only once more, on July 14, 2009, successfully placing a Malaysian Earth-observation satellite called RazakSAT into orbit — a quiet, competent commercial mission that demonstrated the company could deliver for a paying customer. And then Falcon 1 was retired, after five flights and two successes. It had done its job. It had taught SpaceX how to build and launch an orbital rocket, had proven the concept to the world, and had earned the company the credibility and the contract it needed to attempt something far greater. The company's entire focus now shifted to Falcon 9, a rocket many times more powerful, with nine Merlin engines clustered in its first stage instead of one, capable of lifting the heavy payloads that made up the real launch market, and designed from the start with an ambition that the industry still regarded as fantasy: to come back from space and land, so that it could fly again. The desperate years were over. The audacious ones were about to begin.

Chapter 9: Falcon 9

If Falcon 1 was a school, Falcon 9 was the career. The rocket that first flew from Cape Canaveral on the fourth of June 2010 was a different order of machine from the slender vehicle that had struggled off the pad on Omelek — larger, vastly more powerful, and designed not to prove a point but to make a living. Its name announced its architecture: nine Merlin engines clustered at the base of its first stage, burning together to produce the thrust needed to lift heavy payloads off the Earth. Those nine engines were the same Merlins, in an upgraded form, that had powered Falcon 1, and the decision to use nine of them rather than one big engine was one of the shrewdest choices SpaceX ever made. It let the company reuse the engine it had already developed rather than designing a new large one from scratch, saving years and money. And it gave the rocket a property almost no other launch vehicle had: engine-out capability. With nine engines arranged in a three-by- three grid, together generating roughly five thousand kilonewtons of thrust — a little over a million pounds of force — at the moment of liftoff, Falcon 9 could lose one engine during ascent and still complete its mission by burning the others a little longer, exactly as a multi-engine airliner can fly on fewer than all its engines. This redundancy, borrowed from aviation, made the rocket more robust, and it was not a theoretical selling point. It would one day be tested in flight, with the whole world's cargo hanging on whether the idea worked.

The first flight itself was a study in how much can go right and how much can still be learned. The maiden launch had a boilerplate stand-in for the Dragon capsule bolted to the top — a "Dragon Spacecraft Qualification Unit," a mass model deliberately not designed to separate — and an early attempt to fly it was scrubbed when a sensor on one of the first-stage engines read out of range. When the rocket did climb, on the fourth of June, it reached orbit cleanly, but in the final seconds of the second stage's burn the spent stage began to roll, spiraling and venting propellant as it tumbled, a small blemish on an otherwise historic success. For a company that had so recently been failing repeatedly on a much smaller rocket, putting a heavy- lift vehicle into orbit on its first try was a striking display of how much had been absorbed from the pain of Falcon 1. The more important early milestone came six months later, in December 2010, when Falcon 9 launched carrying the first operational Dragon capsule under the NASA COTS program. On that flight, Dragon separated, completed a couple of orbits of the Earth, and then did something no privately built spacecraft had ever done: it reentered the atmosphere, survived the searing heat behind its heat shield, deployed its parachutes, and splashed down intact in the Pacific, where it was recovered. With that flight, SpaceX became the first private company ever to launch a spacecraft into orbit and bring it safely back to Earth — a feat that, until then, only a handful of national governments had achieved. The scorched but whole capsule was proof that the company could do the round trip.

The design of Falcon 9 reflected everything SpaceX had learned and everything it believed. It was built for manufacture, not just for flight: the company intended to produce these rockets in quantity, on something closer to a production line than the artisanal, one-off assembly of traditional launch vehicles, and the design was simplified and standardized to make that possible. It used the same cheap, well-understood kerosene-and-oxygen propellants as Falcon 1. It kept the number of distinct parts and suppliers as low as possible, with SpaceX building most components in-house at Hawthorne. And, most consequentially, it was designed from the beginning with reuse in mind — not achieved at first, but intended. Musk had never lost sight of the original insight from the flight home from Moscow: that the great waste in spaceflight was the throwing away of the rocket, and that the path to radically cheaper launch ran through recovering and reusing the hardware, above all the expensive first stage with its nine engines. Falcon 9 was conceived as the vehicle that would finally make reuse real, even though its early flights would, of necessity, be expendable while the company learned to bring the boosters home.

The engine-out promise was put to the test on the seventh of October 2012, on a mission designated CRS-1, the first operational cargo run to the International Space Station. Seventy-nine seconds into the flight, as the rocket bored through the region of maximum aerodynamic pressure, engine number one suffered a sudden loss of pressure and shut down; a shower of debris, briefly visible in the launch video, prompted alarmed speculation that the engine had exploded, though SpaceX maintained it had simply lost containment and been commanded off. What happened next was the vindication of a design choice made years before. The flight computer, sensing the loss, recalculated the ascent in real time and simply burned the remaining eight engines longer — the first stage firing some twenty-eight seconds past its planned cutoff, the upper stage another fifteen — and Dragon reached its correct orbit and went on to berth with the station without any effect on the cargo mission. It is often told as an unqualified triumph, and for the primary mission it was; but there is an honest asterisk that the celebratory accounts leave out. Riding along as a secondary payload was a prototype Orbcomm communications satellite, and NASA's range-safety rules, written to protect the crewed station, forbade the upper stage from reigniting to raise that satellite unless the odds of a stray collision were vanishingly small. The extra fuel spent saving Dragon had eaten the margin, the restart was disallowed, and the little

Orbcomm satellite was stranded in a uselessly low orbit and reentered days later. Engine-out capability had saved the mission that mattered most and quietly sacrificed the one that mattered least — a lesson, in miniature, in how redundancy buys survival at a price.

It is worth pausing to register how quickly SpaceX was now moving relative to the industry it was challenging. Traditional aerospace programs took decades to move from concept to a mature, operational vehicle. SpaceX had gone from a failing small rocket in 2008 to a successful heavy-lift rocket and a recoverable spacecraft by the end of 2010 — a span of about two years — and by late 2012 it was flying cargo to the station and shrugging off an engine failure in flight. The pace was a direct product of the company's culture and structure: the vertical integration that let it change designs without renegotiating with dozens of contractors, the tolerance for failure that let it test aggressively, the in-house manufacturing that let it iterate hardware rapidly, and the founder's relentless pressure that kept everyone sprinting. That pace would only accelerate. Over the following years, SpaceX would upgrade Falcon 9 repeatedly, stretching its tanks, uprating its engines from the Merlin 1C to the far more powerful Merlin 1D, and steadily working toward the goal that lay behind all of it: bringing the first stage back.

The rocket that flew in 2010 was, in truth, only the first draft of a machine that would be redrawn again and again. That original version, known internally as Falcon 9 v1.0, flew just five times before it was superseded in 2013 by a substantially larger vehicle, Falcon 9 v1.1, whose tanks were stretched, whose nine Merlins were rearranged from the tidy three-by-three grid into a circular pattern SpaceX called the "octaweb" — eight engines ringing a central ninth — and whose upgraded Merlin 1D engines gave it something like sixty percent more thrust. It was this version that carried the SES-8 communications satellite toward high orbit in December 2013, the company's first flight to the distant geostationary transfer orbit where the big commercial birds live, and thus its true debut as a competitor for the world's most lucrative launches. Two years later came the version that would make history by landing, its propellants chilled to a denser, colder state to squeeze out still more performance. The point worth grasping is that "Falcon 9" was never a fixed object but a moving one, a design under permanent revision, each iteration a little more capable than the last — a way of building that had far more in common with successive versions of a piece of software than with the frozen, decade-locked designs of traditional rocketry.

There is a persistent belief that clustering many small engines is a recipe for disaster, and it is not an irrational fear; it has a famous cautionary tale behind it. The Soviet N1, the giant Moon rocket meant to beat Apollo, packed thirty engines into its first stage, and all four of its launches ended in failure, at least one of them in one of the largest non-nuclear explosions ever recorded — a debacle often invoked to argue that more engines mean more ways to fail. SpaceX's answer was that the N1's problem had not been the number of engines but the primitive means of controlling and monitoring them; with modern sensors, fast computers, and software able to sense a failing engine and instantly recompute the flight, a cluster of nine — or, on the later Falcon Heavy, twenty-seven — became not a liability but the very source of the rocket's resilience. The engine-out save on CRS-1 was the proof: the thing skeptics feared as a weakness was, handled correctly, a strength.

The economics driving all this were compelling and, once one grasped them, almost irresistible. Consider the arithmetic in the most concrete terms. A Falcon 9 first stage — the large lower section with its nine engines, its tanks, its avionics, its structure — represented the majority of the cost of the whole rocket, on the order of three-quarters of it. On a conventional expendable launch, that expensive piece of hardware did its job for about two and a half minutes, pushing the upper stage and payload up and toward orbit, and was then discarded, falling back through the atmosphere to smash into the ocean and sink. Every launch threw away a machine worth tens of millions of dollars after a working life measured in minutes, as if an airline scrapped a jet at the end of its first flight. If that first stage could instead be recovered intact and flown again, the savings would be enormous — not immediately, because the first attempts would cost more in development than they saved, but eventually, as reuse became routine and refurbishment cheap. Musk's vision was of a rocket that could be flown, landed, inspected, refueled, and flown again within days, like an airliner, driving the cost per launch down toward the cost of the propellant and the operations rather than the cost of building a new vehicle each time. It was a vision the industry regarded, at best, as decades away and, at worst, as a fantasy that would never pay off. SpaceX set out to prove it was neither.

The skepticism was not foolish. Reusing a rocket is fantastically difficult, far harder than it might sound to a layperson accustomed to the idea that airplanes fly again and again. A rocket's first stage, after doing its job, is traveling at thousands of kilometers an hour, high in the atmosphere, pointed the wrong way, nearly empty of fuel. To recover it, you must somehow slow it down, turn it around, steer it back through the violent heat and buffeting of atmospheric reentry, and set it down gently and precisely on a target — all autonomously, all in a few minutes, all while conserving enough propellant to do the job without stealing too much performance from the actual mission of putting the payload in orbit. Every kilogram of fuel and hardware devoted to landing is a kilogram of payload lost. The engineering challenges were immense, and the history of attempts to build reusable rockets was, as we have seen, a history of expensive failure. NASA's own Space Shuttle had been sold as a reusable vehicle that would make spaceflight cheap and routine, and had instead become one of the most expensive launch systems ever operated, its reusability more nominal than real. Against that backdrop, the promise that SpaceX would make reuse work — and make it cheap — sounded like the same old dream that had disappointed everyone before.

The Falcon 9 that first flew in 2010 is, in a sense, still flying today, though almost nothing about it is the same. The rocket that carries astronauts to the station, launches the great constellations of Starlink, and lofts the satellites of governments and rivals alike is a grandchild of that first vehicle, grown taller, stronger, and — the decisive difference — routinely reusable, its boosters returning to land and flying again and again until the tally on a single first stage runs into the dozens. When a modern Falcon 9 lights its nine engines and climbs off the pad, it is executing the same basic bargain struck in 2010: nine small engines instead of one large one, kerosene and oxygen instead of anything exotic, a design meant to be built by the hundred and, above all, meant to come home. To understand why SpaceX succeeded where others had failed, and why so many smart people were sure it would fail, we need to look closely at the idea of reusability itself: what it demands, why the industry had turned away from it, and how a company of software people and rocket hobbyists set out to do the thing that could not be done.

Chapter 10: Selling Rockets

A rocket that no one buys is an expensive hobby, and one of the least appreciated reasons for SpaceX's success is that it was, from early on, extraordinarily good at selling launches. The engineering triumphs — the landings, the reuse, the astronauts — tend to dominate the story, but underneath them ran a relentless commercial engine that filled the manifest, won the contracts, and generated the revenue without which none of the engineering could have continued. The person most responsible for that engine was Gwynne Shotwell, and the weapon she wielded was price.

Shotwell's arrival at SpaceX was almost accidental, which makes her importance to the company easy to underrate. A mechanical engineer and applied mathematician by training, she had spent years in the aerospace establishment, at the Aerospace Corporation and then at a small satellite company, before a friend, the SpaceX propulsion engineer Hans Koenigsmann, invited her to lunch and introduced her to Musk. She joined in 2002 as employee number eleven, hired to run business development and sales — which, at a company with no rocket that had yet reached orbit, meant selling a promise. She proved spectacularly good at it, and by 2008, the year Falcon 1 finally flew and NASA awarded its first big contract, she had been made President and Chief

Operating Officer, running the company day to day while Musk pushed the engineering. The division of labor became one of the underappreciated engines of the whole story: his ambition, her execution; his willingness to bet everything, her ability to reassure the customer whose satellite was riding on the bet.

When SpaceX entered the commercial launch market, that market was a cozy, expensive oligopoly. A handful of providers — the American United Launch Alliance, the European Arianespace, the Russians, and a few others — sold launches to the operators of communications satellites and to governments at prices that reflected decades of cost-plus habit and limited competition. A launch to geostationary transfer orbit, the destination of the big communications satellites that were the market's staple, might cost a customer well over a hundred million dollars, and customers had little choice but to pay, since the providers were few and the barriers to entry enormous. United Launch Alliance, the Boeing-Lockheed joint venture that held an effective monopoly on American military launch, drew on a supply chain of more than a thousand subcontractors, and the price of putting a sensitive national-security payload into orbit could run past four hundred million dollars. Into this comfortable arrangement SpaceX arrived with a rocket it could build and fly for a fraction of the incumbents' costs, and it did the thing oligopolies fear most: it competed on price, openly and aggressively.

The transparency was itself a form of aggression. Where the launch business had always been conducted in opaque, individually negotiated deals, SpaceX simply posted its prices on its website for anyone to read — a startling gesture in an industry that treated pricing as a closely held secret. By the end of 2013 the advertised figure for a Falcon 9 to low Earth orbit was around fifty-six and a half million dollars, and a standard commercial launch settled near sixty million, numbers so far below the incumbents' that they could not be matched without abandoning the fat margins on which the older companies' cost structures depended. It was, in effect, a public dare. A satellite operator who had spent years grumbling about the cost of getting to orbit could now see, in black and white, exactly how much cheaper the alternative was, and grumbling turned into signed contracts.

Someone had to go first, and the willingness of a blue-chip customer to take the gamble mattered enormously. That customer was SES, the Luxembourg-based operator that was among the largest in the world, which agreed to put one of its valuable communications satellites, SES-8, atop the newly enlarged Falcon 9 for its inaugural flight to geostationary transfer orbit in December 2013. It was a genuine risk — the upgraded rocket had never flown a mission to that demanding high orbit, and a failure would have cost SES a satellite and months of lost revenue — and SES took it in exchange for a price reported to be well under sixty million dollars, a fraction of what a comparable ride on an Ariane or a Proton would have run. When the Falcon 9 delivered the satellite cleanly, the psychological barrier fell. If the industry's establishment could trust SpaceX with a flagship commercial payload, so could everyone else, and the manifest that had been a trickle of the adventurous became a flood of the merely prudent. It is a pattern worth noticing in the diffusion of any disruptive technology: the decisive moment is often not the first flight of the machine but the first time a conservative, credible buyer stakes something real on it.

The effect was seismic. Satellite operators flocked to the newcomer, and as SpaceX proved its reliability, flight after flight, the trickle became a flood, and the company captured an ever-larger share of the global commercial launch market, a share that had once been divided comfortably among the established providers. The incumbents, structurally unable to match SpaceX's prices, found themselves losing business they had taken for granted, forced to respond with cost-cutting, new rockets, and, in some cases, appeals to governments to protect their strategic position. Arianespace lobbied European governments for support; ULA began, belatedly, to design a cheaper vehicle to replace its costly Atlas and Delta. The comfortable oligopoly was shattered, and the shattering was the work not only of SpaceX's cheaper rockets but of its willingness to use them as a commercial weapon, slashing prices and seizing market share with a ruthlessness that the genteel launch business had never seen.

The incumbents' responses, when they came, were themselves a measure of the disruption. Arianespace, which had dominated the commercial market for decades with its reliable but costly Ariane 5, found its order book eroding and went to the European governments that backed it for support, ultimately setting in motion the development of a cheaper successor, the Ariane 6, explicitly to answer the SpaceX threat. United Launch Alliance, the Boeing-Lockheed venture whose Atlas and Delta rockets had never had to compete on price, began designing an all-new vehicle, Vulcan, to replace them, and cut thousands of jobs in a scramble to lower its costs. Even the Russians, whose Proton had long been the budget option, found themselves undercut by a rocket cheaper still. The remarkable thing is how completely the terms of competition had shifted: for half a century the launch business had competed on reliability, on relationships, on national prestige — on nearly everything except price — and within a few years SpaceX had made price the axis on which the whole market turned, forcing every established player to reorganize itself around a game it had never had to play.

Shotwell was the architect of this commercial assault, and her role is easy to overlook amid the attention lavished on Musk. Where Musk provided the vision, the engineering drive, and the willingness to bet everything, Shotwell provided the commercial and operational execution that turned the engineering into a business. She was, by all accounts, a formidable negotiator and a superb operator, respected throughout the industry for her competence and her steadiness, and she provided a crucial counterweight to Musk's volatility — the person who could reassure a nervous customer, manage a government relationship, or hold the organization together through a crisis. It is telling that when SpaceX at last went public, in the June 2026 offering that valued it among the most valuable private companies ever to list, Shotwell — who had privately harbored doubts about an IPO for years, worried it might distort the culture that had made the company work — was the executive investors most wanted to hear from. Many of SpaceX's most important commercial and government contracts bore her mark, and it is no exaggeration to say that without her, Musk's engineering vision might never have been translated into a sustainable, profitable enterprise. Her steadiness mattered most in the moments when the engineering faltered. When a Falcon 9 was lost in flight in the summer of 2015 and another exploded on the pad the following year, it was Shotwell and her teams who kept the customers from fleeing — who explained, reassured, and held the manifest together through the months of grounding, so that when the rocket returned to flight the business was still there waiting for it. A company that could build brilliantly but could not keep its customers calm through a crisis would not have survived those blows; that SpaceX did owed as much to the trust Shotwell had banked over years as to any engineering fix.

The commercial strategy had a logic that went beyond simply winning market share, and it connected directly to the deeper mission. Every launch SpaceX sold generated revenue that funded the company's development — of Falcon 9's reusability, of Dragon, of Starship, of the whole long march toward Mars. And crucially, the commercial launches kept the factory busy and the rockets flying, building the cadence, the manufacturing efficiency, and the operational experience that made the reusable-rocket enterprise work. High cadence was not just a consequence of commercial success but a cause of further success: the more SpaceX flew, the better and cheaper it got at flying, the lower its costs fell, and the more it could undercut the competition and win still more business. And once the first stages began to come home and fly again, the advantage compounded ferociously. Equity analysts watching the December 2015 landing estimated that reuse might eventually cut launch prices by something like forty percent — and that estimate, made before anyone had actually reflown a booster, would in time look conservative. The commercial engine and the engineering engine drove each other, in a virtuous cycle that competitors, lacking either the cheap reusable rockets or the aggressive commercial strategy or both, could not replicate. The scale this eventually reached is hard to overstate, and it is the form in which most people meet the company's commercial dominance today: from a rocket industry that for decades had measured itself in a couple of dozen launches a year worldwide, SpaceX alone climbed past a hundred launches in a single year, flying more often than the rest of the planet — every other company and country combined — put together. A cadence like that is not merely a scoreboard; it is a moat. Each flight lowered the cost of the next, sharpened the factory, and deepened the reliability record that made customers comfortable, and no rival could hope to close the gap without first matching a launch rate that only reusable rockets and a full manifest could sustain. And when Starlink came along, SpaceX gained the ultimate captive customer — itself — guaranteeing a high cadence regardless of the commercial market and completing the self-reinforcing system that made the company so dominant.

The commercial creativity did not stop at undercutting the big operators; it reached down and opened space to customers who had never been able to afford it at all. SpaceX built a rideshare business, packing a single Falcon 9 with dozens or even scores of small satellites for universities, startups, and small nations, and — in keeping with the company's habit of transparency — it published a price for the service, a figure on the order of a million dollars to carry a couple of hundred kilograms to orbit, bookable almost like a parcel. On one such flight it lofted a hundred and forty-three satellites at once, a record, a whole swarm of small craft that would once have had no way to reach space but the years-long wait for a berth on someone else's expensive rocket. This was the same commercial instinct working at the other end of the market: not only to steal the incumbents' richest customers with lower prices, but to conjure entirely new customers into being by making orbit cheap enough to be an ordinary business expense rather than a national undertaking.

The lesson of SpaceX's commercial success is that engineering brilliance, however dazzling, is not enough. Many technically superb ventures have failed because they could not sell what they built, could not win customers, could not translate their innovations into a sustainable business. It is often said that SpaceX simply out- engineered its rivals; in fact it out-sold them just as decisively, and the two victories were inseparable. SpaceX succeeded not only because it built better rockets but because it sold them relentlessly, priced them aggressively, delivered them reliably, and built a commercial operation as formidable as its engineering. The landings and the astronauts and the constellations captured the imagination, but it was the unglamorous work of winning contracts and filling the manifest — the work of Shotwell and her commercial teams — that paid for it all and kept the company alive and growing. Every time a communications company today shops its next satellite to launch and finds that the cheapest, most available, most reliable ride is a reused Falcon 9, it is meeting the legacy of that commercial revolution as surely as any engineering one. In an industry that SpaceX transformed with its engineering, it is worth remembering that the transformation was also, and essentially, a commercial one: the breaking of an expensive oligopoly by a competitor willing to build cheaper and sell harder than anyone had dared. That commercial dominance would soon be tested in the most demanding arena of all — the launch of the nation's most sensitive military and intelligence satellites, a market long closed to SpaceX by an entrenched monopoly that the company would have to fight, in the courts as well as the marketplace, to break open.

Chapter 11: Dragon

The spacecraft that SpaceX built to carry cargo to the International Space Station was named Dragon, and the name carried a private joke with an edge to it. Musk called it after "Puff, the Magic Dragon," the gentle folk song, because, he said, so many people had told him when he founded the company that his plans were pure fantasy, a magic dragon that did not exist. Naming the capsule Dragon was a quiet retort to the skeptics — a way of turning the accusation of fantasy into a badge. By the time the joke was made, of course, the fantasy was becoming hardware, and Dragon would go on to become one of the most important spacecraft of its era, the vehicle that restored America's ability to reach the space station and, later, to carry human beings there.

The task Dragon was built for was harder than it might appear. Delivering cargo to the space station is not simply a matter of launching a box into orbit. The station moves around the Earth at nearly twenty-eight thousand kilometers an hour, and a spacecraft coming to meet it must be maneuvered into precisely the same orbit, then guided through a delicate, patient approach — matching speed and position to within centimeters, closing the final distance at the pace of a slow walk, all while both vehicles hurtle around the planet. Get it wrong and you risk a collision that could damage or destroy a station housing human crews. The rendezvous and approach must be executed with extraordinary precision and with fail-safes at every step, because there is no room for error when the target is irreplaceable and inhabited. For a young company that had only recently learned to reach orbit at all, developing a spacecraft capable of this was a formidable undertaking. It was also a matter of some urgency to the nation: with the Space Shuttle retired in the summer of 2011, the United States had no vehicle of its own capable of reaching the station it had spent tens of billions of dollars to build, and was reduced to buying seats and cargo space from Russia. Dragon was, among other things, a way home.

The first true test came in May 2012, on the flight designated COTS Demonstration 2 — only the third flight of the Falcon 9 rocket, and the first American vehicle to visit the station since the Shuttle's final mission. A Falcon 9 lifted a Dragon capsule to orbit on the twenty-second of May, and over the following days the spacecraft performed the full sequence of maneuvers required to approach the station: raising its orbit, testing its sensors and thrusters, demonstrating that it could abort and back away safely if commanded, and finally closing in on the outpost as the crew aboard watched and, at the critical moment, reached out with the station's robotic arm to grapple the capsule and berth it to a docking port. The astronaut at the controls of the arm was Don Pettit, working from the station's cupola, and as he captured Dragon high over Western Australia he radioed a line to the ground that has entered the lore of the era. "Looks like we got us a Dragon by the tail," he told the capsule communicator in Houston, a fellow astronaut named Megan McArthur. It was the kind of unscripted, faintly cowboy remark that fixes a moment in memory, and it did double work: it marked a milestone in the sober language of a working spacecraft crew, and it made a small joke of the very name that had begun as a retort to the skeptics. The magic dragon that so many had said did not exist was now, quite literally, being held by the tail at the end of a robotic arm, four hundred kilometers above the Earth. With that grapple, SpaceX became the first private company in history to send a spacecraft to the International Space Station, and the first ever to rendezvous and berth one commercial vehicle with another in orbit. The choreography that led to that grapple deserves a moment, because it shows how much trust was riding on the software. NASA did not simply wave Dragon in. The mission plan required the capsule first to fly beneath the station at a distance of a few kilometers, proving that its navigation sensors could lock onto the outpost and track it, then to demonstrate that it could hold station, retreat on command, and abort cleanly if anything looked wrong — a whole graded examination of the spacecraft's judgment before the controllers on the ground and the crew aboard would permit it to creep into the delicate zone close to the station. Only after Dragon had passed each test in turn was it cleared to make its slow final approach into the reach of the robotic arm. It was a measure of NASA's growing confidence that it had allowed SpaceX to fold two planned demonstration flights into this single mission, letting one capsule both prove itself and deliver real cargo — a compression of the schedule that would have been unthinkable under the cautious old way of doing things. For NASA, which had helped fund and guide the development, it was a vindication of the commercial approach. For SpaceX, it was proof that the company could do not just the spectacular but the exacting — that it could be trusted with missions where failure meant more than a lost rocket. That first demonstration Dragon carried a modest load up, about four hundred and sixty kilograms of supplies, and it brought roughly six hundred and twenty kilograms back down, a foretaste of the capability that would soon set it apart from every other vehicle serving the station.

With the demonstration complete, the cargo flights began in earnest. Under the Commercial Resupply Services contract, Dragon started making regular trips to the station, carrying up food, water, equipment, experiments, and spare parts, and — crucially — bringing things back down. This last capability set Dragon apart from the other cargo vehicles serving the station. It is often assumed that all the world's cargo craft were more or less interchangeable delivery trucks; in fact Dragon was, for years, the only one that could make the return trip. The Russian Progress, the European Automated Transfer Vehicle, the Japanese HTV — all could carry supplies up, but none could return cargo to Earth intact; they were designed to be loaded with the station's rubbish and to burn up in the atmosphere at the end of their missions, incinerating themselves and everything inside. Dragon, with its heat shield and its capacity to survive reentry and splash down under parachutes, could bring things home: completed experiments, biological samples that had to reach a laboratory before they degraded, failed equipment engineers wanted to examine, the results that scientists needed in their own hands rather than as data beamed down a radio link. For the research mission of the space station, this "down-mass" capability was uniquely valuable, and it made Dragon indispensable in a way that went beyond simple delivery.

Dragon's design carried a second, quieter cleverness in the form of its "trunk," the unpressurized cylindrical section beneath the capsule that housed its solar panels and, in the volume around them, could carry cargo too large or too rugged to fit inside the pressurized cabin — the bulky external experiments and replacement parts that had to be mounted on the outside of the station and handed over by robotic arm rather than carried in by hand. The trunk was jettisoned before reentry and burned up, so it was a one-way ride, but it meant a single Dragon could deliver both the delicate laboratory samples that travelled in the pressurized volume and the refrigerator-sized hardware that rode outside. Few of the vehicles serving the station could do both, and none could then turn around and bring the pressurized cargo home. It was this combination — up and down, inside and out — that made the homely cargo capsule so much more than a delivery van.

The cargo Dragon flights, which continued year after year, did something for SpaceX that is easy to overlook amid the more dramatic milestones of landings and astronauts: they made the company operationally mature. To fly cargo to the station on a regular schedule, reliably, safely, for a demanding government customer with the lives of astronauts on the line, is to become a serious, disciplined spaceflight organization. It requires processes, quality control, mission assurance, the ability to plan and execute complex missions again and again without the drama of the early years. SpaceX, a company born in a culture of move-fast improvisation, had to add to that culture the sober discipline of routine operations, and the Dragon program is where much of that maturation happened. The flights were not glamorous — a cargo run to the station rarely makes headlines — but they built the muscle memory and the credibility that everything else would rest on. They also built the relationship with NASA into something deep and durable, a partnership of trust that would prove essential when the far bigger prize came into view: carrying not cargo but people.

There were setbacks, and the honest story requires that they be told as plainly as the triumphs. On the twenty-eighth of June 2015, a Falcon 9 carrying a Dragon full of cargo broke apart a little over two minutes into flight, on the mission designated CRS-7 — the first in-flight failure of a Falcon 9 and a jolt to a company whose confidence had been growing with every success. The investigation traced the disaster to a part almost absurdly small in relation to the destruction it caused: a two-foot steel strut, one of hundreds inside the upper-stage tank, whose job was to restrain a bottle of high-pressure helium against the forces of flight. The strut snapped at a fraction of the load it was rated to bear. SpaceX had bought it from a supplier as an industrial-grade component and, as a later NASA review concluded, had trusted the certified rating without independently testing it — a rare lapse of the paranoid rigor that spaceflight demands. When the strut let go, the helium bottle broke free, the tank ruptured, and the vehicle came apart. There is a haunting coda to the loss: the Dragon capsule actually survived the breakup of the rocket, tumbling free and transmitting data all the way down, and might conceivably have been saved had its software known to deploy its parachutes in such a contingency. It did not, and it was destroyed on impact with the sea — a bitter lesson that SpaceX absorbed by writing exactly that emergency logic into later capsules.

The loss stung the more because of what that particular Dragon had been carrying. Bolted inside it was the first of the International Docking Adapters, the ring of hardware that was meant to be fitted to the station to let the coming generation of crew capsules dock automatically — a piece of infrastructure on the critical path to returning American astronauts to orbit, now scattered across the floor of the Atlantic. Alongside it were a pair of Microsoft's experimental HoloLens headsets, bound for an effort to give astronauts holographic guidance as they worked, and more than thirty experiments designed by schoolchildren, from studies of how bees pollinate in weightlessness to a test of a sunlight-blocking plastic. There is a particular cruelty in a rocket failure that is easy to forget when the loss is tallied only in dollars: a child who had spent a year preparing an experiment for space watched it vanish in a streak of vapor over Florida because a two-dollar strut had been trusted a little too far. SpaceX rebuilt the docking adapter and flew a replacement; the lesson about testing every part to destruction, rather than trusting a supplier's paperwork, it did not have to learn twice.

The loss grounded the fleet for months while SpaceX investigated, found the cause, and changed its procedures, and it was followed the next year by another failure, this one on the launch pad, to which we will return, because how the company responded to these blows says as much about it as its successes do. But the larger arc of the Dragon program bent toward reliability. The cargo capsule flew and flew, delivered and returned, and in doing so it established SpaceX as the workhorse of the American space program, the company NASA turned to when it needed something carried to orbit and brought back. When you read today of an experiment returning from orbit — a protein crystal grown in weightlessness, a sample of astronaut blood, a piece of hardware pulled from the station for failure analysis — the odds are overwhelming that it came home in a Dragon, the only Western vehicle for years that could make the journey both ways. The first-generation cargo Dragon flew some twenty missions to the station across the better part of a decade before it was retired in 2020 and replaced by a larger, smarter successor derived from the crew capsule — a vehicle that could dock itself to the station automatically, without waiting for an astronaut to reach out and grab it with the arm. But the humble original had already done the essential work. It had proved that a private company could be entrusted with the supply line of an inhabited outpost, and it had taught SpaceX, mission by mission, the discipline that carrying human beings would require. When astronauts finally rode a Dragon to orbit, they were riding on a foundation laid by years of grapples and berthings, of experiments carried up and samples brought home, of a company slowly earning the right to be trusted with lives.

That foundation of trust, built one unglamorous cargo run at a time, was the platform on which SpaceX would attempt its most audacious feats — beginning with the one the whole industry had declared impossible, the recovery and reuse of the rocket itself.

Chapter 12: The COTS Revolution

Behind the rockets and the landings lay a quieter revolution, one conducted not in the skies over Texas but in the offices and hearing rooms of Washington, and it may have been as consequential as anything SpaceX built. It was a revolution in how the government bought spaceflight — a shift from a decades-old model that had made space expensive and static to a new one that unleashed the commercial sector and made SpaceX possible. Understanding it is essential to understanding the whole story, because SpaceX did not succeed in spite of the government; it succeeded, in large part, because a handful of reformers changed the way the government did business.

For most of the space age, NASA had bought rockets and spacecraft the way it had bought them since Apollo: through cost-plus contracts, in which the government specified in detail what it wanted, and contractors built it while the government reimbursed their costs plus a guaranteed profit. This model had put men on the Moon, and it produced magnificent hardware, but it had a fatal flaw when it came to controlling cost. Because contractors were paid their costs plus a percentage, they had no incentive to be efficient — indeed, the more they spent, the more profit they made — and the government, owning and operating everything, bore all the risk and all the expense. The result was spaceflight that was reliable but ruinously expensive, and a launch industry that had no reason and no ability to become cheaper. The Space Shuttle, built and operated this way, had cost far more than promised. Any new government spacecraft, developed by the same model, would cost billions and take a decade or more — a point made vivid, later, by the Orion crew capsule, developed under a traditional cost-plus arrangement at a price that ran into the tens of billions of dollars and a schedule that stretched across the better part of two decades.

The reformers who challenged this model — figures inside and around NASA who had watched the costs spiral and believed there was a better way — proposed something radical: that for routine tasks like carrying cargo to the space station, NASA should stop being an owner and become a customer. Instead of designing and owning the vehicles, the government would tell private companies what service it needed — so many tons of cargo delivered to the station — and pay them for the service, at a fixed price, for results delivered, leaving the companies to design, build, own, and operate the hardware themselves, and to bear the risk and reap the reward of doing it efficiently. Crucially, the money was paid out not against invoices for hours worked but against milestones achieved: a company earned its next tranche only when it had actually flown a test, passed a review, delivered a result. If it overran its budget getting there, the overrun was its own problem, not the taxpayer's. This was the philosophy behind the Commercial Orbital Transportation Services program, launched in 2006, and its successor, the Commercial Resupply Services program. The idea was to harness the incentives of the market — the drive to control costs, to innovate, to compete — that the cost-plus model suppressed, and to use them to make access to space cheaper.

The reform was fiercely controversial, and its passage was a genuine political fight. The traditional aerospace contractors, and their allies in Congress, had every reason to resist a change that threatened their comfortable cost-plus business, and many in the space establishment were genuinely skeptical that private companies could be trusted with something as demanding as spaceflight. The champions of the commercial approach — among them Lori Garver, who as a senior NASA official fought for it against entrenched opposition — had to battle to fund the programs, to defend them against those who preferred the traditional model, and to keep them alive through the doubts and the setbacks. The program had been created in 2006 under the administrator Michael Griffin, but it was in the years that followed, when Garver served as NASA's deputy administrator and pressed to extend the same commercial logic from cargo to crew, that the fight turned bitter. Powerful members of Congress with traditional contractors and their jobs in their districts preferred a big government rocket built the old way, and the struggle to protect and fund commercial spaceflight against that preference made Garver, by her own later account, something of a pariah within parts of the very agency she served — mocked, undercut, dismissed as naive for believing that upstart companies could be trusted with the nation's astronauts. That she and her allies held on long enough for the results to speak for themselves is a reminder that the COTS revolution was not an inevitability but a near-run political thing, won by a handful of stubborn people against the grain of an establishment that had every reason to want them to fail. The initial COTS purse was a modest half a billion dollars, spread between two competitors, and SpaceX's share of the agreement came to about two hundred and seventy-eight million dollars — a sum trivial by the standards of a cost-plus rocket program, and a bet that many insiders expected to lose. It is worth remembering how genuinely improbable the wager looked at the time: NASA was staking a piece of the station's supply line on a company that had, when the deal was signed, never once reached orbit.

The program's structure was itself a piece of quiet legal ingenuity, and it is often overlooked. Rather than a conventional procurement contract, laden with the thousands of pages of federal acquisition rules that governed cost-plus work, COTS was built on a mechanism called a Space Act Agreement, a flexible instrument that let NASA share the cost and the risk of development while leaving the company to run the project its own way, unburdened by the government looking over every engineering decision. That legal choice was not a technicality; it was what allowed a startup to move at a startup's speed while spending public money. And SpaceX was not the only horse NASA backed. When the first agreements were signed in August 2006, the agency chose two winners: SpaceX, with its roughly two hundred and seventy-eight million dollars, and a company called Rocketplane Kistler, awarded a somewhat larger sum. Kistler, for all its experienced aerospace pedigree, could not raise the private capital the deal required, and NASA — in a demonstration of exactly the discipline the old model lacked — terminated its agreement in 2007 when it missed its funding milestones, and reallocated the money in a fresh competition that brought in Orbital Sciences, whose Antares rocket and Cygnus freighter would become the program's second success. The willingness to cut a failing participant loose, rather than pour good money after bad as cost-plus programs so often did, was part of what made the model work.

It is sobering to consider how close the whole thing came to not happening. In the last week of December 2008, SpaceX was nearly out of money, its founder's personal fortune all but exhausted by the years of failures that had finally ended with Falcon 1's first success only weeks before. It was into that near-collapse that NASA delivered the operational cargo contract worth well over a billion dollars — a lifeline that Musk has said, plainly, saved the company. Had the reformers lost their fight, had the agency stuck to the old way of buying spaceflight, there would have been no such contract to award, and the company that would go on to reshape the space age might simply have run out of cash and disappeared. The revolution in government contracting was not an abstraction to SpaceX; it was, at the decisive moment, the difference between existence and bankruptcy.

The results vindicated the reformers beyond their hopes, and they did so in a way that could be measured. NASA itself commissioned an analysis, using its long-established cost-estimating model, of what it would have cost the agency to develop the Falcon 9 the traditional way. The model returned a figure of about four billion dollars under standard cost-plus assumptions; adjusted to reflect a leaner, more commercial style of development, it still predicted roughly one and a seven-tenths billion. SpaceX, by its own accounting, had developed the rocket for something on the order of three hundred million — a fraction, perhaps a tenth or less, of what the conventional approach would have consumed. The commercial cargo program worked, delivering supplies to the station reliably and cheaply, and it paved the way for the commercial crew program that would restore American human spaceflight. For a total public investment that eventually reached about eight hundred million dollars across the program, the government had helped bring into being two new medium-class launch vehicles and two automated cargo spacecraft — a return that would have been unthinkable under the old model. The success of the commercial approach transformed how NASA and the government thought about buying spaceflight, and it became the template for an ever-widening range of activities — not just cargo and crew but lunar landers, space stations, and more — as the government increasingly turned to the commercial sector for capabilities it had once developed in-house.

It helps to hold two numbers side by side to feel the force of the comparison. To restore its own ability to carry astronauts to the station, NASA in 2014 bought commercial crew transport the same way it had bought cargo, on fixed-price terms, and SpaceX's share of that award was about two and a half billion dollars — for a human-rated spacecraft, its launch escape system, and a series of flights. Set that against the Orion capsule and the giant rocket built to carry it, developed the old cost-plus way at a cost that ran into the tens of billions and a schedule stretched across the better part of two decades. The commercial program did not merely save money at the margin; it delivered a comparable capability for something like a tenth of the price, and delivered it while the traditional program was still working toward its first crewed flight. That the same company could be, at once, the beneficiary of government seed money and the living rebuke to government cost-plus contracting is the paradox at the center of the whole story — and the resolution of the paradox is simply that the two programs, commercial and traditional, were buying spaceflight in fundamentally different ways.

There is an important corrective in this story to the popular image of SpaceX as a triumph of pure private enterprise over government. The truth is more interesting and more instructive. It is often said, by admirers, that SpaceX proved private enterprise could do what a bloated government agency could not; and by critics, that SpaceX was merely a creature of federal subsidy dressed up in libertarian rhetoric. Both readings miss what actually happened. SpaceX succeeded through a partnership with government — a government that had reformed the way it did business precisely in order to harness private innovation. The commercial model was not a rejection of government but a reinvention of its role, from owner and operator to customer and enabler, and it was that reinvention, as much as any private brilliance, that made SpaceX's rise possible. The seed money mattered, but so did something less tangible: NASA's willingness to share its hard-won expertise, to assign experienced engineers to advise the newcomer, to be a demanding but patient customer while a young company learned. The reality was a genuine and productive collaboration, a new model in which public purpose and private execution reinforced each other, and in which the reform of government contracting was as essential to the outcome as the engineering of rockets.

The legacy of the COTS revolution is visible today far beyond the space station's supply line, and far beyond SpaceX. The same fixed-price, service-buying model that carried cargo to orbit now underlies the way NASA is buying rides to the surface of the Moon, the way it is seeking commercial successors to the aging station, the way a dozen young companies have found the footing to attempt things that only governments once dared. When NASA set out to return astronauts to the lunar surface, it did not design and own the lander in the old manner; it bought a landing service, and the vehicle it chose was a version of SpaceX's own Starship, purchased on the same fixed-price logic first tried out on cargo runs to the station a generation earlier. When the agency began to plan for the day the aging station is retired, it turned again to the commercial model, seeding private companies to build the space stations that will replace it and promising to be a customer rather than an owner. The line runs unbroken from a modest cargo contract signed in 2006 with a company that had never reached orbit to the architecture of America's return to the Moon — which is why the COTS revolution, for all that it was fought in hearing rooms rather than on launch pads, deserves to be counted among the most consequential decisions of the era. Every time an agency announces that it will pay for a result rather than reimburse a process — that it will be a customer rather than an owner — it is drawing on the precedent that a handful of reformers fought to establish in 2006. It is a model with lessons far beyond spaceflight, about how governments might harness private innovation to public ends, and it is one of the quieter but more important legacies of the whole enterprise — a legacy that would be tested and extended as the same commercial approach was applied to the far more demanding and dangerous task of carrying human beings, and later to the grandest prize of all, the return to the Moon.

Chapter 13: The Reusability Heresy

To grasp why reusability was treated as heresy — why serious, intelligent people in the aerospace industry regarded SpaceX's plan to land and reuse rockets as either impossible or economically pointless — you have to understand the shadow cast by the Space Shuttle. For thirty years, from 1981 to 2011, the Shuttle was the United States' primary human spaceflight vehicle, and it had been sold to the public and to Congress as the answer to the wastefulness of expendable rockets. It was a reusable spaceplane. The orbiter, that iconic winged vehicle, flew to space and glided back to a runway landing, to be refurbished and flown again. Its twin solid rocket boosters parachuted into the ocean, were fished out, and were reused. Only the great orange external tank was thrown away. The promise had been dazzling: routine, frequent, cheap access to space, flights every couple of weeks, the cost per launch driven down by reuse to a fraction of the expendable alternative. Early projections had spoken of flying the fleet dozens of times a year and of driving the cost of reaching orbit down toward a few hundred dollars a pound.

The reality had been almost the opposite. The Shuttle turned out to be one of the most expensive launch systems ever operated; by the time the full cost of the program was divided across the flights actually flown, each launch had cost on the order of a billion and a half dollars. Its reusability was real in a narrow sense — the orbiters did fly again — but the refurbishment between flights was so extensive, so labor-intensive, so painstaking, that it devoured most of the savings reuse was supposed to provide. Consider the thermal protection system alone. The orbiter was clad in something close to twenty-four thousand heat-shield tiles, most of them individually shaped for their exact position on the airframe and not interchangeable one for another, and after every flight technicians had to inspect each of them, waterproof them, and replace the many that had been chipped, cracked, or lost. The three main engines, among the most sophisticated ever built, had to be largely removed and torn down between missions. The solid boosters, recovered from corrosive seawater, had to be disassembled and rebuilt at great cost. Far from flying every two weeks, the Shuttle fleet managed only a handful of flights a year; across the whole thirty-year program it flew a hundred and thirty-five times, an average of about four and a half missions a year against the fifty or more that had once been promised. And the system was not only expensive but dangerous: two of the five orbiters were lost in catastrophic accidents, Challenger in 1986 and Columbia in 2003, killing fourteen astronauts and grounding the fleet for years each time. By the time the Shuttle was retired in 2011, its lesson had been absorbed throughout the industry as a kind of grim orthodoxy: reusability sounds wonderful in theory, but in practice it costs more than it saves.

Here lies the misconception at the heart of the heresy, and it is worth stating plainly, because SpaceX's whole enterprise turned on correcting it. It is often said that the Shuttle proved reusability does not pay. What the Shuttle actually proved was narrower and more useful: that a particular, baroque way of being reusable does not pay. The vehicle had been made reusable in the most demanding manner imaginable — a delicate winged spaceplane, hand-built, wrapped in fragile tiles, its engines run so close to their limits that they needed rebuilding after every flight — and reuse of that kind required a standing army of thousands of technicians and months of work between missions. The failure was not of the idea but of the implementation. This distinction was lost on much of the industry, which drew from the Shuttle the blanket conclusion that reuse was a mirage, and it was precisely this blanket conclusion that SpaceX set out to overturn.

When Musk and his engineers talked about landing and reusing Falcon 9 first stages, the industry's response was informed skepticism rooted in hard experience. Reusability, the veterans said, had been tried; it did not pay. The refurbishment would eat the savings. The mass penalty — the fuel and hardware needed to bring a stage back — would cripple the rocket's performance, forcing it to carry less payload and undercutting its competitiveness. This last objection was the most serious, and it was true as far as it went: every kilogram of propellant held back for the landing burn, every kilogram of legs and fins and reinforcement, is a kilogram that cannot be spent lofting payload, and flying a stage all the way back to its launch site can cost a rocket a third or more of the mass it might otherwise carry. SpaceX's answer was not to deny the penalty but to build a rocket big enough to swallow it — to make Falcon 9 more powerful than its payloads strictly required, so that even after reserving the fuel and bearing the weight of the recovery hardware, it could still deliver a competitive load to orbit. The margin that looked like overengineering to a rival building an expendable rocket was, in SpaceX's design, the very thing that made reuse affordable without sacrificing the mission. The engineering was fantastically hard. And even if you could land a stage, inspecting and recertifying it for another flight would be so laborious that you might as well build a new one. The smart, experienced, credentialed view was that SpaceX was about to repeat the Shuttle's mistake, chasing a mirage that had already lured the world's most capable space agency into a decades-long, budget-devouring dead end.

SpaceX's answer to this skepticism was not to argue but to build, and its answer contained a crucial insight about how a different approach might succeed. What SpaceX aimed for was something the Shuttle had never been: a rocket rugged and simple enough that it could be recovered and flown again with minimal work — inspected, refueled, and relaunched, ideally in days, without the vast standing army of technicians the Shuttle had demanded. The goal was not merely reuse but rapid, cheap reuse, the kind that actually pays. The prize, if it could be won, was enormous, and a single comparison makes it vivid. The propellant that fills a Falcon 9 — the kerosene and liquid oxygen that it burns on the way up — costs only a few hundred thousand dollars, a rounding error beside the tens of millions the vehicle itself is worth. On an expendable rocket, nearly all of that vehicle cost is thrown away on every flight, so the fuel bill is trivial and the hardware bill is everything. But if the first stage, which is the bulk of the hardware cost, comes home and flies again and again, then over enough flights the cost of building it is spread thin and the price of a launch drifts down toward the cost of the fuel and the labor of turning the rocket around — exactly the arithmetic that makes air travel affordable, where no one scraps the airliner after a single trip and the ticket reflects the kerosene and the crew, not the price of a new jet. That was the promise: not a marginal saving but a change in kind, launch priced like a flight rather than like a fabrication. And the way to get there, SpaceX believed, was to design for it from the start: to build a first stage that could survive reentry and landing without exotic, fragile protection, that used engines robust enough to fly repeatedly, and that could be brought home and turned around quickly. This was the vision behind the propulsive landing — bringing the stage down vertically on its engines, standing it on its legs like a science-fiction rocket, rather than gliding it back like a delicate spaceplane.

The choice of vertical, propulsive landing was itself contentious and difficult, and its central problem was one of arithmetic that most people never think about. A rocket engine sized to heave a fully fueled vehicle off the pad is monstrously overpowered for the same vehicle when it returns nearly empty, its tanks drained, weighing a small fraction of what it did at liftoff. Even a single one of the nine Merlins, throttled as low as it will go, produces more thrust than the empty stage weighs — which means the stage cannot simply hover down gently. It must instead perform what engineers call a "hoverslam" or suicide burn: ignite its engine at precisely the right instant so that the vehicle arrives at the pad at the exact moment its downward speed reaches zero, with no margin to pause and no ability to hang in the air and feel for the ground. Time the burn a half-second early and the rocket climbs away and falls again; time it late and it drives itself into the concrete. The task has been compared to balancing a broom on your hand, except the broom is falling from the edge of space at supersonic speed, is nearly out of fuel, and must be caught, upright, on a target a few meters wide. The vehicle had to steer itself through the thin upper atmosphere and the thick lower one, correcting for wind and error, using aerodynamic surfaces and engine gimbaling and cold-gas thrusters in a carefully choreographed sequence, and all of it had to be done autonomously, by the rocket's own computers, in the space of a few minutes, with no possibility of human intervention. The computational and control challenges were as daunting as the mechanical ones, and here SpaceX's software-heavy culture was an asset: the company was full of people who thought naturally in terms of control algorithms and rapid iteration, and who could write and refine the flight software that would, in the end, make the landings possible.

SpaceX was not, it should be said, inventing the idea of landing a rocket on its tail from nothing. There was a half-forgotten precedent, and the company's engineers knew it well. In the mid-1990s, McDonnell Douglas had built a stubby, cone-shaped demonstrator called the DC-X, or Delta Clipper, that took off and landed vertically in the New Mexico desert, rising a few hundred meters, translating sideways, and settling back down on four legs — even, on one occasion, turning itself around in flight and demonstrating a turnaround between launches measured in hours. The DC-X was never meant to reach orbit; it was a low-and-slow proof that a rocket could be flown like a helicopter, and it worked, until the program was cancelled and the little vehicle destroyed in a landing accident when a leg failed to deploy. That the concept had been shown to be possible two decades earlier cuts against the lazy notion that SpaceX conjured propulsive landing out of thin air; what SpaceX did was not to invent the idea but to carry it, at last, all the way from a desert hop to the recovery of a stage returning from space — the leap that everyone before had found too far.

Before any of this could be attempted from orbit, SpaceX practiced closer to the ground. Beginning in 2012, at the McGregor test site in Texas, the company flew a series of experimental vehicles designed to develop and prove the technology of vertical takeoff and landing. The first was called Grasshopper — a Falcon 9 first-stage tank, some hundred feet tall, fitted with a single Merlin engine, four sturdy steel landing legs, and guidance. Its first outing, on the twenty-first of September 2012, was almost comically modest: it rose about six feet into the air, hovered for a few seconds, and settled back onto the pad. But over the following year each flight went higher and did more. The Grasshopper hovered, translated sideways to prove it could correct its position, flew through a simulated wind, and by its eighth and final flight, in October 2013, climbed to nearly two and a half thousand feet before descending under its own power to land softly, upright, on its legs. To watch the Grasshopper videos from that era is to watch a rocket doing something rockets were never supposed to do: rising into the Texas sky, pausing, and then coming back down to stand where it started, ready to fly again. A successor vehicle, larger and fitted with the steerable grid fins that would guide the real boosters home, extended the envelope further — and, in a fashion that would become a SpaceX signature, ended its own career in a fireball when, on a flight in August 2014, a blocked sensor sent it drifting off course and the range-safety system destroyed it in mid-air over the Texas scrub. SpaceX reported the loss with a shrug, noting that with test vehicles pushing the envelope, something like it was likely to happen again — a sentiment that could serve as a motto for the whole enterprise. The destruction of an experimental rocket was not a scandal to be buried but an expected cost of learning quickly, and the willingness to say so out loud, and to keep flying, was itself part of what set the company apart from an industry that treated every failure as a catastrophe to be avoided at nearly any price. These test flights were not the real thing — landing from a hover a couple of thousand feet up is far easier than landing a stage returning from the edge of space at supersonic speed — but they proved the concept, developed the control software, and gave the team the confidence and the tools to attempt the vastly harder task of recovering a first stage from an actual orbital launch. That attempt, and the string of spectacular near-misses and fiery failures that preceded its success, is where the reusability heresy was finally put to the test of flight.

Chapter 14: Learning to Land

The Grasshopper hops had shown that a rocket could take off and land under its own power. But there is a vast gulf between hovering a couple of thousand feet over a Texas field and recovering a first stage that has just done its real job — pushing a payload toward orbit, then falling back from the edge of space at many times the speed of sound, its tanks nearly empty, pointed the wrong way, tens of kilometers downrange. Crossing that gulf took SpaceX several years and a series of attempts that ranged from encouraging to spectacularly destructive, and that the company documented with an unusual, almost cheerful candor, releasing videos of its rockets exploding on landing under the wry label of "rapid unscheduled disassembly."

The strategy for recovering an orbital first stage came together as a carefully sequenced set of maneuvers, each solving one piece of the problem. After the first stage finished its burn and separated from the upper stage, it would use its cold-gas thrusters to flip around, pointing its engines forward into the direction of travel. Then it would perform a "boostback burn," firing some of its engines to slow its downrange velocity and, for landings back at the launch site, to begin steering itself back toward land. As it fell into the thickening atmosphere, it would perform an "entry burn," firing its engines again to slow down and shield itself from the worst of the aerodynamic heating and stress — using the engine exhaust, in effect, as a buffer against the onrushing air. During the long descent it would steer itself using four grid fins, waffle- like aerodynamic surfaces that deploy near the top of the stage and can be angled to adjust the vehicle's trajectory, correcting for wind and error and guiding it toward the landing target. And finally, in the last moments, it would perform the "landing burn," reigniting one or three engines, throttling them precisely to slow the stage to a gentle stop just as it reached the ground or the deck of a ship, extending its four landing legs, and settling down upright. Each of these maneuvers had to work, in sequence, autonomously, within tight margins, or the stage would be lost — and there was no dress rehearsal, because each attempt rode on the back of a real, paying mission whose primary job had to come first.

The hardware that made this possible was, in places, startlingly rugged for so delicate a task. The grid fins that steered the falling stage were slabs of latticework mounted near the top of the booster, folded flush during ascent and swung out into the airstream during descent, where they could be twisted to bank and guide a hundred-foot vehicle screaming down through the atmosphere; the early aluminum fins ran so hot in the hypersonic flow that they sometimes scorched or caught fire, and SpaceX would eventually forge later versions from single pieces of titanium that could survive reentry untouched and be reused without repair. At the base, four long legs, folded against the body like the limbs of an insect, snapped out and locked in the final seconds before touchdown to catch the stage and hold it upright. None of this bore any resemblance to the fragile, tile-clad delicacy of the Shuttle; it was deliberately, almost crudely, tough — machinery built to be knocked about, brought home, checked over, and flown again, which was the whole point.

SpaceX began testing these maneuvers not by trying to land

on solid ground — too risky, with a real mission's rocket and the danger to people and property if something went wrong — but by attempting "soft landings" on the surface of the ocean. On a series of Falcon 9 flights in 2013 and 2014, after the first stage had done its job and separated, SpaceX had it perform the full landing sequence but aim for a point over open water, touching down (or trying to) on the sea surface, where a failure would harm nothing. These ocean tests were revealing. On some flights, the stage successfully performed the maneuvers, slowed itself, and reached the ocean surface in a controlled vertical descent — demonstrating that the sequence worked — before toppling over and being destroyed by the water, since a rocket cannot float upright on the sea. On others, problems emerged: the stage spun up too fast for its thrusters to control, or ran out of the hydraulic fluid that steered the grid fins, or hit some other snag. But each test taught the team something, and the videos and telemetry showed steady progress. The stages were reaching the surface more and more gently, more and more precisely. The concept was working; what remained was to do it on a solid target the stage could actually land on and survive.

The turning point in that campaign came in April 2014, on the CRS-3 cargo mission, when a Falcon 9 first stage performed the full return sequence and, for the first time, brought itself down to the ocean's surface in a genuinely controlled vertical descent, its engine firing, its speed nulled almost to zero at the waterline. The stage did not survive — the Atlantic swell knocked it over and broke it apart before any ship could reach it, and the recovery was only ever meant to prove the maneuver, not to salvage the hardware — but the telemetry told the story that mattered. A booster returning from the edge of space had flown itself to a soft, upright, essentially pinpoint touchdown; it had simply done so on water, which no rocket can stand on. Musk was blunt about what it meant and what it did not: the data showed the stage had "landed," in every sense but the existence of something solid beneath it. From that point the problem was no longer whether a returning stage could be flown to a gentle stop but only where it could be caught.

Why go to sea at all, rather than always fly the stage home to a pad on solid ground? The answer is a matter of fuel arithmetic, and it is worth working through, because it shaped everything. To turn around and fly all the way back to the launch site, a first stage must cancel its considerable downrange speed and then reverse it — a "boostback burn" that swallows a large share of the propellant it might otherwise have spent lifting payload. For a light satellite bound for a low orbit, the rocket has fuel to spare and can afford the round trip. But for a heavy satellite headed to a high orbit, the first stage burns nearly all its propellant just doing its primary job, and has nothing left to fly home; its only hope of survival is to keep coasting downrange and come down far out to sea, where a much smaller burn can set it on a nearby target. A drone ship parked hundreds of kilometers out in the ocean, directly under the stage's natural path, was therefore not a gimmick but a necessity: without it, reuse would have been confined to the lightest missions, and the heavy, lucrative launches that made up much of the business would have gone on throwing their boosters away.

For landings far downrange, where a stage did not have the fuel to fly all the way back to the launch site, SpaceX needed a target in the middle of the ocean. Its solution was one of the more memorable pieces of engineering in the whole saga: the autonomous spaceport drone ship, a converted barge fitted with powerful thrusters that let it hold its position precisely in the open sea, its deck painted with a landing bullseye, onto which a returning first stage was supposed to descend and land. Musk gave these vessels names borrowed from the novels of the Scottish science-fiction writer Iain M. Banks, whose sprawling "Culture" series he loved — "Just Read the Instructions" and "Of Course I Still Love You," the names of vast, sardonic, sentient starships that appear in the novel "The Player of Games." Banks had died in 2013, and the names were meant as a posthumous homage, a private tribute folded into the hardware; they became part of the company's lore. The vessels themselves were ordinary ocean-going deck barges, the kind that haul cargo up and down the coast, rebuilt for the purpose: a landing platform welded across the deck, blast shielding added, and — the crucial modification — four powerful diesel-driven thrusters bolted on so that the barge could hold its position in the open sea to within a few meters, keeping the bullseye steady even as the swell tried to shove it around. The drone ships were positioned hundreds of kilometers downrange, in the path of the descending stage, and the rocket was tasked with finding this station-keeping ship in the vast ocean and setting itself down on a deck only a little larger than the stage itself. It was, as Musk acknowledged, an absurdly difficult thing to attempt — the stage arriving from eighty kilometers up, still traveling at something like ten times the speed of sound when the grid fins bit into the air — and the first attempts failed in ways that were both instructive and, in the videos, darkly comic.

The first true drone-ship landing attempt came on the tenth of January 2015, on the CRS-5 cargo mission. That the stage found the ship at all, out in the Atlantic, was itself an achievement; but it arrived too hard and at an angle, having run its grid fins dry of the hydraulic fluid that steered them in the final phase of descent, and it struck the deck and was consumed in a fireball. Musk described the result with rueful economy: the rocket had "landed hard," he wrote, but the ship itself was fine, and only some of the support equipment on the deck would need to be replaced. A few months later, on the CRS-6 mission of the fourteenth of April 2015, another attempt came achingly close. The stage descended to the ship, touched down, and stood upright on the deck for a moment — success seemingly in hand — before it became clear it had arrived with a little too much sideways momentum. A throttle valve had stuck, Musk later explained, and the control system could not react fast enough to null out the lateral motion; the tall stage overbalanced, toppled, and broke apart in a slow, almost mournful deflagration that

SpaceX, characteristically, released to the world on video.

Each of these failures was a step forward, and this is the point most easily missed by anyone watching the explosions and concluding that SpaceX could not land a rocket. The stages were arriving at the target, executing the landing sequence, coming down in the right place; the failures were in the final seconds, in details that could be fixed — a fluid reserve too small, a valve that stuck. SpaceX was not failing to solve the problem; it was converging on the solution, closing the gap between near-miss and success with each flight. The company understood this, and communicated it, framing the explosions not as embarrassments but as data, as the visible evidence of a hard thing being learned. It was a framing that some found refreshing and others found like spin, but it was, in the engineering sense, accurate. And the candor was, by the standards of the industry, genuinely radical. Traditional aerospace treated a failure as a scandal to be managed, its footage locked away, its details parceled out grudgingly to investigators; a rocket blowing up was an institutional humiliation. SpaceX did the opposite, releasing its crashes to the world, and in time it would go so far as to stitch its worst moments together into a blooper reel — a cheerfully edited compilation of boosters toppling, exploding, and cartwheeling into the sea, set to jaunty music and titled, with self-deprecating swagger, a guide to how not to land an orbital rocket. To broadcast one's failures as entertainment was unheard of in a business built on the appearance of infallibility, and it did something subtle and important: it reframed the explosions, in the public mind as well as the company's own, as the natural cost of learning fast rather than as evidence of incompetence, and it made everyone who watched a partner in the eventual triumph. Every "rapid unscheduled disassembly" retired a specific unknown; the envelope of what the team understood was expanding with each fireball, even as the public saw only a rocket blowing up on a barge.

The landings were coming. The only question was when the

last piece would fall into place — and whether it would happen on the pitching deck of a ship at sea or on solid ground, where SpaceX had also begun preparing a landing zone a few kilometers from the Cape's launch pads. The sea, in a sense, was the harder target: a barge that heaves and rolls with the swell is a crueler place to balance a returning rocket than a slab of concrete that stays put. Consider what the guidance had to accomplish. A stage that began its fall some eighty kilometers up, still traveling at many times the speed of sound, had to steer itself across that entire descent and arrive, a couple of minutes later, upright and nearly stationary on a spot a few meters wide — a precision, over such a distance and from such speed, comparable to throwing a dart the length of a football field and hitting not the board but a chosen number on it, in a crosswind, with the board itself sliding about on the ocean. That the early attempts came as close as they did — finding the ship, reaching the deck, standing up for a heartbeat before some last small fault undid them — was, rightly understood, astonishing; the failures were failures only against a standard of perfection that nothing in the history of rocketry had ever met. It made a certain sense, then, that the first success would come not on the water but on land. In December 2015, on a cold Florida night, the question was answered, and a rocket came home standing up.

Chapter 15: The Software Rocket

There is a way of describing SpaceX that its engineers sometimes used and that captures something the outside world often missed: it was, in a real sense, a software company that happened to build rockets. This was an exaggeration, but a revealing one. The company's culture, its methods, and much of its advantage came from treating problems the way software people treat them — as things to be iterated rapidly, controlled by code, and improved continuously — and this software sensibility, applied to the hardware-heavy, failure-averse world of rocketry, was one of the deep sources of SpaceX's difference.

Consider the act of landing a rocket. The mechanical feat — the engines, the grid fins, the landing legs — is impressive, but the truly hard part, in a sense, is the control: the split-second calculations, thousands of times a second, that keep a tall, thin, nearly empty stage balanced on a column of thrust as it descends from the edge of space, correcting for wind and error, steering toward a target a few meters wide, all autonomously, with no human in the loop. That is a software problem — a problem of sensing, computation, and control — and SpaceX's strength in software was central to solving it. It is worth being concrete about how the rocket actually thinks. The Falcon 9's flight software is written largely in C and C++ and runs not on one computer but on three, mounted in triplicate aboard the vehicle, each independently computing what the rocket should do next. On every decision the three compare their answers, and only a result on which they agree is allowed to command the engines and the grid fins — a voting scheme, borrowed from the world of fault-tolerant computing, that lets the rocket keep flying even if a stray cosmic ray or a failing chip corrupts one of the three. Beneath the flame and the steel, a landing is that quiet arithmetic of three computers agreeing, hundreds of times a second, on how to stand a falling tower upright.

There is a deeper heresy buried in that architecture. Traditional spacecraft flew on radiation-hardened processors — special chips, engineered to shrug off the cosmic rays and charged particles that can flip a bit of memory in orbit, and costing a fortune precisely because they were made in tiny quantities for a tiny market. The penalty was that a rad-hardened processor was typically many years, sometimes more than a decade, behind the chips in an ordinary laptop; spacecraft flew on computers that would have embarrassed a desktop of the era. SpaceX made the opposite bet. It used cheap, modern, commodity processors of the kind found in ordinary computers — fast and powerful, but vulnerable to radiation — and defended against their vulnerability not with exotic silicon but with that scheme of redundancy and constant cross-checking, running the calculation three times over and discarding any answer the three did not share. If a stray particle corrupted one processor's result, the disagreement was caught and the corrupted value thrown out; the rocket simply carried on. It was a characteristically software way to solve a hardware problem — buy ordinary parts and make them reliable through architecture and code rather than through costly, slow-moving custom engineering — and it gave SpaceX's vehicles far more raw computing power, for far less money, than the traditional approach allowed. The engine-out save on that early cargo flight, when the rocket lost an engine and its flight computer coolly recomputed the entire ascent in the time it takes to blink, was a glimpse of what that abundance of cheap computation bought: a rocket that could think its way out of trouble.

The software sensibility extended to how SpaceX built and improved its rockets. Traditional aerospace, shaped by the extreme cost of failure and the impossibility of repairing a spacecraft once launched, had evolved a culture of exhaustive upfront design, meticulous documentation, and cautious, slow change — get everything right before you fly, because you fly only once and failure is catastrophe. Software culture was different: build something, test it, see how it fails, fix it, and iterate rapidly toward a working product, treating failure not as disgrace but as information, and improvement as continuous rather than final. SpaceX brought this iterative, build-test-fail-improve approach to rockets, and it was the source of both its speed and its willingness to accept the spectacular failures that punctuated its history. The prototypes that exploded at Boca Chica, the boosters that crashed on drone ships, the Starships lost in the sky — these were, in the software framing, not disasters but iterations, each failure a step toward a working design, each explosion a source of data. This approach was possible only because SpaceX had driven down the cost of its hardware — cheap steel prototypes, mass-produced engines — so that it could afford to fail, but it was rooted in a software mentality that saw rapid iteration through failure as the fastest path to success.

Nowhere was this method carried to a fuller extreme than in the development of Starship, the enormous next- generation vehicle SpaceX built on the flat coast of South Texas. There the company put up prototype after prototype in cheap stainless steel and flew them knowing many would be destroyed, learning from each spectacular failure the way a software team learns from a crashed build — an early prototype crumpling under pressure, others exploding on landing or tearing apart in the sky, each loss feeding the next design. To a traditional program, losing a series of full-scale rockets would have been an unthinkable catastrophe; to SpaceX it was the plan, the fastest way to converge on a working vehicle, and by the middle of the decade the approach had carried Starship from a run of fiery prototypes to a cadence of ever-more-successful test flights. The build-test-fail-improve method that had first been proven landing Falcon 9 boosters had become the company's whole way of building, applied at ever-larger and more ambitious scale.

The reliance on software also showed, unmistakably, in the design of the vehicles themselves — and nowhere more than in the spacecraft that carried human beings. When astronauts first climbed aboard Crew Dragon, they found not the dense thicket of switches, dials, and circuit breakers that had defined every crewed spacecraft since Mercury, but three large touchscreens and a small bank of physical buttons reserved for the most critical actions. The interface behind those screens was, remarkably, built with the ordinary tools of the modern web — a Chromium browser running a JavaScript application — talking to flight-control software written in C++ atop the open-source Linux operating system. The capsule flew and docked itself, autonomously, with the crew as monitors rather than pilots; but the same touchscreens let them take manual control if they had to, nudging a multi-ton spacecraft toward the station with swipes and taps. To a veteran of the Space Shuttle, whose cockpit bristled with more than a thousand switches, it was a startling inversion: the spacecraft as an app, the pilot as a supervisor of code. The astronauts who first flew Crew Dragon were themselves men who had flown the Shuttle, trained for years on its dense analog panels, and they had to learn a wholly new relationship to their vehicle — one in which the machine flew itself and their job was mostly to watch it do so, intervening by touchscreen only if something went wrong. Some of them admitted to a certain unease at first, a test pilot's instinctive discomfort at surrendering the stick to software; but the capsule flew its rendezvous and docking so smoothly, so precisely, that the unease gave way to something closer to respect. The cockpit of the future, it turned out, looked less like an aircraft's and more like a browser tab. Starlink, too, was an enormous software system as much as a hardware one, its thousands of satellites orchestrated into a coherent, self-managing network, each one flying itself and dodging debris on instructions from the ground. It is worth dwelling on the scale of that, because it is the largest expression of the software sensibility the company ever built. Starlink grew into a constellation of many thousands of spacecraft — more satellites than all the rest of the world had launched in the whole history of the space age — and no room full of controllers could possibly fly them one by one. Instead the constellation runs itself, largely as a piece of software: each satellite computes its own maneuvers, autonomously steering clear of debris and of other spacecraft thousands of times a day, while the network as a whole routes internet traffic among orbiting nodes and ground stations in the manner of a distributed computer system rather than a fleet of individual craft. Even the dish on the customer's roof is not really an antenna in the old sense but a flat panel of electronics that steers its beam by software, with no moving parts, tracking satellites that streak overhead by doing arithmetic rather than by turning to follow them. A telecommunications system of that size, managed that way, would have been inconceivable to the aerospace culture that treated each satellite as a precious, hand-tended individual; it was possible only to a company that thought of spacecraft as nodes in a program. And the rockets flew on software that was updated and improved continuously, so that a Falcon 9 could fly better on its hundredth flight than its first not only because the hardware had improved but because the code had learned. SpaceX treated its vehicles as platforms to be continuously improved through software, in the manner of a technology company updating its products, rather than as fixed designs frozen at the moment of their creation. The contrast with tradition here is sharp and worth naming. In the older world, a rocket's configuration was locked down long before flight and changed only through a laborious formal process, because every alteration threatened to invalidate years of analysis; the vehicle you flew was, as much as possible, the vehicle you had certified, unchanged. SpaceX instead refined the flight software between missions the way a software house ships a new release, folding in what it had learned from the last flight so that each booster flew a little smarter than the one before — a better landing algorithm here, a more graceful response to a wind gust there. The hardware improved too, of course, but the software could improve faster and almost for free, and because so much of what a modern rocket does is really a matter of sensing and deciding, an upgrade to the code could make a physically identical rocket meaningfully more capable. The vehicle was less a finished artifact than a living product under permanent revision.

It is worth pausing on a misconception that this culture invites. It is often said that SpaceX succeeded because it wrote better software, as though code were a magic solvent that dissolved the hard physics of spaceflight. In fact the software mattered precisely because it was married to deep, old-fashioned engineering competence — to the metallurgy of engines, the fluid dynamics of turbopumps, the structural design of tanks that must hold cryogenic propellant under enormous load. The touchscreen did not land the booster; the control law running behind it did, and that control law was worthless without engines that could throttle and gimbal and relight on command. What distinguished SpaceX was not software instead of hardware but the fusion of the two: the habit of treating a rocket as a system that could be sensed, computed, and refined like a program, without ever forgetting that it was also a bomb that had to be built with exquisite physical care.

There was, too, a cultural and human dimension to the software sensibility, one that shaped who SpaceX hired and how it worked. The company drew heavily on the talent and the ethos of Silicon Valley as well as of aerospace, recruiting software engineers and people comfortable with the fast-moving, iterative, failure-tolerant culture of the technology world, and blending them with the traditional disciplines of rocketry. This blend — of aerospace rigor and software speed, of hardware craft and iterative improvement — was distinctive, and it was not always comfortable; the two cultures had different instincts about risk, documentation, and the acceptability of failure, and reconciling them was part of what made SpaceX's culture intense and demanding. But the blend was also, at its best, extraordinarily productive, marrying the deep engineering competence of aerospace with the speed and adaptability of software, and producing an organization that could build hardware with a velocity the traditional industry could not approach.

The deeper lesson is that SpaceX's transformation of spaceflight was not only a matter of better hardware but of a different way of thinking about hard engineering problems — iterative, software-inflected, failure-tolerant, continuously improving. This way of thinking, imported from the technology world and applied to the most demanding of physical challenges, was one of the company's genuine innovations, and it has begun to spread beyond SpaceX, influencing how a new generation of engineers and companies approach the building of complex hardware — electric-car makers, drone builders, rocket startups founded by SpaceX alumni who carried the method out with them. The "software rocket" — the vehicle designed, controlled, and continuously improved through code, built by an organization that iterated through failure rather than trying to avoid it — was, in its way, as important an innovation as the reusable booster, and the two were deeply connected, for it was the software sensibility, as much as anything, that made the mastery of reusability, and of everything that followed from it, possible. To understand SpaceX is to understand that it built rockets like software, and that this, as much as any single technology, was the secret of its speed.

Chapter 16: The Night the Rocket Came Home

On the evening of December 21, 2015, a Falcon 9 stood on a pad at Cape Canaveral carrying eleven small communications satellites for a company called Orbcomm. It was, in one sense, an ordinary commercial mission, the kind SpaceX now flew regularly. But it carried an extraordinary hope. This flight would be the first attempt to bring a Falcon 9 first stage all the way back to a landing on solid ground, at a site SpaceX had prepared a few kilometers from the launch pad and named, with deliberate echoes of the Apollo era, Landing Zone 1 — a patch of the Cape that had once been Launch Complex 13, where Atlas rockets flew in the early space age, now repurposed to catch rockets rather than launch them. After the string of near-misses on the drone ships, after the toppled stages and the fireballs, SpaceX would try to set a returning booster down not on a pitching barge but on a fixed concrete pad on the Florida coast, standing it upright on its legs, intact and reusable. It was also the return to flight after a disaster — the CRS-7 failure that summer had grounded the fleet for nearly six months — so the launch carried the additional weight of proving that Falcon 9 was flying safely again. The rocket itself was a new, upgraded version, its tanks stretched and its propellant chilled to a denser, colder state to wring out extra performance, flying for the first time. The stakes, technical and emotional, were high.

Even the details of the countdown reflected how much the landing mattered. The flight had been the debut of a substantially upgraded Falcon 9, a version whose propellants were loaded unusually cold — the liquid oxygen chilled far below its normal temperature so that it grew denser and more of it could be packed into the same tanks, wringing out the extra performance the rocket needed to do two jobs at once: fling eleven satellites toward orbit and still hold back enough fuel to fly the first stage home. And SpaceX had allowed itself an extra day, slipping the launch from the twentieth to the twenty-first of December, a delay it explained partly in terms of improving the odds of the landing by a meaningful margin — a small but telling choice, a company willing to wait a day not to protect the payload, which was ready, but to give the experimental recovery its best chance. That the return of a booster that would ordinarily have been discarded could shape the schedule of a real commercial mission was itself a sign of how seriously SpaceX now took the thing everyone else still called a stunt.

The launch itself was flawless. The Falcon 9 rose off the pad on its column of flame, arced out over the Atlantic, and delivered its upper stage and the eleven satellites toward orbit exactly as intended. But the attention of everyone who understood what was being attempted was on the first stage, which, having done its job and separated, now began the sequence the company had practiced and failed at so many times. It flipped around. It performed its boostback burn, reversing its downrange motion and steering itself back toward the Cape. It fell through the upper atmosphere, deployed its grid fins, and guided itself down through the dark toward the coast. And then, in the final moments, its engines reignited for the landing burn, and the long slender stage descended out of the night sky, slowing, slowing, its exhaust lighting the darkness, until it settled down onto the concrete of Landing Zone 1 and stood there, upright, whole, its engines shutting off, its legs planted on the ground. The rocket had come home.

There is a small, human detail in the story of that night that has become part of its telling, and it belongs to Musk himself. He had been watching from the launch control area, but as the booster came screaming back toward the Cape he ran outside to see it with his own eyes — and what he heard, in the seconds after the landing, was the double crack of the sonic boom the returning stage had dragged down through the atmosphere behind it. Standing in the dark, hearing that concussion roll across the Cape, Musk assumed the worst: that the rocket had struck the ground and exploded, another failure to add to the year's collection. It was only when he hurried back inside and saw, on the monitors, the stage standing serene and intact on its pad that he understood what had actually happened. He ran back toward the control room, by one account not quite shouting, "It's standing up! It's standing up!"

The reaction at SpaceX was pandemonium. In the company's headquarters in Hawthorne, where hundreds of employees had gathered to watch, the control room and the surrounding floor erupted in a roar of disbelief and joy so loud and sustained that it drowned out the mission commentary. People screamed, embraced, wept, pumped their fists, chanted "U-S-A." For a moment the launch webcast, meant to be narrating the deployment of the satellites, was overwhelmed by the sheer noise of a company watching its impossible dream become real. There is a small irony in that noise, worth savoring: the rocket's actual primary mission, the delivery of eleven paying satellites to orbit, was still unfolding overhead, and the commentators trying dutifully to describe it could not be heard over the pandemonium below. The paying job had been eclipsed, in the emotions of everyone present, by the test that rode along on top of it — the clearest possible sign of where the company believed its future lay. The engineers who had spent years on the problem — flying the Grasshopper, attempting the ocean soft-landings, watching stages explode on drone ships — stood among their colleagues and understood that they had just done something that would be remembered as long as people flew to space. An orbital-class rocket had launched a payload toward orbit and then returned itself to Earth and landed, standing up, ready in principle to fly again. It had never been done before. "Welcome back, baby!" Musk tweeted, a few words that carried a decade of vindication.

Here a distinction has to be drawn carefully, because it was the subject of a small, sharp public quarrel at the time and it is still muddled today. It is often said that SpaceX was the first company to land a rocket. It was not; that honor, by a month, belonged to Jeff Bezos's Blue Origin, whose New Shepard vehicle had risen just past the hundred-kilometer edge of space over West Texas on the twenty-third of November 2015 and set its booster back down intact. When Falcon 9 landed a few weeks later, Bezos took to Twitter with a needling note of welcome: "Congrats @SpaceX on landing Falcon's suborbital booster stage. Welcome to the club!" The barb turned on that word "suborbital," and it cut both ways, because the difference between what the two vehicles had done was not a matter of degree but of kind. New Shepard had gone straight up and come straight down, reaching space but not orbit, topping out around Mach three. Falcon 9's first stage had been flung downrange at something closer to Mach ten while accelerating a heavy upper stage and payload toward the twenty-five-times-the-speed-of-sound velocity that orbit actually demands, and then had to turn around, survive a far more violent reentry, and fly itself home across a great arc of sky. Because the energy of motion rises with the square of speed, the gap between reaching space and reaching orbit is not a little larger but roughly a hundredfold larger. Both landings were real achievements, and both companies had earned their celebrations; but they were not the same feat, and conflating them flatters the easier one at the expense of the harder.

It is worth being precise about what the landing did and did not prove, because in the excitement the distinctions can blur. Landing the stage did not, by itself, demonstrate that reuse was cheap or even that this particular stage would ever fly again. That first landed booster was, in fact, never reflown; SpaceX judged it too historically valuable to risk, and after inspecting and test-firing it, the company kept it as an artifact, eventually standing it upright outside its Hawthorne headquarters, where it remains — a scorched, hundred-foot monument to the night the idea stopped being a theory. What the landing proved was that the recovery was possible: that the maneuvers worked, that a first stage could survive the return and touch down intact and undamaged enough to be, at least in principle, reusable. To confirm that last point, SpaceX did something revealing in the days after the landing: it hauled the recovered booster to a launch pad, clamped it down, and fired its engines in a static test, to see whether a stage that had flown to space and back could light again. It could — though the company noted one outer engine showed signs of trouble, exactly the kind of finding that made the difference between a rocket that had merely landed and a rocket that could be trusted to fly a second time. That was the real work reuse would demand: not the drama of the landing but the patient forensic labor of taking a flown stage apart, learning what the journey had done to it, and proving, piece by piece, that it was sound. The enormous economic promise of reuse still had to be realized through the far less glamorous work that would follow — taking landed stages, inspecting them, understanding what reflight actually required, refurbishing them as little as possible, and proving that a flown booster could fly again safely and cheaply. The landing was the beginning of that work, not the end. But it was the indispensable beginning, the moment that turned reusability from a theory the industry had dismissed into a demonstrated fact the industry now had to reckon with.

The significance rippled outward in the days and years that followed. Competitors who had confidently declared rocket reuse impractical or impossible now had to explain a video of a SpaceX booster standing on its pad in Florida. It is often forgotten how recently the consensus had run the other way; only weeks before, a rival's executives had been happy to explain why landing an orbital booster on its tail was a stunt with no economic future. The economic logic that had always underpinned Musk's vision — that recovering the most expensive part of the rocket could eventually slash the cost of launch — now had a proof of concept in hardware. Investors, customers, and governments began to reassess what SpaceX might become and what the launch market might look like if one company could fly its rockets over and over. And within SpaceX, the landing lit a fire under the entire reuse program. If a stage could be recovered, then the next questions became urgent: How quickly could you fly it again? How much refurbishment did it really need? How many times could a single booster fly? The answers to those questions, worked out over the following years, would transform not just SpaceX but the entire economics of spaceflight, and would eventually give the company a cost advantage over its rivals so large that it came to dominate the global launch market. But that dominance lay in the future. On that December night in 2015, the achievement was simpler and purer: a rocket had gone to the edge of space and walked back, and a room full of people who had been told it could not be done had watched it happen.

There is a final measure of what that night meant, and it lies in how quickly the astonishing became routine. On the twenty-first of December 2015, a rocket landing was a thing that stopped a company's work, filled a webcast with screaming, and sent its founder running into the dark convinced he had just heard an explosion. Within a few years, Falcon 9 boosters were returning from space and settling onto their pads and their drone ships so reliably that the landings scarcely rated a mention, individual boosters flying not once or twice but a dozen and then a score of times, each return a small routine miracle that no one bothered to marvel at any longer. That is perhaps the truest sign of a revolution: not the disbelief of the first success but the boredom of the hundredth. The thing the entire industry had sworn could not be done became, in less than a decade, the most unremarkable part of a launch — and the tall, scorched first stage still standing outside the Hawthorne factory is there precisely to remind everyone who walks past it that it was once impossible.

There was, of course, more to do. Landing on solid ground near the launch site was possible only for missions where the first stage had enough leftover fuel to fly back — typically lighter payloads to lower orbits. Many missions, especially the heavy satellites bound for high orbits, used so much of the first stage's fuel that it could not return to land; for those, the stage would have to be recovered downrange, on a drone ship in the middle of the ocean, the very thing SpaceX had been failing to do all year. The land landing was a triumph, but the drone ships were still waiting, still un-conquered, and until SpaceX could land reliably at sea, reuse would be limited to a fraction of its flights. The next chapter of the reusability story would be written not on the Florida coast but hundreds of kilometers out in the Atlantic, on the pitching deck of a robot barge with a whimsical name, where in the spring of 2016 a rocket would finally do the thing that had eluded it all along.

Chapter 17: Ships at Sea

The drone ship was one of those ideas that sounds like a joke until you realize the people proposing it are entirely serious. To recover a first stage that has flown too far downrange to return to the launch site, SpaceX would land it on a boat — specifically, on an autonomous barge holding its position in the open ocean, hundreds of kilometers from shore, its deck painted with a target, waiting for a rocket to fall out of the sky and stand itself up on the pitching steel. The engineering challenge was, if anything, harder than landing on solid ground. The ship, however precisely it held station, was a small, moving target in a large, moving sea, subject to swells and wind, and the returning stage had to find it, aim for it, and touch down on a deck only modestly larger than the rocket's own footprint, all while the deck itself rose and fell with the waves. It was, Musk conceded, an almost comically difficult thing to attempt, and the early attempts had produced the spectacular failures already described: the stage that arrived and exploded, the stage that landed and tipped over.

Even the vessels themselves carried a whiff of the absurd, and the whimsy was deliberate. The barges were not called "landing platform one" or "recovery vessel A" but christened, at Musk's insistence, after sentient starships from the science-fiction novels of the late Scottish author Iain M. Banks — "Just Read the Instructions" and "Of Course I Still Love You," names lifted from Banks's 1988 novel The Player of Games, in which vast, godlike artificial minds pilot planet-sized ships across the galaxy and answer to exactly such droll appellations. Banks had died in 2013, and Musk, a devoted reader of his Culture novels, meant the names as a posthumous homage; a later addition to the fleet would be dubbed "A Shortfall of Gravitas," a nod to a ship in another Banks book. It was a characteristic gesture — a hundred-million-dollar industrial program with a literary in-joke stenciled on its hull — and it captured something true about the enterprise, which was that the people building it took the work with deadly seriousness and themselves not entirely so. The first vessels were not purpose-built spacecraft platforms at all but ordinary ocean barges — flat-decked hulls of the kind that haul gravel and fuel up the Mississippi — retrofitted in a Louisiana shipyard with steel decks, station- keeping thrusters, and, after early mishaps, a steel blast wall to shield equipment from a bad landing. The deck measured roughly the length of a football field, which sounds enormous until one imagines a fourteen-story tower descending onto it from the edge of space, in the dark, with the whole platform heaving on the Atlantic.

Why bother with something so hard? Because the economics demanded it. A large share of SpaceX's commercial business consisted of launching heavy communications satellites to high orbits — geostationary transfer orbits, far above the Earth — and these missions consumed so much of the first stage's propellant that the stage had no fuel to spare for a boostback burn all the way to the launch site. For such missions, the only options were to expend the stage, throwing it away as in the bad old days, or to recover it downrange, on a ship positioned in its natural descent path far out at sea. If SpaceX wanted to reuse its boosters on its most lucrative missions — the ones that paid the bills — it had to master the drone-ship landing. It is often said that the land landings were the real breakthrough and the sea landings a lesser variation on the theme; in fact the reverse is closer to the truth. Land landings were the more dramatic achievement, the ones that drew gasps because a rocket stood upright on concrete where a rocket had never stood before, but sea landings were the more important one commercially, because they made reuse possible on the flights that mattered most to the bottom line — the heavy geostationary payloads that no return-to-launch-site profile could accommodate. The spectacle belonged to the land; the money was at sea.

It is worth pausing on just how thin the geometry was, because it explains why the sea landing resisted the company for so long. A Falcon 9 first stage returns from the edge of space falling faster than sound, and it does not slow itself with parachutes or wings but with its own engines, in a single braking maneuver — the so-called suicide burn or hoverslam — lit at the last possible moment so that the rocket arrives at the deck with its velocity hitting exactly zero at exactly the height of the landing legs. Light the engines a beat too early and the stage stops in mid-air and falls back; a beat too late and it drives itself into the steel at speed. The margin, engineers liked to say, was measured in fractions of a second and a handful of meters, and all of it had to be computed and executed autonomously by the rocket's own flight computer, with no human in the loop, onto a target that was itself sliding and tilting on the ocean. A booster that could put itself down within a meter or two of a painted bullseye after a journey of hundreds of kilometers was performing a feat of precision closer to threading a needle from across a city than to anything in the history of aviation.

That precision had been hard-won through a sequence of near-misses that read, in retrospect, like a study in failing productively. In January 2015 a stage reached the drone ship but ran out of the hydraulic fluid that steered its grid fins and came in at an angle, striking the deck and detonating. That April another arrived almost gently but with too much lateral velocity, tipped over, and was consumed in a fireball. A landing in January 2016 was very nearly perfect — the stage touched down upright and on target — only for one of the four legs to fail to lock, so that the booster leaned, toppled, and exploded a moment after arriving, a heartbreak caught on camera and shared to the world as a lesson in how much could go right and still end in flame. Each failure was mined for data, each fix folded into the next attempt, and the company published the wreckage footage rather than hiding it — a deliberate posture that treated a crash not as an embarrassment to be buried but as a step on a curve that everyone could watch bending toward success.

The breakthrough came in April 2016. On the eighth of that month a Falcon 9 lifted off from Cape Canaveral carrying a Dragon cargo capsule bound for the space station — the CRS-8 mission, hauling more than three thousand kilograms of supplies, including an experimental inflatable habitat called BEAM that would be bolted to the station to test whether a spacecraft could one day be, in effect, blown up like a balloon. The flight sent the first stage far enough downrange that recovery required the drone ship, which was stationed in the Atlantic roughly three hundred kilometers off the Florida coast under the name "Of Course I Still Love You." After the launch and separation, the first stage performed its now-familiar sequence — the flip, the burns, the guided descent through the atmosphere — and came down toward the waiting ship. This time everything held. The grid fins steered it true, the landing burn slowed it precisely, the legs deployed, and about nine minutes after liftoff the stage settled onto the deck of the drone ship and stood there, upright and stable, riding the swells in the middle of the ocean. SpaceX had landed a rocket at sea. In the company's Hawthorne headquarters the roar was deafening; engineers who had watched four previous attempts end in wreckage now watched a booster simply stand there on a barge, and the achievement completed the reusability picture. Now the company could recover its boosters both on land, for lighter missions, and at sea, for the heavier ones, which meant that reuse could be applied across essentially its entire manifest. The last technical barrier to routine recovery had fallen.

With both land and sea landings in hand, SpaceX entered a period of rapid consolidation and improvement, and the videos of returning boosters — standing on Florida concrete, riding barges through Atlantic swells — became almost routine, which was itself the point. What had been a miracle in December 2015 was, within a couple of years, an expected part of nearly every Falcon 9 flight, so ordinary that the company's webcasts would show a booster landing in one window while the mission continued in another, and viewers would cheer and then move on. This normalization of the extraordinary is one of the quiet themes of the SpaceX story: again and again, the company took something that had been impossible or unprecedented and made it routine, and in doing so changed what the world expected of spaceflight. The first landing was history; the hundredth was a Tuesday. By the middle of the following decade the drone ships had between them caught boosters some two hundred times, and the sight of a stage riding home on a barge — an image that would have been mistaken for a film effect a decade earlier — had become so unremarkable that it barely rated a mention in the press. There is no better measure of a revolution than the speed with which its wonders come to be taken for granted.

Recovering the boosters was only half the reuse equation, however. A landed stage is not a reused stage. The economic promise of reusability depended on being able to take a recovered booster, prepare it for another flight quickly and cheaply, and fly it again — ideally many times, with minimal work in between. This was where the Shuttle had failed so badly, its reuse consumed by months of costly refurbishment. SpaceX now had to prove it could do better: that a Falcon 9 booster, once landed, could be turned around and reflown without the army of technicians and the fortune in labor that had doomed the Shuttle's economics. The recovered stages were shipped back to be inspected in minute detail, their engines examined, their structures scrutinized, the company learning, flight by flight, what actually needed to be done to make a booster ready to fly again and what could be left alone. The goal, always, was to do less — to discover that the rocket was more robust than feared, that reflight required inspection more than rebuilding, that the dream of aircraft-like turnaround was achievable if the vehicle was designed and operated for it.

The first reflight of a used booster came in March 2017, and, fittingly, the stage that made the attempt was the very one that had touched down on "Of Course I Still Love You" eleven months earlier — the CRS-8 booster, recovered, inspected, and returned to the flight line. On the thirtieth of the month SpaceX launched a communications satellite for the Luxembourg operator SES on a Falcon 9 whose first stage had already flown and landed nearly a year before. The stage flew again flawlessly, delivered its payload, and — for good measure — landed again, becoming the first orbital-class rocket booster in history to fly, land, and fly a second time. It was another line crossed, another impossibility made real: not just recovering a rocket, but actually reusing one, closing the loop that the whole reusability project had been driving toward. SES, the customer willing to fly its expensive satellite on a used rocket, took a calculated risk and was rewarded with a successful mission and, reportedly, a discount. Martin Halliwell, the company's chief technology officer, was refreshingly candid about the uncertainty of it all — asked how cheap reused launch might eventually become, he admitted, "we'd like it to be as cheap as possible, but realistically where it can get to I don't know" — and he noted, tellingly, that flying on a flown booster had not raised the satellite's insurance rates "in any material way whatsoever." That detail mattered more than any marketing line, because the insurers were the hard-nosed accountants of the space business, and their indifference to the booster's history was a quiet verdict that reuse was not, in fact, more dangerous. The reflight proved that customers would accept flown boosters and that the boosters could be trusted with real payloads.

From there, the pace of reuse accelerated rapidly, and SpaceX began the work of refining the Falcon 9 into a vehicle optimized not just for occasional reuse but for frequent, rapid, cost-effective reflight — the final version of the rocket that would come to dominate the launch industry. The drone ships, meanwhile, settled into their strange vocation as the least glamorous and most indispensable vessels in the fleet, plodding out into the Atlantic before each heavy launch to sit at a computed point in the empty ocean and wait for a rocket to arrive from space — robot barges named for the starships of a dead novelist, catching fourteen-story towers on the open sea as a matter of ordinary business. You meet their descendants today every time a Falcon 9 sends a heavy satellite to orbit and its booster comes down not on Florida but on a deck a hundred kilometers offshore, a routine so reliable that the webcast graphics have a countdown clock for it. But before we follow that story of triumph, honesty requires a detour into failure, because the road to routine reuse was not smooth, and two disasters — one in flight, one on the pad — nearly derailed the whole enterprise and revealed both the fragility of the achievement and the character of the company that had built it.

Chapter 18: The Turnaround

Landing a rocket is a spectacle; reusing one is a business, and the difference between the two is where the Shuttle failed and SpaceX succeeded. To land a booster proves that recovery is possible. To make reuse pay requires something far less glamorous and far harder to sustain: an industrial system for taking a recovered rocket, preparing it for another flight quickly and cheaply, and doing it again and again, hundreds of times, without the costs of refurbishment eating the savings. This is the turnaround problem, and solving it was the real achievement, the one that transformed reusability from an impressive stunt into an economic revolution.

The nightmare that haunted the effort was the Space Shuttle, whose reusability had been undone precisely at this point. The Shuttle flew again, but preparing it to fly again took months and an army of technicians, tearing down and rebuilding the engines, inspecting and replacing thousands of fragile heat-shield tiles, refurbishing the solid boosters recovered from corrosive seawater. The scale of the labor is easy to understate. Each orbiter was covered in roughly twenty-four thousand thermal tiles, very nearly unique in shape, that had to be individually inspected and, in the hundreds, replaced between flights; the main engines were partially disassembled after every mission; the recovered solid rocket casings had to be fished out of the Atlantic, where saltwater had gone to work on them, and shipped across the country to be cleaned, reloaded, and reassembled. The refurbishment was so extensive and so costly that it devoured the savings reuse was supposed to provide, and the Shuttle ended up more expensive per flight than the expendable rockets it was meant to replace — a program that had promised airline-like operations at a fraction of the old cost and delivered neither. The lesson the industry drew was that reuse did not pay. The lesson SpaceX drew was different: that reuse did not pay if you did it the Shuttle's way, but that it might pay handsomely if you designed the vehicle and the process for cheap, fast turnaround from the start. The whole point of the Block 5 Falcon 9 was to be a rocket that could be flown again with inspection and refueling rather than rebuilding, and the whole point of the turnaround operation was to make that vision real.

The work began the moment a booster landed. A recovered stage — whether it had touched down on land at a landing zone or on a drone ship at sea — was transported back to be inspected and prepared for its next flight. In the early days, this process was cautious and thorough, as SpaceX learned what actually happened to a booster during flight and reentry, what wore, what cracked, what needed attention, and what came through unscathed. The company studied its recovered rockets obsessively, tearing them down, examining every component, building up a detailed understanding of how a Falcon 9 aged with each flight. And what it learned, over time, was encouraging: the rockets were more robust than feared, the wear less severe, the refurbishment required less extensive than the Shuttle's had been. The most revealing measure of that progress was money. In the program's early years the company reportedly spent something on the order of thirteen million dollars to ready a flown booster for its next mission; within a few years, as the process matured and the fear gave way to data, that figure is thought to have fallen toward a single million — a better-than-tenfold collapse in the very cost that had strangled the Shuttle. As confidence grew, the process streamlined. Inspections that had once been exhaustive became targeted. Components that had once been replaced were found to be reusable. The turnaround time shrank — from months, at first, toward weeks, and the ambition was to shrink it further still, toward the aircraft-like turnaround of days that was the ultimate goal.

The trajectory of that shrinking cadence can be traced in specific, almost sporting records. For years a good turnaround meant a couple of months between a booster's flights. Then, in early 2021, SpaceX flew the same first stage twice in twenty-seven days; not long after, another booster halved the gap again, flying, landing, and flying once more in just twenty-one days. Average turnaround across the fleet settled into a matter of a few weeks, with the fastest boosters coming back to the pad in under three. It is easy to let such numbers slide past, but the human meaning of them is worth holding onto: a machine that had journeyed to the edge of space, endured the fire of reentry, and fallen back to a pinpoint landing could be checked over, refueled, and sent up again before the month was out — something no rocket in history had ever approached, and something the Shuttle, for all its promise, had never come within a factor of a hundred of achieving. Individual boosters flew not once or twice but ten times, then fifteen, then twenty and beyond, each flight demonstrating that the vehicle could indeed be reused far more than originally promised, and the certified flight limit was raised again and again as the boosters proved their durability. When Musk had first unveiled Block 5 he had set the target at ten flights with no significant refurbishment between them and perhaps a hundred with periodic maintenance — a figure that struck many in the industry as fantasy, and that the fleet would, in time, quietly begin to make look conservative.

This industrialization of reuse required building infrastructure and processes that had never existed. SpaceX had to develop the facilities to receive, inspect, refurbish, and re-stack recovered boosters; the fleet of drone ships and recovery vessels to retrieve stages from the ocean and bring them home; the systems to track and manage a fleet of reusable boosters flying in constant rotation; and the manufacturing to keep producing the expendable upper stages — for the Falcon 9's second stage was still thrown away on each flight — and the components that reuse consumed. It was, in effect, the creation of an airline-like operation for rockets, a system for flying the same vehicles over and over on a regular schedule, and it demanded a kind of operational discipline and industrial capability quite different from the one-off, artisanal production of traditional rocketry. It is often assumed that reusability let SpaceX slow down its factory, since it no longer had to build a new booster for every flight; in fact the opposite pressure took hold. The bottleneck simply moved. With boosters flying again and again, the constraint on launch cadence became the upper stage, thrown away every time, and the fairings, and the sheer throughput of the refurbishment lines — so that the company found itself building second stages and Merlin engines at a pace, and running a launch tempo, that the old expendable model could never have sustained. SpaceX built it, and in building it, transformed itself from a company that launched rockets into a company that operated a reusable fleet, with all the efficiency and cadence that implied.

The booster was the largest prize, but it was not the only piece of the rocket worth saving, and the pursuit of reuse spread outward to the humbler hardware as well — most visibly to the payload fairing, the clamshell nose cone that shrouds the satellite during ascent and splits in two once the rocket is above the atmosphere. The pair of fairing halves was no trifle; it cost something on the order of six million dollars, close to a tenth of the price of an entire booster, and in the old way of doing things it was simply dropped into the sea and abandoned. SpaceX set out to recover it, and the effort produced one of the more improbable spectacles of the whole program. Each half was fitted with cold-gas thrusters to orient itself and a steerable parafoil to glide down under control, and the company dispatched fast boats — christened, with the same deadpan whimsy as the drone ships, Ms. Tree and Ms. Chief — rigged with enormous nets slung on arms above their decks, sent racing across the ocean to position themselves beneath a falling fairing and catch it out of the air before it ever touched the water. It was, as an engineering proposition, absurd, and it mostly did not work; the nets scored the occasional catch amid a long string of maddening near-misses, the fairing splashing down just beyond the rim. In the end SpaceX concluded that the nets were more trouble than they were worth and simply let the fairings land gently in the water to be fished out and dried off, having discovered — as with the boosters — that the hardware was tougher than anyone had feared, and that a fairing dunked in seawater could be cleaned, inspected, and flown again. The lesson repeated at every scale: reuse was less a matter of protecting fragile hardware than of learning, empirically, how much abuse the hardware could actually shrug off.

The payoff was the cost structure that underlay SpaceX's dominance. By reusing its boosters many times, SpaceX amortized the cost of building each one over dozens of flights, so that the marginal cost of a launch fell toward the cost of the expendable upper stage, the propellant, the refurbishment, and the operations, rather than the cost of a whole new rocket. Musk himself once put rough numbers to it, suggesting that the marginal cost of a Falcon 9 launch, once the booster was flying often, might be on the order of fifteen million dollars — against a sticker price to customers that had long hovered above sixty. Precisely how low SpaceX's internal cost per launch fell was a closely guarded number, but it was clearly far below what any competitor building expendable rockets could achieve, and it gave SpaceX a commercial weapon of extraordinary power — the ability to underprice its rivals while still making a profit, and to fly at a cadence limited not by how fast it could build new rockets but by how fast it could turn around the ones it had. The turnaround operation, unglamorous as it was, was thus the hinge on which the whole economic transformation turned. The landings got the attention, but it was the quiet, industrial work of recovering, inspecting, and reflying the boosters — over and over, faster and cheaper each time — that actually delivered the promise of reusability that the Shuttle had failed to keep.

To grasp the size of the shift, it helps to carry the aviation analogy all the way through, because it makes concrete what the abstractions of amortization obscure. Imagine an airline that built a brand-new wide-body jet for every single flight, flew it once across the ocean, and then pushed it into the sea on arrival. No fare any passenger could plausibly pay would cover the cost of a fresh aircraft per trip; air travel as we know it would not exist, and crossing the Atlantic would remain the privilege of governments and the very rich. That was, almost exactly, the economics of spaceflight for its first sixty years — a new rocket, built at vast expense, discarded after a single use. What SpaceX did was, in effect, to persuade the rocket to behave like an airliner: to fly, land, be checked over, and fly again, spreading the enormous fixed cost of its manufacture across ten or twenty or thirty missions instead of one. It is often said that reusability simply "halves the cost of launch," as though the first stage were half the rocket and saving it saved half the bill; the reality is both messier and more dramatic. The upper stage was still expended every flight, the propellant and range services and refurbishment and recovery fleet all still had to be paid for, so the saving on any one launch was bounded — but because the same booster could absorb those savings again and again, the cumulative effect over a booster's career was to drop the amortized cost of the most expensive single component toward a rounding error, which no amount of belt-tightening on an expendable rocket could ever match.

There was a broader significance in this, one that pointed toward the future SpaceX was trying to build. The whole vision of cheap, frequent access to space — of mega-constellations, of missions to the Moon and Mars, of a genuine expansion of human activity beyond the Earth — depended on making launch not a rare and precious event but a routine, high-cadence operation, like air travel. And that depended, in turn, on mastering the turnaround: on being able to fly a rocket, recover it, and fly it again quickly and cheaply, over and over. The Falcon 9's reusability, and the industrial operation that made it real, were the proof of concept for that vision, the demonstration that rockets could indeed be operated like aircraft, at least approximately, and that the cost of access to space could be driven down not by tenths but by factors. You meet the consequence of it every time a phone in a remote valley connects to the internet through a Starlink satellite that could only be in orbit because launch had become cheap enough to loft thousands of them — a constellation that would have been economically unthinkable at the old prices, and that the turnaround made routine. And it was the template for the far more ambitious reusability of Starship, which aimed to make not just the booster but the entire vehicle fully and rapidly reusable, and thereby to carry the cost revolution to its ultimate conclusion. The turnaround problem, solved for Falcon 9, was the foundation on which the whole edifice of the new space age was built — and the model that Starship would have to surpass if the grandest ambitions were ever to be realized.

Chapter 19: Fire on the Pad

Success in the launch business is always provisional, and SpaceX learned this twice in the space of fourteen months, in ways that grounded its fleet, rattled its customers, and tested whether the company's move-fast culture could survive contact with catastrophe. The two failures — one in flight in June 2015, one on the launch pad in September 2016 — are essential to an honest account, both because they were serious and because the way SpaceX responded to them reveals as much about the company as any landing.

The first came on June 28, 2015. A Falcon 9 lifted off from Cape Canaveral carrying a Dragon capsule loaded with cargo for the space station, on the mission designated CRS-7. For the first two minutes the flight was normal, the rocket climbing through the atmosphere exactly as it should. Then, about two minutes and nineteen seconds into the ascent, moments before the first stage was to hand off to the second, the vehicle abruptly broke apart, disintegrating in a cloud of vapor and debris high over the Atlantic, the Dragon capsule and its cargo lost. It was the first in-flight failure of a Falcon 9, after eighteen straight successes, and it was a shock — the more so because it came at what should have been a placid stretch of the flight, the rocket coasting smoothly upward with the hardest aerodynamic stresses already behind it. The investigation, which combed through data from more than three thousand telemetry channels along with video and recovered debris, traced the cause to a small, almost trivial component: a steel strut, one of many inside the upper-stage propellant tank, that held down a bottle of high-pressure helium. The strut had failed at a fraction of the load it was rated to bear — designed to withstand some ten thousand pounds of force, it had snapped at around two thousand, barely a fifth of its rating — allowing the helium bottle to break free inside the tank. The loose bottle and the escaping helium overpressurized the tank, which ruptured, and the rocket came apart. A part that should have held a load of several tons had given way at a small fraction of that, and a mission was lost.

What made the strut failure so unsettling was its ordinariness, and Musk was blunt about the puzzle it posed. "We tested several hundred struts," he told reporters. "On the outside they looked normal. But inside there was a problem." A later, independent NASA review drew a sharper and less flattering conclusion than SpaceX's own: it found not merely a rogue bad part but a design error, the company having specified an industrial-grade rod rather than a certified aerospace-grade one, and having trusted the supplier's paperwork instead of screening each piece under load. It is often remembered, in the tidied-up telling, that CRS-7 was simply a case of a faulty component slipping through — bad luck rather than bad practice. The truth was more instructive, and the company took the harder lesson to heart: as a rocket grows more complex and flies more often, the number of small parts that can fail multiplies, and a single weak component, anywhere in the vast assembly, can destroy everything. SpaceX responded by changing how it sourced and tested such struts, inspecting them individually rather than trusting supplier certifications, and by adding software that would, in a future emergency of that kind, command the Dragon capsule to deploy its parachutes and save itself — a capability that, the company realized to its chagrin, had very nearly been available on CRS-7 itself, since the capsule survived the rocket's breakup intact and might have been recovered had it only been programmed to try. The fleet was grounded for nearly six months while the investigation ran and the fixes were implemented, and the return to flight came in December 2015 — the very mission, as it happened, that produced the first land landing. The company had turned its return from failure into its greatest triumph to date, folding recovery and redemption into a single flight.

There was a bitter irony folded into the cargo that CRS-7 carried to its destruction. Among the supplies lost when the Dragon fell into the Atlantic was the first International Docking Adapter — a ring of hardware meant to be fitted to the space station so that commercial spacecraft could dock with it automatically, the very piece of infrastructure on which the coming era of private crewed flight would depend. The rocket that failed had been carrying, in effect, part of the doorway through which SpaceX's own future astronauts were meant to arrive, and that doorway now lay scattered on the ocean floor. A replacement had to be built and flown on a later mission, delaying the station's readiness for the crewed vehicles then in development, so that a single snapped strut rippled outward to touch not just one lost cargo run but the timeline of American human spaceflight itself. It was a small, sharp lesson in how tightly the pieces of the enterprise were coupled, and how a trivial component's failure could cost far more than the mission it destroyed.

The second failure was stranger and, in some ways, more unsettling, because it did not happen during a launch at all. On September 1, 2016, a Falcon 9 stood on its pad at Cape Canaveral, being fueled in preparation for a routine pre-launch test firing of its engines — not even a launch, just a standard ground test conducted days before the planned mission. The rocket was carrying, atop it for the test, a valuable communications satellite called AMOS-6, owned by the Israeli operator Spacecom and worth some two hundred million dollars, intended to expand internet access over Africa and, by arrangement, to serve a connectivity project of the social-media company Facebook. During the fueling, with no warning, the rocket erupted in a massive explosion, a fireball that destroyed the vehicle, the satellite, and much of the launch pad, sending a column of smoke over the Cape. No one was hurt — the pad was cleared for the hazardous fueling operation — but a rocket and an expensive satellite were gone in an instant, and the cause was, at first, a genuine mystery. Rockets are not supposed to explode while sitting still on the ground being fueled. Mark Zuckerberg, touring Africa at the moment his company's payload went up in flame, wrote that he was "deeply disappointed to hear that SpaceX's launch failure destroyed our satellite," a rare public wince from one titan at another's misfortune. Musk, for his part, would call the episode the "most difficult and complex failure" SpaceX had ever had in its fourteen years — and he did not say it lightly, for this was a man who had watched three of his first rockets fail in a row.

The investigation into the AMOS-6 explosion led SpaceX into some of the subtlest and most unexpected physics it had ever confronted, and for a time the mystery was deep enough that the company chased genuinely exotic possibilities — at one point it went so far as to ask its rival United Launch Alliance whether it could examine the roof of a nearby building, having detected an odd sound and glint that raised, however briefly, the outlandish question of whether something had struck the rocket from outside. The prosaic answer, when it came, was stranger than sabotage. The culprit lay in the interaction between the super-chilled liquid oxygen SpaceX used, the high-pressure helium bottles submerged within the oxygen tank, and the carbon-fiber-and-metal construction of those bottles — the composite-overwrapped pressure vessels, or COPVs. SpaceX chilled its propellants to unusually low temperatures to pack more of them into the tanks and boost performance, and at those extreme temperatures the liquid oxygen could, under certain conditions, freeze solid. The lining of a COPV could buckle in tiny folds, and liquid or even solid oxygen could seep into the gaps between that liner and the carbon-fiber overwrap; when the bottle was pressurized, the trapped solid oxygen, pressed against carbon fibers, could ignite through friction — a spark struck not by any flame but by the sheer mechanical intimacy of frozen oxidizer and fiber. In effect, the very technique that gave the rocket extra performance had created a rare and unforeseen failure mode, a way for the quiet act of fueling to trigger an explosion. It was the kind of deep, non-obvious problem that lurks at the edges of complex engineering, invisible until it destroys something, and finding it required painstaking detective work, including SpaceX deliberately firing projectiles at pressurized bottles to reproduce the conditions in the laboratory. Once understood, the problem could be addressed through changes to the bottles and to the fueling procedures, and the Falcon 9 returned to flight in January 2017, but the episode was a humbling reminder that even a mature, successful rocket could harbor lethal surprises.

The AMOS-6 explosion also carried a hidden cost that the fireball made vivid: it destroyed not only a rocket and a satellite but a launch pad. Space Launch Complex 40, the workhorse Cape Canaveral pad from which most Falcon 9 flights departed, was wrecked by the blast and the fire, its ground equipment, propellant lines, and the great strongback that raised the rocket all badly damaged, and it would take the better part of a year and a substantial rebuild before it could host a launch again. In the meantime SpaceX leaned on its newly refurbished pad at the Kennedy Space Center — historic Launch Complex 39A, from which the Apollo missions and many Shuttles had flown — and it was from the West Coast, at Vandenberg, that the Falcon 9 finally returned to flight in January 2017, lofting the first batch of satellites for the Iridium communications network. That the company could absorb the loss of a primary launch pad and keep flying from others within months was itself a quiet demonstration of how far it had come from the fragile outfit that had once staked its survival on a single rocket. It is sometimes said that the AMOS-6 vehicle "blew up" in the sense of a bomb going off; in fact investigators were careful to characterize it not as a detonation but as a very fast fire, a conflagration that tore the vehicle apart in seconds — a distinction that mattered technically, because it pointed the inquiry away from the engines and toward the fueling system and the frozen oxygen lurking in the pressure vessels.

The shadow of AMOS-6 fell longest over an argument that had nothing to do with cargo. To wring maximum performance from the Falcon 9, SpaceX loaded its super-chilled propellants late, in the final half hour before liftoff, a procedure nicknamed "load and go" — and once the rocket was destined to carry astronauts, that meant the crew would be strapped into their seats atop a rocket that was still being fueled, the very operation during which AMOS-6 had exploded. Members of a NASA safety advisory panel found the prospect deeply uncomfortable and argued that convention held it safer to fuel first and board the crew afterward. SpaceX countered that its approach, combined with the always-armed launch escape system, was in fact the safer choice, since the crew could blast clear at any instant. The debate ran for months and was resolved only after the redesigned pressure vessels had been tested exhaustively and the fueling procedure demonstrated over dozens of uncrewed flights — a direct, traceable line from the pad fire of 2016 to the caution that would surround the first astronauts to ride a Falcon 9. It was, in its way, the clearest proof that the company's failures were being converted into information: a rare and subtle explosion had reshaped the very procedures by which SpaceX would one day launch human beings.

The two failures grounded the fleet, delayed missions, cost the company and its customers dearly, and provoked the perennial criticism that SpaceX moved too fast, cut corners, and traded safety for speed. That criticism deserves to be taken seriously, and the failures gave it real ammunition; the NASA finding on CRS-7, in particular, suggested a company that had let a genuine design shortcut through, and a firm that flies frequently and iterates aggressively will, other things equal, encounter more failures than one that moves slowly and conservatively. But the fuller picture is more nuanced. In each case, SpaceX found the root cause, often a subtle one, and fixed it, and the Falcon 9 that emerged from the two investigations was more robust than the one that had entered them. The company's willingness to fly often meant that when problems existed, they surfaced and were caught and corrected, rather than lurking undiscovered — and it is worth noting that the AMOS-6 failure mode, buried in the interaction of subcooled oxygen and composite fiber, might well have gone undetected for years in a program that flew only a handful of times a year, waiting to strike a crewed mission instead of an uncrewed test. And crucially, both failures occurred without human beings aboard — a fact that would weigh heavily as SpaceX turned toward its next great challenge, carrying astronauts, where a failure like CRS-7 or AMOS-6 would mean not lost cargo but lost lives. The company knew that the standard for human spaceflight was different in kind, not just degree, and the hard lessons of 2015 and 2016 fed directly into the rigor it would bring to Crew Dragon — right down to the COPV redesign that NASA would scrutinize for years before it would let its astronauts sit atop a fueled Falcon 9. Failure, at SpaceX, was supposed to be information. The test of that philosophy would come when the payload was people.

Chapter 20: Flight Proven

SpaceX had a phrase for a rocket that had already flown: "flight-proven." It was a marketing term, coined to counter the instinctive worry that a used rocket must be more dangerous than a new one, and it inverted that worry with a certain logic. A brand-new rocket, the argument went, is an unknown quantity; it has never flown, and its first flight is its riskiest. A rocket that has already flown, landed, and been inspected has, in a sense, been tested in the most demanding way possible — by actually going to space and coming back — and its successful flight is evidence of its soundness. Rather than "used," which sounded shabby, SpaceX would say "flight-proven," which sounded reassuring, even superior. The instinct behind the phrase was not pure spin; it echoed a hard fact of aviation and rocketry alike, the so-called bathtub curve of reliability, in which failures cluster at the very beginning of a machine's life, when latent manufacturing flaws reveal themselves, and again at the very end, when things wear out, with a long, safe valley in between. A booster on its fifth or tenth flight was, by this reasoning, sitting comfortably in the valley — past its infancy, nowhere near its old age. Whether the framing was wholly persuasive was another matter, and early customers still needed convincing and discounts, but as the flown boosters accumulated flights without incident, the phrase came to seem less like spin and more like description. By the 2020s, flying on a flight-proven booster was not a risk customers accepted grudgingly but the normal, expected way to launch, and the boosters with the most flights were, if anything, the most trusted — a reversal so complete that a customer assigned a brand-new first stage might quietly wish for a seasoned one instead. It is worth noting how strange this would have sounded to the aerospace establishment a decade earlier, when the very phrase "reusable rocket" was, for most serious engineers, a synonym for the Shuttle's broken promises; that a used booster could become the safer, preferred option was not merely a commercial victory but a genuine inversion of the industry's settled wisdom.

The pursuit of practical reusability drove SpaceX to a final, definitive version of the Falcon 9, introduced in 2018 and known as Block 5. Where earlier versions had been landed and, in some cases, reflown, they had not been optimized for the rapid, repeated, low-cost reuse that the economics ultimately demanded. Block 5 was. It incorporated hundreds of changes aimed at making the booster more durable, more reusable, and easier to turn around: a more robust heat shield at the base to protect the engines during reentry, more capable and reusable engines, and — a small detail that captures the whole philosophy — titanium grid fins in place of the earlier aluminum ones. The original fins, cast in aluminum and coated for protection, had a habit of catching fire on the way down as they steered the booster through the searing heat of reentry, and had to be refurbished or replaced between flights; the new titanium fins, larger and forged from a metal that could shrug off the heat, came home unscathed and ready to fly again, sometimes still bearing the scorch marks of a dozen reentries like a well-used tool. There were countless such refinements, learned from the recovered stages of earlier versions and folded into the design. The goal was ambitious: Musk described a booster meant to fly ten times with no refurbishment beyond inspection between flights, and as many as a hundred times with periodic overhaul — a claim that sounded, at the 2018 unveiling, like the familiar Musk habit of promising the impossible. Block 5 was, in effect, SpaceX's answer to the Shuttle's failure — a reusable vehicle engineered from the start so that reuse would actually be cheap, so that turning a booster around would be a matter of inspection and refueling rather than months of rebuilding.

The results, as Block 5 matured over the following years, vindicated the vision more completely than even many supporters had expected. Individual boosters flew not ten times but fifteen, twenty, and beyond, each flight demonstrating that the vehicle could indeed be reused far more times than originally promised, with the company steadily raising the certified flight limit as the boosters proved their durability. By the mid-2020s, some Falcon 9 boosters had flown more than two dozen times, and by 2026 a single first stage had reached its twenty-ninth flight — one booster that had gone to space and come home twenty-nine times, a feat that would have seemed like fantasy a decade earlier, when the entire industry threw its rockets away after a single use. It is worth dwelling on that number for a moment, because it is easy to let it wash past. A traditional expendable rocket is a matchstick: struck once, spectacularly, and discarded. Twenty-nine flights from one booster means twenty-eight rockets that did not have to be built, twenty-eight sets of nine engines that did not have to be cast and assembled and tested, a whole squadron of vehicles collapsed into a single hardworking stage. The turnaround times shrank as well, from months to weeks, as SpaceX refined its inspection and refurbishment processes and built the infrastructure to handle a fleet of reusable boosters flying in constant rotation. The dream of aircraft-like reuse, while not fully realized — a Falcon 9 still required more work between flights than an airliner between hops — had come remarkably close, close enough to transform the economics of the business.

If any single machine embodied the whole improbable story, it was the booster designated B1058, and its life deserves telling because it makes the abstraction of "reuse" into something with a beginning and an end. B1058 made its debut in May 2020 carrying the first astronauts SpaceX ever flew, and for that occasion it wore, near its base, NASA's sinuous red "worm" insignia — a logo the agency had retired back in 1992 and revived for the moment — so that the booster became a small flying monument, the only one of its kind so marked. It then went to work, and it worked relentlessly: the first stage to reach a fourteenth flight, then a fifteenth, a sixteenth, a seventeenth, an eighteenth, a nineteenth, hauling satellites and Starlinks and cargo to orbit and coming home each time. Its end, when it came, was almost gentle and strangely moving. After its nineteenth landing, in December 2023, the booster was standing on the deck of a drone ship being towed back toward port when it met high winds and heavy seas, and the tower that had survived nineteen trips to the edge of space and nineteen fiery returns simply toppled over on the pitching deck and broke apart — undone at last not by the violence of spaceflight but by the ordinary fury of the Atlantic. SpaceX salvaged what engines it could, meaning to learn from them what nineteen flights had done to the hardware, wringing information even from the wreck. A single booster had flown as many missions as some entire national rocket programs manage in a decade.

And transform them it did. The cost savings from reuse, once the development investment had been recouped and the boosters were flying many times each, were enormous. SpaceX did not have to build a new, expensive first stage for every launch; it could amortize the cost of a booster over dozens of flights, paying mainly for propellant, refurbishment, and operations on each mission. This drove SpaceX's internal cost per launch far below that of any competitor still building expendable rockets, and it gave the company a commercial weapon of extraordinary power. SpaceX could offer launches at prices its rivals could not match while still making a profit, and it could fly at a cadence — a frequency of launches — that no one else could approach, because it was not limited by the rate at which it could build new rockets but only by the rate at which it could recover, refurbish, and refly the ones it had. The reusability that the industry had dismissed as impractical had become the foundation of a dominance that would, within a few years, see SpaceX launching more mass to orbit than the rest of the world's rockets combined — a single company, on some reckonings, lifting more tonnage in a year than every other launch provider on Earth, state and commercial, put together.

It is worth dwelling on the strategic dimension of this, because it is easy to treat the landings as a technical spectacle and miss their economic and geopolitical weight. Reusability did not merely make SpaceX's launches cheaper; it changed the structure of the entire industry. Launch had always been a scarce, expensive resource, rationed by cost and by the limited number of rockets the world could build each year. By driving the cost down and the cadence up, SpaceX effectively expanded the supply of launch, making it possible to put far more into orbit for far less — which in turn made possible things that had been economically unthinkable, above all the mega-constellations of thousands of satellites, of which SpaceX's own Starlink would be the first and largest. Here the argument closed into a loop that few outsiders appreciated: the cheap, reusable rocket made Starlink possible, and Starlink, by supplying a bottomless internal demand for launches, in turn justified flying the boosters often enough to drive their per-flight cost down still further — the constellation feeding the rocket that fed the constellation. The reusable rocket was thus not just a cheaper way to do the old business but an enabler of entirely new ones. It also gave the United States a launch capability so cheap and so abundant that it reshaped national-security space, commercial space, and the global competitive landscape, concentrating an extraordinary share of the world's access to orbit in a single private company — a concentration that would raise its own difficult questions, to which we will return.

There is a psychological dimension to the flight-proven story that is easy to miss beneath the engineering, and it concerns how quickly a fear can invert into a preference. When SpaceX first proposed reflying boosters, insurers, satellite operators, and government customers alike treated a used stage as a liability to be discounted for and hedged against; the burden of proof lay entirely on the rocket. Within a few years the burden had flipped. Operators began to understand that a booster on its tenth flight had, by definition, survived nine launches and nine landings without failing, whereas a factory-fresh stage had survived none, and that every flown booster was a data point in SpaceX's favor rather than a gamble against it. By the mid-2020s the company's most experienced boosters — its "fleet leaders," the individual stages pushing the flight count ever higher — were not the ones customers avoided but the ones engineers watched most closely and trusted most, precisely because they had revealed, flight after flight, exactly how a Falcon 9 ages. The certified limit on flights, which had begun life at that cautious figure of ten, was raised and raised again as the fleet leaders kept coming home intact, until the original ceiling looked less like a limit than a first guess. The word "flight-proven," coined as a defensive euphemism, had become a plain statement of fact, and then a boast.

The Falcon 9, in its mature Block 5 form, became the most reliable and most flown rocket in the world, a workhorse that launched satellites, cargo, and eventually astronauts with a regularity and a success rate that the industry had never seen. It flew dozens of times a year, then more than a hundred times a year, an astonishing cadence for a single vehicle, and it did so while landing and reusing its boosters on nearly every flight — so that the very act of a booster returning to Earth, an event that had stopped the world's breath in December 2015, had become the least remarked-upon part of a launch, a footnote to the mission rather than the headline. The bare cadence is worth pausing on, because it is the kind of number that beggars the imagination until one does the arithmetic: more than a hundred launches in a year means a Falcon 9 leaving the Earth, on average, better than once every three days, week in and week out, each one carrying its booster back for another turn — a tempo the entire rest of the world's launch industry, every nation and company combined, could not come close to matching. Much of that traffic was Starlink, the constellation the cheap rocket had made possible, and the two grew together in a spiral that fed itself. You meet the end of that spiral in the most ordinary places now: a fishing boat far past the reach of any cell tower, a farmhouse at the end of a dirt road, a disaster zone where the ground networks have failed, all of them reaching the internet through satellites that could only be up there because a rocket had learned to come home and fly again. It was, by any measure, one of the most successful launch vehicles in history, and it was the vehicle that made SpaceX not just a scrappy disruptor but the dominant force in global spaceflight. But even as Falcon 9 was conquering the market, SpaceX was building something bigger and stranger, a demonstration of raw power that would strap three Falcon 9 cores together and hurl a sports car toward Mars — a spectacle that was part serious heavy-lift rocket and part audacious piece of theater, and that announced to the world, in the most flamboyant way imaginable, that the company had arrived.

Chapter 21: Starman

On February 6, 2018, SpaceX launched the most powerful operational rocket in the world, and it did so carrying, instead of a satellite, a cherry-red electric sports car with a mannequin in a spacesuit at the wheel. The rocket was Falcon Heavy, and the payload was Elon Musk's own Tesla Roadster, dispatched on a trajectory that would carry it beyond the orbit of Mars, its sound system nominally playing David Bowie, its driver — a dummy the company named "Starman," after the same Bowie song — gazing out through the windshield at the receding Earth. It was, all at once, a serious feat of engineering, an outrageous marketing stunt, and a piece of performance art, and it captured the peculiar character of SpaceX at the moment it announced itself as a giant: technically formidable, gleefully theatrical, and entirely willing to do something absurd if the absurdity served the mission of getting the world excited about space again. "It's kind of silly and fun," Musk said of the whole enterprise, "but I think silly, fun things are important" — a sentence that could serve as an epigraph for the entire day.

Falcon Heavy was, in essence, three Falcon 9 first stages strapped together — a central core with two additional boosters attached to its sides — their combined twenty-seven engines generating more than five million pounds of thrust at liftoff, enough to make it, at its debut, the most powerful rocket flying and the most powerful launched since NASA's Saturn V had sent astronauts to the Moon nearly half a century earlier. That comparison invites a common misconception worth correcting: for all the fanfare, Falcon Heavy did not rival the Saturn V, which had produced some seven and a half million pounds of thrust and stood taller and lifted more; it was the most powerful rocket in operation, and the most powerful since Apollo, but it was not the most powerful ever built. The concept, in any case, sounded simple — take a rocket that works and bolt three of them together — and in practice it was fiendishly difficult. The three cores had to be structurally integrated to withstand the tremendous stresses of launch, their engines had to fire in concert, and the aerodynamic and vibrational environment of the clustered vehicle turned out to be far more complex than anticipated. Musk himself later admitted that Falcon Heavy had been much harder to build than expected, that the company had naively assumed it would be a straightforward combination of existing hardware and had instead had to redesign a great deal, and that at several points he had doubted it would work at all. He was so uncertain, in fact, that he told the public before the flight that he would consider it a win if the rocket merely cleared the pad without destroying it — his private odds of success, he suggested, were little better than even. The development stretched for years past its original schedule, and by the time it flew, the ever-improving Falcon 9 had grown so capable that it could handle most of the missions Falcon Heavy had originally been meant for, leaving the bigger rocket a somewhat niche vehicle for the heaviest payloads.

But if the business case for Falcon Heavy had narrowed, its debut was nonetheless a spectacle for the ages, and SpaceX played it for everything it was worth. The choice of payload was pure Musk. A rocket's first flight is risky, and customers are reluctant to put a valuable satellite on an unproven vehicle, so first flights often carry a mass simulator — a block of concrete or steel, dead weight to stand in for a real payload. Musk found the idea of launching a lump of concrete boring, and decided instead to launch something that would, in his words, make people excited, settling on his own midnight-cherry Tesla Roadster, the first car his company had made, with a spacesuited mannequin strapped into the driver's seat. The car was dressed for the occasion with a hacker's sense of humor: a copy of Douglas Adams's The Hitchhiker's Guide to the Galaxy tucked in the glovebox alongside a towel and a sign on the dashboard reading "DON'T PANIC!"; a Hot Wheels miniature of the Roadster mounted on the dash with a tiny Starman of its own at the wheel; and a plaque bearing the names of the thousands of SpaceX employees who had built the rocket, etched with the words "Made on Earth by humans." Cameras were mounted to capture the surreal images of a car floating in space against the blue curve of the Earth. It was self-indulgent, it was brilliant marketing for both his companies, and it was, undeniably, unforgettable. The launch drew one of the largest live audiences in the history of spaceflight webcasts — millions watching in real time as the countdown ran — and the crowds that had gathered along the Florida beaches and causeways to see it in person rivaled those of a Shuttle launch, a turnout no commercial rocket had ever commanded. The images of Starman cruising through the void, the Earth reflected in his visor and the dashboard sign still cheerfully warning him not to panic, became instantly iconic, shared around the world, imprinting the SpaceX brand on the global imagination in a way no conventional launch ever could. It is worth remembering, too, that the two side boosters were not new but refurbished veterans of earlier Falcon 9 flights, pressed back into service to help launch the car — so that the spectacle doubled as a quiet advertisement for reuse, the whole audacious stunt riding partly on hardware that had already been to space and come home.

The flight itself delivered a moment of pure engineering theater that rivaled the stunt. As the rocket ascended and the two side boosters completed their work and separated, they flipped around and flew back toward Cape Canaveral, descending in tandem toward two landing zones side by side

— and touched down, together, almost simultaneously, two towering boosters settling onto their pads in near-perfect synchrony, their sonic booms cracking across the Florida coast in a rapid double report that startled onlookers for miles. The image of the two boosters landing side by side, upright and identical, was so precise and so science-fictional that many first-time viewers assumed it was computer animation. It was not. It was two real rockets, having just helped launch a car toward Mars, returning to Earth and landing in unison, a demonstration of reusability and control so polished that it looked staged. The central core, aiming for the drone ship "Of Course I Still Love You" far out at sea, was less fortunate: it ran out of the igniter fluid needed to relight enough of its engines for the landing burn, so that only its center engine fired for the touchdown, and it struck the water beside the ship at some three hundred miles an hour, throwing up a plume of shrapnel that knocked out two of the ship's thrusters — a reminder, amid the triumph, that the margins remained thin. But the two synchronized landings, and the car sailing off toward the asteroid belt, made the day an unambiguous sensation.

There was one final, unscripted flourish. The upper stage was meant to coast through the radiation of the Van Allen belts and then fire a last time to send the Roadster onto a trajectory reaching out to the orbit of Mars. The burn overperformed. Instead of settling into an orbit that grazed the red planet, the car was flung onto a longer ellipse that carries it, at its farthest, out past Mars toward the inner edge of the asteroid belt, some one and a half times Earth's distance from the Sun. It is often said that the Roadster is "headed for Mars," or even that it might one day strike the planet; in fact its orbit is tilted just enough relative to Mars's own that the two can never collide, and the car is not going to Mars at all but circling the Sun on its own lonely heliocentric loop, a red sports car that has become, in effect, a small artificial asteroid. Barring some improbable gravitational nudge, it will keep making that circuit for millions of years, long after everyone who watched it launch is gone — the first automobile in deep space, and very possibly the longest-lasting artifact of the whole affair.

Curiosity about the car's ultimate fate was strong enough that a team of astronomers actually ran the numbers. Within weeks of the launch, the researcher Hanno Rein and two colleagues published a study modeling the Roadster's chaotic, planet-crossing orbit forward across the ages, and their Monte Carlo simulations produced a result at once reassuring and faintly comic: a roughly six percent chance that the car would collide with Earth over the next million years, and about a two-and-a-half percent chance of hitting Venus, with the odds slowly mounting over longer spans, so that the vehicle had perhaps an even chance of surviving intact for a few tens of millions of years before some eventual impact or the slow scouring of solar radiation did it in. There is something bracing in the thought that the likeliest long-term destination of Elon Musk's sports car is, after all, a fiery return to the planet it came from. It is worth dispelling one small myth while we are here: the Roadster's radio was often described as endlessly "playing" David Bowie into the void, and Musk did loop "Space Oddity" and "Life on Mars?" through the car's sound system during the early hours of the flight, but sound requires air to travel through, and in the vacuum of space the music plays to no one and carries nowhere — the car sails on in perfect silence, the gesture pure symbol, which was rather the point.

Beneath the spectacle, Falcon Heavy's debut carried real significance. It demonstrated that SpaceX could build and fly a genuinely heavy-lift rocket, expanding the range of missions the company could bid for, including lucrative national-security launches of the largest and most sensitive military and intelligence satellites, which required exactly this kind of capability. It further cemented the company's technical credibility and its reputation for doing things no one else could. And it served the mission Musk had articulated from the start — reigniting public enthusiasm for space — more effectively than any number of routine launches could have. The image of a sports car in orbit did something that data sheets and cost figures never could: it made spaceflight feel exciting, romantic, and open again, a domain of wonder rather than just of government contracts and communications satellites. Whatever one thought of the extravagance, it worked. A generation that had grown up regarding space as a settled, unglamorous business suddenly saw a rocket land two boosters in unison and a car floating past the Earth, and paid attention. You can meet the residue of that day even now, in the online trackers that still plot, years later, exactly where in the solar system the Roadster is — a small piece of internet folklore kept alive by people who were, briefly, made to feel that space belonged to them again. That feeling, more than any tonnage or thrust figure, was the true payload of the mission — the sense, half-forgotten since Apollo, that the frontier overhead was open and inviting rather than closed and bureaucratic.

Falcon Heavy would go on to fly a modest number of missions over the following years, handling the heaviest and most demanding payloads — a specialized tool in SpaceX's kit rather than a frequent flyer. Its second flight, in April 2019, carried a real paying customer's satellite at last, the Arabsat-6A communications spacecraft, and this time all three cores came home, the two side boosters touching down together as before and the center core riding the drone ship in — only to topple over on the deck in rough seas on the voyage back to port, an eerie rehearsal of the fate that would one day befall the fleet's most celebrated Falcon 9. From there the big rocket settled into a career of rare but consequential flights: it lofted some of the military's heaviest and most secret satellites; it hurled NASA's Psyche probe toward a distant metal asteroid in 2023; and in 2024 it sent the Europa Clipper on its long journey to Jupiter's ice-crusted moon to search for the conditions of life — the kind of flagship deep-space science mission that, a generation earlier, would have flown only on a government-built rocket. Falcon Heavy became, in other words, exactly what its debut had promised: a heavy-lifter kept in reserve for the payloads nothing else could carry. In the larger arc of the company's history it was something of a bridge: a demonstration of heavy-lift capability that pointed toward the far more ambitious vehicle already taking shape in Musk's mind, a fully reusable super-heavy rocket that would dwarf even Falcon Heavy and that was meant, at last, to make the Mars mission real. But before we follow SpaceX to the stainless-steel behemoth rising on the Texas coast, we must turn to the achievement that, more than any rocket landing or any car in orbit, changed how the world saw the company and restored a capability the United States had lost: the return of American astronauts to orbit, launched from American soil, on a spacecraft built by a private firm. For nine long years after the Space Shuttle's retirement, the world's leading spacefaring nation had been unable to launch its own people into space. SpaceX was about to end that humiliation.

Chapter 22: The National Security Launch

Of all the markets SpaceX set out to conquer, none was more tightly closed, more jealously guarded, or more strategically important than the launch of the United States' national-security satellites — the reconnaissance spacecraft, the communications and navigation and early-warning satellites on which the nation's defense depended. For years this market had been the exclusive preserve of a single provider, the United Launch Alliance, a joint venture of Boeing and Lockheed Martin formed in 2006 that held an effective monopoly on launching the government's most sensitive and valuable payloads, flying reliable but extremely expensive expendable rockets — the Atlas V and the Delta IV — under long-term contracts that guaranteed its business and its profits. The scale of the premium is worth stating plainly: government analyses put the average cost of a national-security launch under this arrangement at something approaching a half a billion dollars, several times what SpaceX said it could do the job for. Breaking into this closed market was one of SpaceX's most consequential battles, and it was fought not only in the marketplace but in the courts and the halls of Congress — a campaign that pitted a young company against two of the largest defense contractors in the world, the Air Force procurement bureaucracy, and decades of accumulated institutional habit, all of them invested, one way or another, in keeping the arrangement exactly as it was.

The national-security launch market was closed to SpaceX for reasons both legitimate and self-serving. The legitimate reasons had to do with the extraordinary stakes: the satellites in question were worth billions and were critical to national defense, and the government had every reason to insist that they be launched only by a provider it fully trusted, with a proven record of reliability. A single launch failure could destroy an irreplaceable asset and blind or deafen the nation's defenses, so caution was warranted. But the self-serving reasons were real too. The incumbent monopoly had every interest in keeping the market closed, in maintaining the certification requirements and contracting arrangements that protected its position, and in resisting the entry of a competitor that could launch the same payloads for a fraction of the price. The result was an arrangement in which the government paid enormous sums for national-security launches to a single provider, with no competition to discipline the cost, and in which a well-established set of rules made it very difficult for a newcomer, however capable, to break in.

There was, moreover, a strategic irony buried in the incumbent's rockets that SpaceX would exploit to devastating effect. The workhorse of the fleet, the Atlas V, was powered at liftoff by the RD-180, a superb engine — but a Russian one, built by NPO Energomash and imported from a country with which relations were rapidly souring. When Russia annexed Crimea in 2014 and Washington answered with sanctions, the spectacle of the United States launching its most secret spy satellites atop Russian engines became suddenly, acutely awkward, and Musk pressed the point relentlessly, arguing that the nation's defense should not depend on the goodwill of Vladimir Putin's government. It was an argument that fused SpaceX's commercial self-interest with a genuine question of national security, and it landed hard on Capitol Hill.

SpaceX challenged the whole arrangement with characteristic aggressiveness, and when the normal channels proved too slow, it went to court. In April 2014 SpaceX sued the U.S. Air Force to challenge a large, sole-source "block buy" — a bulk contract for dozens of rocket cores awarded to the incumbent without competition — arguing that it was capable of launching national-security payloads at far lower cost and that it should at least be allowed to compete for the work. Musk made the accusation personal and pointed, suggesting that Air Force officials responsible for the arrangement seemed to be angling for future jobs at the very contractors they were protecting. The lawsuit, and the broader pressure SpaceX brought to bear, forced the issue into the open. In January 2015 the two sides settled, with the former U.S. attorney general John Ashcroft serving as mediator, and the Air Force agreed to open more launches to competition and to help move SpaceX's certification along. That certification arrived later in 2015, after the Falcon 9 had demonstrated through a long record of successful flights that it met the exacting reliability standards the government demanded, and SpaceX began winning national-security launch contracts — breaking the monopoly and introducing, for the first time in years, genuine competition into the launch of the nation's most sensitive satellites. The savings to the government were substantial, and the introduction of competition disciplined the whole market, forcing the incumbent to cut its own costs and to develop a new, cheaper rocket, the Vulcan, built around an American engine, precisely so that it could shed its dependence on Moscow and survive the price war SpaceX had started.

The fight over the Russian engine did not stay confined to SpaceX's lawyers; it moved to the Senate, where it found an unlikely ally in Senator John McCain, who seized on the RD-180 as a symbol of everything wrong with the cozy arrangement — a national-security program dependent, at its heart, on the industrial goodwill of a hostile power, its costs shielded from competition. McCain pressed to cap and ultimately end the military's use of the Russian engine through the annual defense-authorization bills, and the resulting political squeeze on the incumbent gave SpaceX's commercial argument a sharp geopolitical edge it might otherwise have lacked. The convergence of a lawsuit, a price differential of hundreds of millions of dollars per launch, and a genuine national-security embarrassment proved impossible for the system to resist, and the door that had been sealed for a decade swung open.

Once through it, SpaceX moved with dispatch. In 2017 the Falcon 9 flew its first mission for the National Reconnaissance Office, the agency that operates the country's spy satellites, and in 2018 it launched the first of a new generation of Global Positioning System satellites for the Air Force — the same GPS constellation on which navigation, banking, and precision weapons the world over silently depend. The savings were concrete and enormous: where a national-security launch under the old monopoly had averaged something close to a half a billion dollars, SpaceX won GPS launches at roughly a sixth of that price, and later, having demonstrated that it could do so without loss of reliability, it even persuaded the government to let it fly some of those missions on previously flown, flight-proven boosters — reusing, on the nation's most exacting payloads, the very hardware the establishment had once insisted must be pristine. By the time the Space Force ran its next great procurement, the Phase 2 competition of 2020, the question was no longer whether SpaceX could be trusted with national-security launch but merely how the work would be divided; the answer split the missions between the two providers, with roughly sixty percent going to the incumbent and forty percent to SpaceX — a company that six years earlier had been suing for the right to bid at all.

It would be a mistake, though, to imagine that winning this market was simply a matter of underpricing the incumbent, for the technical and bureaucratic barriers were real and expensive to clear. Certification alone demanded a long series of successful flights and an exhaustive baring of the rocket's design to government engineers, who reviewed the Falcon 9 to a depth no commercial customer ever required. And some of the military's most demanding payloads imposed physical requirements that the SpaceX way had never accommodated: where the company built and integrated its rockets horizontally, laying the vehicle on its side to mate the payload before rolling it out and raising it upright, certain sensitive national-security satellites were designed to be stacked vertically and could not tolerate being tipped on their sides at all. Meeting that requirement forced SpaceX to design and build new vertical-integration infrastructure at its pads — a significant investment in facilities that served no commercial purpose whatsoever, undertaken solely to satisfy the peculiar needs of the defense customer. The lesson was that breaking into the national-security market cost SpaceX real money and real years even after the legal victory, and that its dominance, when it came, was earned against genuine obstacles and not merely handed over on a spreadsheet of prices.

The opening of the national-security market was significant not only commercially but strategically. It gave the United States a second, independent provider of launch for its critical defense payloads, ending a dangerous dependence on a single company and introducing the resilience that comes from having more than one option. It saved the government enormous sums, freeing resources for other purposes. And it demonstrated that the commercial revolution SpaceX had wrought extended even to the most demanding and most guarded corner of the launch business, that a private company could be trusted with the nation's most sensitive assets, and that competition and reusability could deliver the same strategic capabilities the government needed at a fraction of the traditional cost. It is worth noting how completely this inverted the received wisdom of the industry, which had long held that reusable, commercially developed rockets might be fine for cargo and communications satellites but could never meet the punishing standards of national-security work — the heaviest payloads, the most exotic orbits, the zero tolerance for failure. SpaceX not only met those standards; it eventually flew some of the nation's most sensitive spacecraft on flight-proven boosters, reusing the very hardware the establishment had once insisted must be pristine. As SpaceX proved itself, it won an ever-larger share of national-security launches, becoming an indispensable provider of the government's access to space for defense as well as civilian purposes, and cementing its position at the center of American spaceflight.

There was, of course, a shadow side to this success, one that grew more prominent as SpaceX's dominance of national-security launch deepened. The very achievement of becoming an indispensable provider of the government's most critical launches created a new dependence — on SpaceX itself — that mirrored the old dependence on the monopoly it had displaced. As SpaceX came to launch a large and growing share of the nation's defense payloads, the government found itself relying heavily on a single private company, controlled by a single individual, for a strategically vital capability, precisely the kind of concentration that the opening of the market had been meant to remedy. The introduction of competition had broken one monopoly, but SpaceX's overwhelming cost and capability advantage threatened, over time, to create another, with SpaceX in the place the incumbent had once held. And this dependence was compounded by the concerns, examined elsewhere in this book, about the concentration of so much critical capability in one individually controlled firm — concerns sharpened by the fact that the same individual also controlled Starlink, a communications network that had itself become woven into military operations, so that a single person's decisions could bear on both how the nation's satellites reached orbit and how its forces communicated once there. The government, aware of the risk, worked to maintain competition and to cultivate alternative providers — the incumbent's new Vulcan, and eventually Blue Origin's New Glenn — precisely to avoid becoming as dependent on SpaceX as it had once been on the monopoly SpaceX had broken.

Underlying the whole contest was a doctrine the national-security establishment held almost sacred, known as assured access to space — the principle that the nation must never be left unable to launch its most vital satellites because a single rocket had been grounded by a failure. For decades this doctrine had been the incumbent's best argument, the justification for paying a premium to keep two distinct rocket families, the Atlas and the Delta, alive under one roof. SpaceX turned the doctrine against its authors. The surest guarantee of assured access, it argued, was not a single protected provider flying two expensive rockets but genuine competition between rival companies, each with its own vehicles, each disciplined by the other's prices and kept honest by the other's existence. When a Falcon 9 anomaly or a Vulcan delay grounded one provider, the other could carry the load — and the government, for the first time in a generation, had somewhere else to turn. The argument proved persuasive precisely because it aligned the nation's strategic interest with the introduction of the competitor the incumbent least wanted, and it recast SpaceX not as a risk to assured access but as its guarantor.

The story of the national-security launch market thus captures, in miniature, the larger arc of SpaceX's rise and the ambivalence it provoked. SpaceX broke open a closed, expensive, monopolized market, introducing competition, slashing costs, and giving the nation a capable new provider of its most critical launches — a genuine public good, achieved through a combination of engineering excellence, commercial aggressiveness, and legal persistence. But in doing so, it rose toward a dominance that recreated, in a new form, the very concentration it had displaced, making the nation dependent on a single company for a vital strategic capability and raising anew the concerns about concentration and control that shadowed all of SpaceX's success. It was a pattern that would repeat across the whole of the new space age: SpaceX, by being better and cheaper than anyone else, won market after market, and in winning them, accumulated a dominance that was both a triumph and a danger — a source of enormous value and enormous risk, indispensable and worrying in equal measure, and nowhere more so than in the domain of national security, where the stakes were highest and the concentration most fraught.

Chapter 23: The Gap

In July 2011, the Space Shuttle Atlantis touched down at the Kennedy Space Center, and with it the United States' ability to launch human beings into space came to an end. It was a strange and, for many Americans in the space community, a painful moment. The nation that had put the first men on the Moon, that had built and flown the Shuttle for thirty years, that had led human spaceflight for half a century, would now have no way to send its own astronauts to orbit. The Shuttle was retired — its aging fleet increasingly costly and, after two fatal accidents that had killed fourteen astronauts, increasingly hard to justify — with no American replacement ready to take its place. For the foreseeable future, the only way for American astronauts to reach the International Space Station, the enormous orbiting laboratory the United States had spent tens of billions of dollars to build and lead, would be to buy seats aboard Russian Soyuz spacecraft, launched from the steppe of Kazakhstan, at the pleasure and the price of a foreign government.

There was a quiet finality to that last landing that those who witnessed it did not soon forget. Atlantis came home before dawn, its final mission — STS-135, a four-person crew sent up to stock the station against the lean years everyone knew were coming — ending in darkness on the Kennedy runway, and as the orbiter rolled to its stop its commander, Christopher Ferguson, radioed a benediction of sorts, noting that the vehicle had earned its place in history and that, with its wheels stopped, an era had ended. Thousands of the workers who had spent their careers on the program stood watching, many of them about to lose their jobs, and the scene had the air of a retirement and a wake at once. For all its flaws, the Shuttle had been a national institution for three decades, the machine on which two generations of Americans had learned to picture spaceflight, and its passing left not only a capability gap but an emotional one — a sense, keenly felt in the space towns of Florida and Texas, that the country had let something slip through its fingers without a replacement in hand.

This period came to be known simply as "the gap," and it lasted nine years. During that time, every American astronaut who went to the space station flew on a Russian rocket, in a Russian capsule, under the control of Russian mission planners, having trained in the closed city of Star City outside Moscow and learned enough Russian to operate a Soyuz — for the spacecraft's displays and checklists were in Russian, and a non-Russian-speaking astronaut was a hazard to the crew. The United States paid Russia for the privilege, and the price rose steadily and inexorably over the years, from around twenty million dollars per seat in the early days to, by the end of the gap, something on the order of ninety million dollars for a single ride — a sum that reflected both inflation and Russia's shrewd awareness that it held a monopoly on the only service in town and could charge whatever the traffic would bear. The arrangement was, on one level, a remarkable testament to international cooperation in space, a partnership that endured even as relations between the two countries deteriorated on Earth, and there is something genuinely moving in the fact that American and Russian crews kept flying together, and depending on one another for their lives, through years when their governments were barely on speaking terms. But it was also, for the United States, a strategic vulnerability and a national embarrassment. The world's leading space power could not launch its own people, and depended for its access to its own space station on a geopolitical rival whose cooperation could not be guaranteed.

The dependence was not merely theoretical, and it was thrown into humiliating relief in 2014. As Washington imposed sanctions on Moscow over the annexation of Crimea, Dmitry Rogozin, the Russian deputy prime minister overseeing the space industry, taunted the Americans with a suggestion that, if they disliked Russian policy, they might ferry their astronauts to the station "by means of a trampoline." The jibe stung precisely because it was true: for the moment, the United States had no answer, and both sides knew it. There was no American rocket standing by, no capsule waiting on a pad; the world's foremost spacefaring nation could do nothing but keep writing checks to Moscow and hope the seats kept coming. That it was a Russian official mocking American helplessness in the one arena — spaceflight — where America had for decades been supreme made the barb land all the harder, and it would be remembered, and repaid, six years later, when an American rocket finally carried an American crew to orbit and Musk could not resist observing, dryly, that "the trampoline is working."

The decision to retire the Shuttle had not been made lightly, nor was it a sudden loss of nerve; it was the delayed consequence of catastrophe. In February 2003 the orbiter Columbia disintegrated over Texas during reentry, killing all seven aboard, the second time in seventeen years that the program had lost a crew and a vehicle. The investigation that followed concluded not only that a piece of foam had doomed Columbia but that the Shuttle was, at bottom, an aging and inherently risky design that the nation had come to operate with a dangerous complacency. In response, the administration of George W. Bush announced in 2004 a new direction: the Shuttle would be flown only long enough to finish building the space station and then retired, and a new generation of vehicles would be developed to carry astronauts beyond low Earth orbit. The retirement, in other words, was scheduled years in advance and for sound reasons of safety and cost — the fault lay not in ending the Shuttle but in the failure to have its successor ready in time, a failure compounded when an independent review panel convened in 2009 found the follow-on Constellation program so far over budget and behind schedule as to be effectively unexecutable within any plausible funding. The gap, then, was less an accident than a slow-motion policy failure, foreseen by many and prevented by no one.

The financial tally of the dependence, when it is added up, is startling. Across the years of the gap the United States paid Russia something approaching four billion dollars for rides to a station it had largely built and paid for — a steady transfer of hard currency to the space program of a strategic rival, at prices that rival set and raised at will. Every dollar of it was, in a sense, a measure of American self-neglect: money spent not on any new capability but simply on renting access to orbit because the country had allowed its own means of getting there to lapse. And the vulnerability was not merely financial. A single Soyuz mishap, a diplomatic rupture, a decision in Moscow to withhold seats, could at any moment have stranded American astronauts on the ground or, worse, aboard the station with no ride home — a dependence on the goodwill of a government with which Washington was, through much of the period, openly at odds.

The gap had its roots in decisions made years earlier, and in the perennial difficulty of replacing a spacecraft as complex and as entrenched as the Shuttle. The program meant to succeed the Shuttle, an ambitious effort called Constellation that aimed to return astronauts to the Moon with a new capsule and a pair of new rockets, had run over budget and behind schedule and was largely cancelled around 2010, leaving a hole where the next American human-spaceflight vehicle should have been. Into that hole stepped a policy idea that was, at the time, genuinely radical: rather than have NASA design, build, and operate its own new crew vehicle in the traditional way — a process that would take many years and cost many billions — the agency would do for crew what it had begun doing for cargo, turning to private companies to develop spacecraft and then buying rides as a customer. This was the Commercial Crew Program, and it extended to human spaceflight the same philosophy that had produced Dragon and the cargo flights: fixed-price contracts, private ownership of the hardware, NASA as a purchaser of services rather than an owner of vehicles. It was controversial. Many in Congress and in the traditional aerospace world were deeply skeptical that private companies could be trusted to safely carry human beings, the most precious of payloads, and Congress underfunded the program in its early years, appropriating well below what NASA requested — a shortfall that, by the agency's own reckoning, added years to the gap and hundreds of millions of dollars in Soyuz payments to Russia, a self-inflicted wound whose bill was paid in rubles.

The path to the two 2014 contracts ran through several years of gradual winnowing, for NASA did not simply hand the work to SpaceX and Boeing out of the blue. Beginning around 2010 the agency seeded a wider field of contenders with small development awards, and at various stages the competition included not only SpaceX's Dragon and Boeing's Starliner but also Sierra Nevada Corporation's Dream Chaser — a graceful little winged spaceplane, a miniature echo of the Shuttle, that would have landed on a runway rather than splashing into the sea. When the final selection came down to two, Sierra Nevada was left out, and the company protested the decision, briefly halting the program while the government auditors reviewed it; the protest was denied, and the Dream Chaser was eventually redirected to fly cargo instead of crew. The episode was a reminder that the outcome we now take as inevitable — SpaceX and Boeing, the upstart and the giant — was in fact one branch of several that history might have taken, and that a spaceplane came within a whisker of carrying American astronauts back to orbit instead of a capsule.

The stakes of the Commercial Crew Program were enormous, and not only for national pride. If it succeeded, it would restore American human spaceflight, end the dependence on Russia, and prove that the commercial model could handle even the most demanding and safety-critical missions — carrying astronauts. If it failed, or if a commercial crew vehicle killed its passengers, it would discredit the entire approach and set human spaceflight back for years. NASA, aware of these stakes, structured the program to have two competitors, following the sound principle that redundancy protects against the failure of any single provider. In September 2014, the agency made its choices: it awarded a contract worth about four and two-tenths billion dollars to Boeing, the storied aerospace giant, to develop a crew capsule called Starliner, and a contract worth about two and six-tenths billion to SpaceX, the upstart, to develop a crewed version of its Dragon capsule. Boeing received the larger sum — some sixty percent more for the same job — a reflection of its stature, its long history with NASA, and, perhaps, the agency's greater confidence in the established firm over the disruptor. Few observers at the time doubted that Boeing, with its decades of experience building spacecraft, would be the more reliable partner. There was a certain logic to the confidence: Boeing and its heritage companies had built spacecraft since the dawn of the space age — the Apollo command and service modules, much of the Shuttle, great portions of the space station itself had passed through those hands — whereas SpaceX had never carried a human being and had, only recently, been losing rockets. The conventional wisdom held that SpaceX was the riskier bet, more likely to stumble, and that Boeing would set the standard; some in Congress had questioned whether the scrappy newcomer belonged in the competition at all, and to bet on it to fly people first would have struck almost any sober analyst of the day as reckless.

The conventional wisdom, as it turned out, had it almost exactly backward, and the divergent fortunes of the two Commercial Crew providers would become one of the most instructive stories in modern aerospace — a natural experiment, almost, in the merits of the old way and the new, with the same requirement, the same customer, and the same destination handed to two very different companies at the same moment. But that divergence lay in the future. In 2014, what mattered was that the program was underway, that the money and the mandate existed to end the gap, and that SpaceX now faced its most demanding challenge yet. Carrying cargo was one thing; a failure meant lost supplies and a bad day. Carrying astronauts was another entirely. A crewed spacecraft must protect its occupants against every conceivable failure, must be able to abort and carry its crew to safety at any moment from the pad to orbit, must maintain a survivable environment, must be tested and scrutinized and certified to a standard far beyond anything required for cargo. The margin for error, always thin in spaceflight, shrank to nothing when lives were aboard. SpaceX, a company that had built its culture on moving fast and tolerating failure — that had, only recently, lost two rockets in fourteen months — now had to prove that it could also move carefully, that it could meet the exacting standard of human spaceflight without losing the speed and ingenuity that made it special. The development of Crew Dragon would test whether the SpaceX way could be reconciled with the demands of carrying people — and, on a spring day in 2020, it would end the gap and return American astronauts to space from American soil.

Chapter 24: Commercial Crew

Building a spacecraft to carry human beings is a fundamentally different undertaking from building one to carry cargo, and the difference is not merely one of adding seats and life support. It is a difference of philosophy, of standards, of the entire posture an engineering organization must take toward risk. A cargo spacecraft can fail; the cost is money and inconvenience. A crewed spacecraft that fails can kill, and the history of human spaceflight is punctuated by the names of those it has killed — the crews of Apollo 1, of Challenger, of Columbia, of Soyuz 1 and Soyuz 11. Every organization that has ever put people into space has done so under the shadow of those losses, and NASA, having endured three fatal accidents of its own, imposed on the Commercial Crew providers a regime of scrutiny, testing, and certification designed to drive the probability of catastrophe as low as human effort could make it. The agency even set a number on the intolerable: the odds of a fatal accident over a mission were to be held to no worse than roughly one in two hundred and seventy, a figure that looks reassuring until one remembers that the Shuttle, across its career, had actually killed its crews at a rate closer to one flight in sixty-eight. For SpaceX, this meant learning to layer onto its fast-moving culture the slow, exhaustive rigor that human-rating demands — to hold, in the same organization, both the instinct to build fast and break things and the discipline to test a valve ten thousand times before trusting a life to it. It was, in a sense, a test of whether the company could grow up without slowing down, whether the culture that had made it could survive the demands of carrying people.

Crew Dragon — formally Dragon 2 — was a substantial evolution of the cargo Dragon that had been flying to the station for years. It had to carry up to four astronauts in a shirtsleeve environment, maintaining breathable air, comfortable temperature, and pressure through launch, orbit, and reentry. It needed displays and controls the crew could use — and here SpaceX made a characteristic break with tradition, replacing the hundreds of switches, dials, and circuit breakers that had cluttered every previous spacecraft cockpit with a spare bank of large touchscreens, a decision that struck some veteran astronauts as unnerving until they flew with it. It was designed to fly autonomously, docking itself to the station without the manual piloting or robotic-arm capture that cargo Dragon required, so that the crew were passengers and supervisors more than pilots. It needed seats that could absorb the shock of landing, a life-support system with redundancy upon redundancy, and communications to keep the crew in contact with the ground. And, most critically, it needed a launch escape system — a means of pulling the capsule and its crew rapidly away from the rocket in the event of a failure during ascent, the capability that the Space Shuttle had conspicuously lacked, and whose absence had doomed the crew of Challenger, who are believed to have survived the initial breakup only to have no way to escape the falling crew cabin. SpaceX's solution was elegant and characteristically integrated: rather than a separate escape tower that is bolted to the nose and then jettisoned and wasted a few minutes into every flight, as most capsules use, Crew Dragon carried eight SuperDraco engines built into its own walls, powerful thrusters that could fire to blast the capsule away from a failing rocket at any point from the pad to orbit, and that were, in principle, reusable and even capable of setting the capsule down on land under rocket power — though SpaceX ultimately chose, for simplicity and safety, to land the capsule under parachutes in the ocean for crewed flights.

Before any astronaut would trust the vehicle, it had to prove itself empty, and in March 2019 SpaceX flew the first full dress rehearsal — Demo-1, an uncrewed Crew Dragon launched to the space station and brought safely home. In the commander's seat rode a spacesuited test dummy bristling with sensors, named Ripley in honor of Sigourney Weaver's character in the Alien films, there to record the forces a human body would feel. And floating beside Ripley was a touch of the whimsy that had by now become a SpaceX signature: a plush toy of the Earth, round and blue and wearing an expression of cartoon astonishment, tucked aboard to serve as the flight's "zero-gravity indicator," the homely object whose sudden drifting would confirm that the capsule had reached weightlessness. Demo-1 went nearly flawlessly, the capsule docking itself to the station autonomously and returning to a clean splashdown, and for a few weeks it looked as though SpaceX might be only months from flying people. Then the same capsule, being readied for a later test, exploded on the ground.

The development was long, difficult, and marked by setbacks that tested the company's commitment to the higher standard. Human-rating a spacecraft involves demonstrating, through analysis and test, that it can survive an enormous range of failure scenarios and protect its crew throughout. SpaceX had to prove its parachutes would work reliably even if one of the four should fail, which meant a long campaign of drop tests, some of which revealed problems and forced a redesign of the parachute system to a stronger material and pattern. It had to prove its launch escape system worked, which it demonstrated in dramatic fashion with an in-flight abort test in January 2020, deliberately triggering an escape at the most violent moment of ascent — the point of maximum aerodynamic pressure, just after the rocket had gone supersonic — and showing that the capsule could pull itself clear of the rocket, which was allowed to break apart in the airstream behind it, and descend safely to the sea. And it had to weather a serious accident. In April 2019, during ground testing, a Crew Dragon capsule — and not just any capsule, but the very one that had already flown a successful uncrewed demonstration flight to the station two months earlier — was destroyed in an explosion during a test of its SuperDraco escape thrusters. The blast was total; the capsule was obliterated, and only the fact that it happened on a test stand, with no one aboard, kept it from being a catastrophe. The cause, when investigators ran it down, was subtle and alarming: a slug of liquid oxidizer had leaked past a check valve into a high-pressure helium line, and when the system was pressurized, that slug was driven into a titanium component at such speed that the impact itself ignited the metal. SpaceX's fix was telling in its bluntness — it replaced the check valves, which could leak, with burst discs, one-use barriers that either hold completely or rupture completely and cannot dribble. That the failure happened during testing, on the ground, without a crew aboard, was exactly the point of testing: to find such failures before they could kill anyone. SpaceX fixed the problem, and the program pressed on.

Meanwhile, Boeing's Starliner was traveling a rockier and, ultimately, far more troubled road, though this was not yet fully apparent. Boeing brought to the task all the advantages of an established aerospace prime: decades of experience building crewed spacecraft, deep institutional knowledge, a long and trusted relationship with NASA. It also brought the disadvantages that had come to afflict the traditional aerospace model — a slower, more bureaucratic development style, higher costs, and, as would become painfully clear, quality and engineering problems that its reputation did not prevent. Starliner's uncrewed test flight in December 2019, meant to demonstrate the capsule could fly to the station and back before carrying astronauts, went badly wrong: the spacecraft's onboard clock had been set incorrectly, so that Starliner believed it was at a different point in its flight than it actually was and fired its thrusters to burn fuel at the wrong time, stranding itself in the wrong orbit, unable to reach the station. The mission had to be cut short and the capsule brought home without ever docking. Worse, a subsequent review found not just the one error but a pattern: a second software flaw, caught only by luck during the flight, could have caused the capsule to collide with its own discarded service module during descent — a potentially fatal blunder that had gone undetected through all of Boeing's vaunted testing until the vehicle was already in space. Boeing had to essentially redo the entire uncrewed test flight, which it eventually managed successfully in 2022, but the troubles foreshadowed deeper problems — stuck valves, weak parachutes, flammable tape, leaking thrusters — that would continue to plague the program for years.

The contrast between the two providers was becoming stark, and it inverted the expectations of 2014. SpaceX, the risky upstart, was moving steadily toward crewed flight, its setbacks — the parachute problems, the ground explosion — surfacing during testing and being corrected. Boeing, the trusted giant, was falling behind, beset by software failures and quality issues that spoke of an engineering culture that had, in the view of many observers, lost some of the discipline that had once made it great — the same erosion that, in those very years, was manifesting in far grimmer form in the company's troubled commercial airliners. The Commercial Crew Program, intended as a competition between two providers to ensure redundancy, was turning into a demonstration of how far the new model had outpaced the old. It would be SpaceX, not Boeing, that first carried astronauts under the program, and by a margin measured not in months but in years. The company that NASA had funded with the smaller contract, the company the conventional wisdom had judged the riskier bet, was about to end the nine-year gap and return American human spaceflight, while the storied incumbent struggled to get its capsule to work at all.

Even the look of the thing announced that something had changed. Where generations of astronauts had climbed aboard their spacecraft in bulky orange or white pressure suits studded with hoses and hardware, the two who would fly Crew Dragon wore sleek, tailored white-and-black suits that SpaceX had developed with an eye as much to aesthetics as to function — designed in part with the help of a Hollywood costume designer, and meant to look like the future rather than the past. The suits plugged into the seat through a single umbilical, the gloves were touchscreen-capable so the crew could work the cabin's displays, and the helmets were printed and custom- fitted. It was a small thing beside the engineering, and yet it was not nothing: part of Musk's stated purpose had always been to make spaceflight feel inspiring again, and a generation of children who might have found a NASA pressure suit forbidding looked at the Crew Dragon suits and saw something they wanted to wear. The medium was, as ever with SpaceX, part of the message.

On May 30, 2020, then, with the world watching and a pandemic raging on the ground, two NASA astronauts — Robert Behnken and Douglas Hurley, veteran fliers and close friends who had trained together for years — climbed into a Crew Dragon they had named Endeavour and launched from Launch Complex 39A, the same Florida pad that had sent the Apollo missions to the Moon and the Shuttle to orbit. There was a small, human grace note aboard that captured the moment better than any speech: a plush toy dinosaur, an Apatosaurus named Tremor, chosen jointly by the astronauts' two young sons and tucked into the capsule to serve as the flight's "zero-gravity indicator" — the object whose sudden floating would tell the crew, and the watching world, that they had reached orbit. When Tremor drifted up from its tether a few minutes after liftoff, it marked the first time in nearly a decade that Americans had reached orbit from American soil, the first human beings ever launched to orbit by a private company, and the quiet closing of a wound that had festered since Atlantis rolled to its stop nine summers before. Watching from the ground, Musk allowed himself the line he had been saving for six years, a reply to a Russian official's old taunt: the trampoline, he said, was working.

Demo-2 was a test flight, but it was quickly followed by the real thing. That November the first operational mission, Crew-1, carried four astronauts to the station for a full half-year tour, and from then on Crew Dragon became the workhorse of American human spaceflight, ferrying crew after crew to orbit on a regular rotation and even carrying the first entirely private astronauts to fly with no professional government pilot at the controls — a routine so reliable that, within a few years, an American launching to space aboard a commercially built capsule had ceased to be news at all. The same capsule that flew Behnken and Hurley, Endeavour, would itself fly again and again, reused like the boosters beneath it. Boeing's Starliner, meanwhile, would not carry its first crew until years later, and even then under a cloud of the same technical troubles that had dogged it from the start, so that the redundancy NASA had paid handsomely to secure rested, in practice, on the provider it had once considered the long shot. It was a vindication of the commercial model so complete that it reframed the whole debate that had preceded it: the question was no longer whether private companies could be trusted to carry human beings safely, but why anyone had ever doubted it — and how a storied incumbent, given the larger contract and every advantage, had been so thoroughly outrun by the upstart it was meant to eclipse.

Chapter 25: Demo-2

The two men who would end the gap were named Bob Behnken and Doug Hurley, veteran NASA astronauts and test pilots, both of whom had flown on the Space Shuttle before its retirement. They had been chosen for the first crewed flight of Crew Dragon, a mission designated Demo-2, the final demonstration that the vehicle could safely carry human beings to the space station and back before it entered regular service. There was a particular poignancy in the choice of Hurley, for it had been Hurley who piloted STS-135 in July 2011, the very last flight of the Space Shuttle, the mission that lowered the curtain on American human spaceflight and opened the nine-year gap; now he would be at the controls of the vehicle that closed it, bookending an era in a single career. The two men had trained for years, working closely with SpaceX engineers, learning the new spacecraft with its touchscreen controls and its autonomous flight profile, so different from the switch-and-dial cockpit of the Shuttle they had known. They wore sleek white pressure suits designed not by a traditional aerospace contractor but by the Hollywood costume designer Jose Fernandez, whose credits included superhero films, and they rode to the pad not in the boxy silver Astrovan of the Shuttle era but in a gullwing-doored Tesla Model X, a detail that captured, in miniature, how thoroughly the aesthetic and the ownership of American spaceflight had changed. And they would launch, fittingly, from Launch Complex 39A at the

Kennedy Space Center — the same pad from which Apollo 11 had departed for the Moon in 1969, and from which many Shuttle missions had flown, now leased and refurbished by SpaceX. The symbolism was rich: the historic pad, dormant since the Shuttle's retirement, brought back to life to send Americans to space once more, this time on a rocket and capsule built by a private company.

The launch was scheduled for late May 2020, and it took place against a backdrop that gave it an almost surreal weight. The COVID-19 pandemic had swept the world that spring, shutting down economies, filling hospitals, and confining much of humanity to their homes. The United States was in the grip of the crisis, and the launch, which would ordinarily have drawn hundreds of thousands of spectators to the Florida coast, proceeded under pandemic restrictions, NASA urging the public to stay home and watch from afar even as President Trump and Vice President Pence traveled to the Cape to witness it. There was something poignant about it — a moment of national achievement and hope arriving in the middle of a season of fear and loss, a reminder that human beings could still do extraordinary things even as an invisible enemy raged. The first attempt, on May 27, was scrubbed by bad weather with the astronauts already strapped in, the launch called off with less than seventeen minutes on the clock, a lightning-laced cold front refusing to clear. Three days later, on May 30, the weather cooperated, and the countdown proceeded to its end. Before liftoff the astronauts performed a small ritual that had become a SpaceX tradition: they released their zero-gravity indicator, a plush toy that would float when the spacecraft reached weightlessness. This one was an Apatosaurus, chosen by Behnken's and Hurley's young sons and named "Tremor," a sequined dinosaur that would drift up on its tether to tell two of the most experienced test pilots alive that they had, at last, reached orbit.

At 3:22 in the afternoon Eastern time, the Falcon 9 carrying Crew Dragon and its two astronauts lifted off from Pad

39A, rising on its column of flame into the Florida sky, and the United States launched human beings into orbit from its own soil for the first time since July 2011. The ascent was clean. The first stage did its work and separated — and then, in a detail that captured how far spaceflight had come, that first stage, a booster designated B1058, flew itself back down and landed on a drone ship named Of Course I Still Love You in the Atlantic, recovered for reuse even on this most historic of missions, a routine miracle folded into a singular one. It is often said that the astronauts flew this new spacecraft, but in truth their piloting was largely a matter of supervision: Crew Dragon flew itself, and although Behnken and Hurley did take manual control briefly during the approach to test the vehicle's handling with the touchscreen thrusters, the rendezvous and docking were autonomous by design. Over the following nineteen hours the capsule caught up with the space station, performed its automated approach, and docked itself to the outpost, the two astronauts monitoring the process as the new spacecraft demonstrated the hands-off docking that was one of its signature capabilities. When the hatch opened and Behnken and Hurley floated into the station to be greeted by the crew already aboard — and when, following another tradition, they revealed the name they had chosen for their capsule, Endeavour, in honor of the Space Shuttle on which both had first flown — the gap was over. American astronauts had reached the station on an American spacecraft, and a private company had, for the first time in history, launched humans into orbit.

The ride to orbit, the two men later reported, was both familiar and strange. The Falcon 9's first-stage ascent was smoother than the bone-shaking violence of the Space Shuttle, whose twin solid rocket boosters had rattled their crews so hard that reading an instrument panel could be a struggle; the Falcon rose, by comparison, on a long, controlled push. But when the second stage lit and its single engine pressed the astronauts back into their seats, the acceleration built with an insistent, growing weight, and the small cabin that had been so calm on the pad filled with noise and force. Then, abruptly, the engine cut off and the weight vanished, and Tremor the dinosaur floated up on its tether to confirm what the instruments already said. From there the machine did almost everything itself, and the veteran test pilots — men whose whole careers had been built on the premise that a pilot flies the vehicle — spent the coast to the station largely as monitors and passengers, watching a spacecraft navigate itself toward a target four hundred kilometers up. It was a strange demotion and a genuine marvel at once, and it foreshadowed the shape of the new age: the human aboard less a pilot than a supervisor, present to intervene if the automation faltered, and trusting, most of the time, that it would not.

It is worth lingering on the two men themselves, because the mission the world watched as an abstraction — America returning to space — was, up close, an intimate human affair. Behnken and Hurley were not only crewmates but close friends of long standing, and both were married to fellow astronauts, Behnken to Megan McArthur and Hurley to Karen Nyberg, a pair of spacefaring couples for whom the dangers of the launch were anything but theoretical. The country came to know the two men simply as "Bob and Doug," the affectionate shorthand catching the unglamorous, workmanlike decency the veteran test pilots projected — a reassuring ordinariness at the controls of something wholly new. NASA had branded the whole effort "Launch America," and although the agency had begged the public to stay away for fear of the virus, tens of thousands drove to the causeways and beaches around the Cape anyway, unwilling to miss the first American human launch in nine years, so that the event became a strange communal gathering in a season of enforced isolation. And there was a quiet coda that captured how thoroughly the new era folded family into history: the following spring, on the first operational rotation flight, Megan McArthur flew to the station aboard the very same capsule, Endeavour, and strapped into the very same seat her husband had occupied on Demo-2 — a private symmetry tucked inside a public triumph, and a small sign of how a reusable spacecraft accrues a history of its own, seat by seat and flight by flight.

There is a detail in that first stage's later fate that says as much about the new era as the launch itself. The booster that carried Bob and Doug, designated B1058, did not retire to a museum as a relic of a historic flight; it was refurbished and flown again, and again, becoming one of the hardest-working rockets in the fleet, launching satellites and cargo and Starlinks in a relentless rotation until, having flown nineteen times, it was finally lost when it toppled on its droneship in heavy seas while being brought back to port at the end of 2023. That the very rocket which had inaugurated American commercial human spaceflight should end its life not in a fireball or on a display pedestal but as a well-used industrial tool, worn out by ordinary service, was a fitting emblem of what SpaceX had done to the business: it had turned the singular into the routine, the monument into the machine.

The achievement was celebrated across the country and around the world, and rightly so. It ended nine years of dependence on Russia for access to the station, years in which NASA had paid ever-rising prices for Soyuz seats, climbing toward ninety million dollars apiece on the final contracts — where a seat aboard Crew Dragon would run the agency closer to fifty-five million, so that the commercial vehicle was not only American but markedly cheaper, the savings a direct dividend of the reusability and competition the program had been built to encourage. It restored a capability the United States had lost, and did so through the commercial model that many had doubted. It proved that a private company could meet the most demanding standard in spaceflight — carrying human beings — and meet it successfully. And it validated, in the most consequential way possible, the long bet that NASA and SpaceX had made together, stretching back to the COTS program and the cargo flights and the near-bankruptcy of 2008. The company that had once been unable to afford a fourth Falcon 1 launch was now trusted with the lives of astronauts, and it had delivered. It is worth correcting a common impression here, one the sleek Tesla and the touchscreens encouraged: Demo-2 was not SpaceX going it alone, a private triumph over a plodding government. It was the fruit of a public–private partnership in which NASA had provided billions of dollars, decades of accumulated expertise, and a small army of engineers who scrutinized every weld and line of code. The genius of the arrangement was that it combined the agency's rigor with the company's speed, and the result belonged to both. For NASA, the mission was a triumph of the commercial approach, evidence that turning to private industry could restore American leadership rather than surrender it. For SpaceX, it was perhaps the proudest moment in the company's history to that point, the culmination of eighteen years of work, and a demonstration that the upstart had become an indispensable pillar of the national space enterprise.

Behnken and Hurley spent two months aboard the station before returning home, and their return provided the final test of Crew Dragon: reentry and splashdown. On August 2, 2020, they undocked, and Endeavour carried them down through the atmosphere, its heat shield glowing against the fire of reentry, its parachutes deploying in a carefully choreographed sequence to slow the descent, until it splashed down gently in the Gulf of Mexico off the coast of Pensacola and was recovered by a SpaceX ship. It was the first crewed water landing by American astronauts since the Apollo era — since the Apollo-Soyuz mission of July 1975, forty-five years earlier, when the last of the ocean-recovery generation had come home. The moment, however, produced an unscripted and faintly alarming coda. Within minutes of splashdown, before the recovery vessel could establish a cordon, some two dozen private pleasure boats came racing across the water and clustered around the bobbing capsule, curious onlookers filming on their phones as recovery crews worked to secure a spacecraft that still vented toxic propellant vapors. The scene, broadcast live, was a vivid reminder of a new reality: a splashdown off the Florida coast was no longer a cordoned-off government operation on the open sea but a public spectacle in busy recreational waters, and NASA and SpaceX would tighten their exclusion procedures for every flight thereafter. Still, the full round trip had been completed successfully, and Crew Dragon was now a proven vehicle for carrying human beings to orbit and back. With Demo-2 complete, the program moved from demonstration to operation, and Crew Dragon began flying regular rotation missions to the station, carrying crews of astronauts — American and international — on six-month tours, replacing the Soyuz as the primary means by which the United States and its partners reached the outpost. What had been an extraordinary achievement in May 2020 became, within a year or two, a routine service, another instance of SpaceX taking the unprecedented and making it ordinary.

The success of Crew Dragon transformed SpaceX's standing and its business. It was now not just a launcher of satellites and cargo but a provider of human spaceflight, one of only a handful of organizations in history — and the only private company — capable of carrying people to orbit. That capsule, Endeavour, would go on to become the most-flown vehicle in the Dragon fleet, carrying crew after crew and private mission after private mission, its pressure hull a piece of living history that anyone tracking a launch schedule might still meet on any given flight. And the capability the mission proved opened new possibilities that went beyond ferrying NASA astronauts to the station. If a private company could safely carry human beings to space, then human beings other than government astronauts might fly — wealthy individuals, private researchers, citizens of nations without their own space programs, adventurers. The door to private human spaceflight, long imagined and often promised, was finally, genuinely open, and SpaceX would be the one to walk through it, launching missions that carried not professional astronauts on government business but private crews on missions of their own devising. Among the first through that door would be a billionaire entrepreneur who bought an entire flight, filled it with ordinary people, and used it to prove that space was no longer the exclusive preserve of governments — and who would go on to push the boundaries of what private citizens could do in orbit, culminating in a feat that not even NASA had attempted in decades.

Chapter 26: Inside Crew Dragon

To carry human beings safely to orbit and back is to solve a thousand problems that carrying cargo never poses, and the spacecraft SpaceX built to do it — Crew Dragon — repays a closer look, both for its engineering and for what it reveals about how the company approached the most demanding task in its business. From the outside, Crew Dragon looked much like its cargo cousin, a rounded capsule atop a cylindrical trunk. Inside and beneath the skin, it was a different machine, engineered around the single overriding imperative of keeping its occupants alive. Before a human being ever rode it, that imperative was rehearsed with a stand-in: the uncrewed Demo-1 flight of March 2019 carried to the station a sensor-laden test dummy that SpaceX had named Ripley, after Sigourney Weaver's character in the film Alien, its instrumented body recording the forces a real astronaut would feel. The joke was characteristic of the company's culture, but the data were deadly serious, and the mannequin's ride was the first full end-to-end proof that the vehicle could do what it was built to do.

The most visible break with tradition was the interior. Where earlier spacecraft cockpits bristled with hundreds of switches, dials, and gauges — the analog controls of the Apollo and Shuttle era — Crew Dragon presented its astronauts with three sleek touchscreen displays, a clean, modern, almost minimalist cabin that looked more like the interior of an electric car than a spaceship. The design reflected SpaceX's software-centric culture and Musk's aesthetic sensibility, and it was more than a style choice: the spacecraft was designed to fly itself, to navigate, rendezvous, and dock autonomously, with the astronauts monitoring rather than manually piloting. This was a significant departure from the tradition of the astronaut as hands-on pilot, and it drew some skepticism from veterans of the old school, who wondered aloud what a test pilot was supposed to do with a glass panel and a gloved fingertip when a spacecraft went wrong. It is often assumed that the touchscreens left the crew helpless in an emergency, mere passengers behind a locked pane of glass, but in fact SpaceX retained a bank of physical buttons beneath the displays for the handful of the most critical and time-sensitive actions — firing the escape system, releasing the parachutes — precisely because there are moments in flight when a pilot must be able to act by feel, without hunting through a menu. The autonomy reflected the reality that computers could fly the delicate approach and docking more precisely and reliably than human hands, and that the astronauts' role had shifted from pilot to supervisor, intervening only if something went wrong; the touchscreens and the automation were emblematic of a new generation of spacecraft, and they worked, Crew Dragon docking itself to the station, flight after flight, with a precision that vindicated the approach.

The clash of cultures was real, and it played out in the training. Astronauts drawn from the military test-pilot tradition were accustomed to sticks and throttles and the muscle memory of hand-flying; now they practiced spacecraft rendezvous by dragging their fingers across glass. On Demo-2, Behnken and Hurley deliberately took manual control for a short stretch during the approach, flying the capsule by touchscreen to prove that a human could step in if the automation ever failed, and they reported that it handled crisply — but the very fact that this was treated as a test, a contingency to be demonstrated rather than the normal way of flying, measured how far the role of the astronaut had shifted. The old hands' skepticism was not mere nostalgia; it reflected a genuine question about whether stripping the pilot of hands-on control made a vehicle safer or more brittle, whether a crew that never truly flew its ship would be ready to seize it in the seconds when everything depended on them. SpaceX's answer, borne out flight after flight, was that a well-designed autonomous system, watched over by a trained crew who could intervene, was more reliable than human hands alone — but it was an answer the profession had to be persuaded, over years, to accept.

Beneath the modern cabin lay the systems that made human spaceflight possible. Life support maintained a breathable atmosphere, a comfortable temperature, and the right pressure through launch, orbit, and reentry, with the redundancy that human-rating demanded — backups upon backups, so that the failure of any single system would not endanger the crew. The seats were engineered to cushion the astronauts against the forces of launch and the shock of splashdown. And at the heart of the safety design was the launch escape system, the capability to pull the capsule and its crew rapidly away from the rocket in the event of a failure during ascent — the very capability whose absence had doomed the Challenger crew, who rode a vehicle with no way to separate from its failing boosters. SpaceX's solution was distinctive and elegant: rather than a separate escape tower mounted atop the capsule and jettisoned early in flight, as most spacecraft from Mercury to Apollo to Orion had used, Crew Dragon had eight SuperDraco engines built into the capsule's own walls, each capable of some sixteen thousand pounds of thrust, powerful enough that firing all eight together could accelerate the capsule from a standstill to well over a hundred miles per hour in seconds, blasting it clear of a failing rocket. These were not the small Draco thrusters, ranged around the vehicle for gentle maneuvering; the SuperDracos were their outsized cousins, and their integration into the hull meant the escape system covered the entire ascent, from the pad to orbit, not just the early phase that a jettisoned tower could protect. It had the added virtue of being carried all the way to orbit rather than thrown away. SpaceX had originally hoped these same engines might allow Crew Dragon to land propulsively, settling gently on land under its own power like a science-fiction lander, but the company ultimately chose the more conservative path of parachute splashdown for crewed flights, judging that certifying a propulsive landing to the standard human life demanded, on the timeline NASA required, was a risk not worth taking.

The SuperDracos themselves embodied the company's willingness to break with aerospace habit. Their combustion chambers were not machined and welded from forgings in the traditional way but 3D-printed in a nickel-based alloy, grown layer by layer by a laser fusing metal powder — the first time a printed rocket engine had ever flown on a spacecraft rated to carry human beings. Printing let SpaceX produce the intricate internal cooling passages of a high-thrust engine as a single seamless part, faster and cheaper than casting and with fewer joints to fail, and it was of a piece with the company's whole approach: treat the hardest components as manufacturing problems to be solved with modern industrial methods rather than as artisanal one-offs. The same instinct showed in the capsule's heat shield, a version of a NASA-developed ablative material called PICA that SpaceX had reformulated as its own PICA-X and manufactured in-house, a shield engineered to survive not merely the fire of a return from low orbit — where the outer surface glows at some 1,900 degrees Celsius — but, in principle, the far fiercer heat of a return from the Moon or Mars, so that the same basic technology could serve missions far beyond the station. It was a habit of mind worth noticing: even in building a capsule for a government contract to reach an orbiting laboratory, the company reached for materials and methods sized to the journey it really cared about.

The development of these systems was long and hard, and

it included the kind of failure that human-rating exists to catch. On April 20, 2019, at Cape Canaveral, the very capsule that had flown the successful Demo-1 mission a few weeks earlier — the vehicle that had carried Ripley to the station — was being put through a ground test of its SuperDraco escape thrusters when, in the final moments before ignition, it was destroyed in an explosion, a plume of orange smoke rising over the Florida coast in footage that soon leaked. The investigation traced the disaster to a subtle and almost fiendish cause: during earlier processing, a small quantity of liquid oxidizer, nitrogen tetroxide, had leaked past a seal and into a high-pressure helium line, and when the system was rapidly pressurized to arm the escape thrusters, that slug of liquid was driven at high speed into a titanium check valve, whose failure under the impact was enough to ignite the titanium in the oxidizer-rich environment and detonate the vehicle. Had it occurred in flight with a crew aboard, it could have been catastrophic. That it happened on the ground, during testing, without anyone aboard, was precisely the point of testing: to find such failures before they could kill. SpaceX investigated, understood the cause, and made a design change that spoke to the company's engineering instincts — it replaced the check valves with burst disks, components that seal completely and open only once, when deliberately ruptured by pressure, eliminating the very flow path that had allowed the accident. Then it pressed on. The company also had to prove its parachutes would deploy reliably even if one of the four failed, which required many drop tests and a redesign to a stronger chute canopy, and it demonstrated the launch escape system in the most convincing way imaginable: on January 19, 2020, it flew a Falcon 9 with an uncrewed Crew Dragon atop it and, about a minute and a half into the ascent, at the moment of maximum aerodynamic stress, deliberately triggered an abort, the SuperDracos firing to yank the capsule clear as the abandoned rocket, its balance destroyed, tore itself apart in a fireball behind it. The capsule descended safely under its parachutes. Each of these tests and failures was part of the exhaustive process of human-rating, the driving of the probability of catastrophe as low as human effort could make it, and it required SpaceX to layer onto its fast-moving culture the slow, meticulous rigor that carrying human lives demanded.

Behind all of this stood a single, sobering number. NASA's human-rating requirement for the Commercial Crew Program set a target for the probability of a catastrophic loss of crew over a mission of no worse than one in two hundred and seventy — a figure meant to represent a marked improvement on the Space Shuttle, whose actual record, two orbiters and fourteen lives lost across a hundred and thirty-five flights, worked out to roughly one chance of disaster in every sixty-eight. To satisfy that standard, every system had to be analyzed, tested, and backed up until the arithmetic of risk fell below the line, and much of the delay that pushed Crew Dragon's first crewed flight years past its early hopes came not from the glamorous hardware but from precisely this grinding work: the micrometeoroid shielding that had to be added when analysis showed the orbital debris threat was worse than assumed, the parachute margins that had to be proven, the escape-system redesign, the endless reviews in which NASA and SpaceX engineers argued line by line over whether the numbers were good enough. Human-rating is, at bottom, a discipline of institutionalized pessimism, an insistence on imagining every way a machine might kill the people inside it and then engineering that possibility down toward zero, and it was the part of the enterprise least visible from outside and most consuming from within.

The astronaut experience aboard Crew Dragon reflected both the advances and the enduring realities of spaceflight. The astronauts who flew it praised its comfort, its clean design, its smooth operation, and the confidence they had in it — a marked contrast, some noted, to the aging systems they had known before. But spaceflight remained spaceflight: the violence of launch, the strangeness of weightlessness, the fire of reentry, the jolt of splashdown, the ever-present knowledge that they were riding a controlled explosion to a place that would kill them instantly if their fragile machine failed. Crew Dragon made the journey safer and more comfortable than it had been, but it did not make it safe or ordinary, and the astronauts who flew it, whether NASA professionals or private citizens, were still doing one of the most dangerous things human beings do. What SpaceX had achieved was to make that dangerous thing routine and reliable, to fly crew after crew successfully, to build a spacecraft that astronauts trusted with their lives and that earned that trust flight after flight. By the mid-2020s Crew Dragon had become the most-flown crewed spacecraft of its era and the backbone of Western human access to orbit, its four capsules cycling between the launch pad and the ocean with a cadence no crewed vehicle before it had matched — a working fleet rather than a museum of one-off machines. Its success, engineered through years of development, testing, failure, and rigor, was one of SpaceX's proudest achievements, the demonstration that the company could meet the highest standard in the business, the standard where lives were at stake, and meet it not once but again and again. And it is worth remembering, when the touchscreens and the sleek suits make the whole thing look effortless, that the ease was purchased with the smoke over Cape Canaveral, the shredded booster over the Atlantic, the mannequin named Ripley — with a deliberate campaign to break the vehicle in every way its engineers could imagine, so that it would not break when it mattered. The measure of that campaign's success was a negative one, the kind that never makes a headline: crew after crew flew and came home, the escape engines never had to fire in anger, the parachutes always opened, and the machine that had once exploded on a test stand became a vehicle astronauts strapped into with the quiet confidence of commuters — which is, in the end, the highest praise an engineered thing can earn.

Chapter 27: A Routine Miracle

Once Crew Dragon had proven itself, something remarkable happened, and its very unremarkableness was the point: launching astronauts became routine. Beginning with the mission called Crew-1 in November 2020 — the first operational flight, which carried NASA's Michael Hopkins, Victor Glover, and Shannon Walker together with the Japanese astronaut Soichi Noguchi in a capsule the crew christened Resilience, a name chosen in the teeth of the pandemic year — SpaceX started flying regular crew-rotation missions to the International Space Station, each carrying a crew of four astronauts for a tour of roughly six months, after which another Crew Dragon would arrive with their replacements and carry the previous crew home. These missions, designated Crew-2, Crew-3, and onward, followed one after another with a regularity that would have seemed astonishing a decade earlier, when the United States could not launch its own people at all. The astronauts aboard were a mix of Americans and international partners — Europeans, Japanese, and others — reflecting the multinational character of the station, and Crew Dragon became the primary vehicle by which the Western partners reached orbit, the Soyuz relegated to a secondary role. It is worth correcting the common assumption that Crew Dragon simply replaced the Soyuz and severed the tie to Russia. In fact, even at the depths of the geopolitical rupture that followed Russia's invasion of Ukraine, NASA and Roscosmos preserved a seat-barter arrangement, quietly flying American astronauts on Soyuz and Russian cosmonauts on Crew Dragon, so that if either vehicle were grounded, both nations would still have someone aboard the station to keep their segments running — a small, stubborn island of cooperation in an ocean of hostility, and a reminder that in orbit the physics of mutual dependence outlasts the politics of the ground.

The crews themselves carried quiet milestones that measured how the meaning of an astronaut was changing. Aboard Crew-1, Victor Glover — a Navy test pilot flying his first mission — became, six decades into the space age, the first Black astronaut to join the International Space Station as a long-duration crew member rather than a short-term visitor, a first that arrived startlingly late and that Glover met with characteristic understatement, more interested, he said, in the work than the label. His crewmate Soichi Noguchi, meanwhile, entered a small pantheon of his own, becoming one of the very few human beings to have reached orbit aboard three different kinds of spacecraft — the American Space Shuttle, the Russian Soyuz, and now the Crew Dragon — a personal span that traced the whole arc of the transition these missions embodied, from the government monoliths of the old order to the commercial vehicle of the new.

The normalization of crewed spaceflight was, in its own quiet way, one of SpaceX's most significant achievements. Spaceflight had always been rare, precious, and dangerous, each crewed launch a national event freighted with risk and significance. SpaceX did not make it safe — nothing makes spaceflight safe — but it made it regular, reliable, and, in an operational sense, almost ordinary, flying crews to the station on a schedule, bringing them home, and doing it again, without the drama and the years-long gaps that had characterized the enterprise before. Consider the contrast in numbers. Across three decades, the Space Shuttle flew 135 missions with a fleet of orbiters that each took months to refurbish between flights, and it lost two of them and fourteen lives; within a few years of Demo-2, a Crew Dragon fleet of just four small capsules was delivering crews to orbit two, three, four times a year, cycling each capsule back to the pad in a matter of months and, by the mid-2020s, having carried dozens of people to the station without losing a single one. This regularity had real consequences. It meant that the space station, the enormous and expensive orbiting laboratory, could be fully staffed and used for research without the uncertainty of relying on a single provider or a foreign government. It meant that the cost of reaching the station came down, as commercial competition and reusability worked their effects. And it meant that human presence in orbit became a continuous, dependable fact rather than an intermittent achievement, a foundation on which more ambitious things could be built.

The economics made the shift concrete. In the years of dependence, NASA had bought Soyuz seats at prices that climbed toward eighty-six and then ninety million dollars each; a seat aboard Crew Dragon cost the agency roughly fifty-five million. Multiply that gap across four astronauts to a flight and several flights a year and the savings ran to hundreds of millions of dollars annually — money that now stayed within the American economy rather than flowing to a geopolitical rival, and that flowed, in part, back to a company reinvesting it in the rockets of the future. The commercial model had promised to make spaceflight cheaper as well as sovereign, and here, in the plain price of a seat, was the promise kept, the abstraction of "commercial crew" reduced to a number a budget officer could point to.

Among those more ambitious things was the arrival of private astronauts on the station — people who flew not as government employees on national missions but as private individuals, their seats purchased commercially. A company called Axiom Space, founded in 2016 by a former manager of the International Space Station program to build commercial space stations and, in the interim, to broker private missions to the existing one, arranged a series of flights on Crew Dragon that carried private crews to the station for stays of a week or more. The first, launched on April 8, 2022, and designated Ax-1, flew aboard the same capsule, Resilience, that had opened the operational era with Crew-1, and it was commanded by Michael López-Alegría, a retired NASA astronaut and veteran of the station who now worked for Axiom, leading a crew of three paying customers who had each reportedly spent on the order of fifty-five million dollars for the experience. They were not idle names. Larry Connor, an Ohio real-estate entrepreneur and accomplished aerobatic pilot, served as the mission's pilot, the first private citizen to fly in that role on an orbital spacecraft; Mark Pathy was a Canadian investor and philanthropist; and Eytan Stibbe was a former Israeli Air Force pilot flying as only the second Israeli ever to reach space. Stibbe's presence carried a private grief and a public tribute, for he had been a squadron mate and friend of Ilan Ramon, Israel's first astronaut, who had died aboard the Space Shuttle Columbia in 2003, and Stibbe carried mementos and conducted experiments in Ramon's memory, so that the mission became, among other things, an act of remembrance. These were not tourists in the trivial sense; they conducted research, followed a demanding schedule, and had trained extensively, and their planned ten-day visit stretched to seventeen days when bad weather at the splashdown site kept them in orbit. But they were private citizens, in orbit on their own initiative and their own money, and their presence marked a genuine shift in who could go to space. The station, built and operated by governments, was now hosting private missions brokered by a private company and flown on a private spacecraft. The commercialization of human spaceflight, long a matter of speculation and promise, was becoming a reality, mission by mission.

The private missions, moreover, did not stop at visiting the station. As the decade wore on they grew stranger and more adventurous, pushing into corners of orbit that even the government programs had never explored. In the spring of 2025 a Crew Dragon carried a crew assembled and led by the entrepreneur Chun Wang on a mission called Fram2 — named for the Norwegian polar exploration ship — into an orbit that passed directly over both of the Earth's poles, a true polar path no human crew had flown in more than sixty years of spaceflight, the astronauts gazing straight down on the white expanses of the Arctic and Antarctic that no previous crew had ever seen from that angle. It was a small but telling illustration of what a genuinely commercial spaceflight capability made possible: not merely wealthy tourists retracing familiar routes, but private crews buying their way to firsts that the national agencies, bound by their own priorities and budgets, had simply never bothered to attempt.

It would be a mistake, though, to picture these private crews as unsupervised joyriders let loose in orbit. NASA kept a firm hand on every mission that visited its station: it required that each Axiom flight be commanded by a professional astronaut, almost always a veteran of its own corps, and it vetted the paying crew members, dictated their training, and folded their activities into the station's tightly scheduled routine. The private astronauts arrived as guests in a working laboratory, not as owners of it, and the arrangement reflected a careful hybrid — commercial money and private ambition operating within the guardrails of a government program, the agency lending its hard-won expertise and its exacting standards even as it opened its outpost to visitors who paid their own way. That hybrid, awkward and endlessly negotiated, was the real texture of the new space age: neither the clean triumph of the private over the public that the boosters proclaimed, nor the sellout of public assets that the critics feared, but an intricate partnership in which each side needed the other and neither could yet do the job alone.

It is worth pausing on the significance of this, because it is easy to be cynical about wealthy individuals buying trips to space, and the cynicism is not entirely misplaced. The people who could afford a private orbital flight were, necessarily, among the richest in the world, and there is something uncomfortable about the spectacle of billionaires and multimillionaires purchasing access to the ultimate luxury while so many on Earth lacked so much. But the deeper trend the private missions represented was one that, over time, could democratize rather than concentrate access to space. Every new market begins by serving the wealthy — the first automobiles, the first airplanes, the first computers, the first cellular telephones, which in the 1980s cost thousands of dollars and weighed as much as a brick, were all playthings of the rich before they became tools of the many — and the logic of falling costs, driven by reusability and competition and scale, pointed toward a future in which spaceflight, like air travel, might eventually become accessible to a far broader range of people. The private astronauts of the early 2020s were the leading edge of that possibility, the first of a kind that might, if the costs kept falling, someday number not in the handfuls but in the thousands. Or they might not; the future is not guaranteed, and space may remain expensive and rare. But the private missions opened a door that had been firmly shut, and what came through it would depend on whether the cost revolution SpaceX had begun could be carried far enough.

The surest sign of the transformation was the fading of attention. The Demo-2 launch had been a national event, carried live for hours, narrated with all the breathless significance the return of American human spaceflight deserved. Within a few years, a Crew Dragon launch to the station rated a brief item, if that — the boosters landing themselves, the crews cycling up and down on schedule, the whole apparatus receding into the background hum of infrastructure, the way scheduled air travel long ago stopped being news. This was not indifference so much as trust, the quiet confidence a society extends to systems that simply work; and it was, paradoxically, the highest tribute the enterprise could be paid. To take the crossing of the most dangerous frontier human beings routinely attempt and render it something the public could safely take for granted was an achievement of a peculiar and almost invisible kind — unnoticed precisely because it was so nearly complete.

The steady rhythm of crewed flights also deepened SpaceX's role as the indispensable partner of the American space program and its allies. By the mid-2020s, Crew Dragon had carried dozens of astronauts to orbit across many missions, becoming the most-flown crewed spacecraft of its era and the backbone of human access to the station for the Western world. The reliability it demonstrated — mission after mission flown successfully, crews delivered and returned safely — stood in ever-sharper contrast to the continuing struggles of its intended counterpart, Boeing's Starliner, which remained mired in problems and delays even as Crew Dragon flew routinely. The Commercial Crew Program had been designed with two providers precisely so that the failure or grounding of one would not leave the nation stranded, but in practice the redundancy ran mostly one way: it was Crew Dragon that carried the load, and Starliner that the program could not yet rely upon. This imbalance would be dramatized in the most vivid way possible when Starliner finally flew astronauts and could not safely bring them home, and Crew Dragon had to be pressed into service to rescue a mission gone wrong — a story we will come to. But the private missions had one more chapter to write first, a chapter in which a single billionaire, not content with visiting the station, would fund a series of increasingly ambitious flights aimed at pushing the frontier of what private human spaceflight could achieve, and would, in the process, reach higher than any human had gone since the days of Apollo and open the hatch of a spacecraft to the vacuum of space in the first spacewalk ever conducted by private citizens.

Chapter 28: The Overview Effect

Amid the engineering and the economics, it is easy to lose sight of the human experience at the center of it all — what it actually means for a person to leave the Earth, to see the planet from above, to float in the vacuum that would kill them in an instant. The new space age, for all its talk of markets and constellations and reusability, was ultimately in the business of sending human beings to space in growing numbers, and it is worth pausing to consider what that experience does to those who have it, because the human meaning of spaceflight is part of why it matters, and part of what SpaceX, in opening space to more people, was changing.

Astronauts have long reported a profound psychological and even spiritual response to seeing the Earth from space, a phenomenon that the space writer and philosopher Frank White named the "overview effect," first in a poster he prepared for a conference in 1985 and then in a 1987 book of that title, built on interviews with some fifty people who had been to orbit and beyond. Looking down at the planet — a fragile blue sphere hanging in the black void, its atmosphere a thin, luminous film, its continents and oceans laid out without the borders that divide them on maps — many astronauts describe a shift in perspective so deep that it changes them permanently. They speak of a sudden, overwhelming sense of the Earth's unity and fragility, of the artificiality of the divisions between nations, of the preciousness and vulnerability of the only home humanity has. Edgar Mitchell, the sixth man to walk on the Moon, put it in words that have become the classic statement of the experience: "You develop an instant global consciousness, a people orientation, an intense dissatisfaction with the state of the world, and a compulsion to do something about it." Mitchell spent the rest of his life pursuing the implications of that moment, founding an institute to study consciousness. The origin of the whole idea is often traced to a single photograph: the image called Earthrise, taken by the Apollo 8 astronaut Bill Anders on Christmas Eve 1968, showing the blue Earth rising over the gray lunar horizon. Anders later distilled the paradox at its heart into a sentence: "We came all this way to explore the Moon, and the most important thing is that we discovered the Earth." That photograph, seen by hundreds of millions, is widely credited with helping to catalyze the modern environmental movement, the first Earth Day following little more than a year later.

The overview effect is not a mere aesthetic response; for many who experience it, it is a genuine transformation, a reordering of priorities, a new and lasting sense of connection to the planet. And it has, on occasion, redirected entire lives. Rusty Schweickart, who piloted the lunar module on Apollo 9 and had a few minutes alone in space with nothing scheduled — a stuck camera left him simply floating and looking — came home so struck by the Earth's fragility that he devoted decades of his later life to the danger of asteroid impacts, co-founding an organization dedicated to planetary defense, as though the man who had seen how small and exposed the world was could not stop working to shield it. To grasp the specific perception at the root of all this, it helps to hold onto one concrete image the astronauts return to again and again: the atmosphere, the entire breathable envelope on which every living thing depends, seen edge-on from orbit, is not the vast sky it appears from below but a startlingly thin blue-white band hugging the curve of the planet, no thicker, in proportion, than the skin on an apple or a coat of varnish on a schoolroom globe. To see the whole of the air that keeps you alive rendered as a fragile rind is, many report, the moment the abstraction of "spaceship Earth" becomes a physical fact.

The testimony spans the whole history and geography of spaceflight, which is part of what makes it persuasive. Yuri Gagarin, the first human to orbit the Earth, radioed down in 1961 a report as plain as a child's — "I see Earth. It is so beautiful" — and reflected afterward, "Orbiting Earth in the spaceship, I saw how beautiful our planet is. People, let us preserve and increase this beauty, not destroy it." A quarter of a century later, Sultan bin Salman, a Saudi prince and the first Arab and first Muslim to fly in space, described the crew's sense of belonging widening day by day aboard the Space Shuttle: "The first day or so we all pointed to our countries. The third or fourth day we were pointing to our continents. By the fifth day, we were aware of only one Earth." That the same shift should be reported, in almost the same words, by a Soviet air-force pilot, a Saudi prince, and an American who had walked on the Moon — men divided by nation, faith, and era — is itself a kind of evidence that the perception is rooted in something real about the vantage rather than in any one culture's expectations of what a person ought to feel.

There is an extreme version of the perspective that pushes the point further still. Michael Collins, who orbited the Moon alone in the command module while Armstrong and Aldrin walked below, became on each pass around the far side the most isolated human being who had ever lived, cut off even from radio contact with Earth, and he wrote afterward with unusual tenderness of how small and fragile the home planet looked from lunar distance, a thing he felt he could hide behind a raised thumb. The most distant portrait of all was made not by human eyes but by a machine. In 1990, at the urging of the astronomer Carl Sagan, the Voyager 1 spacecraft turned its camera back from beyond the orbit of Neptune, some six billion kilometers out, and photographed the Earth as a single pale dot less than a pixel across, suspended by chance in a band of scattered sunlight. "That's here. That's home. That's us," Sagan wrote of the image. "On it everyone you love, everyone you know, everyone you ever heard of, every human being who ever was, lived out their lives." The overview effect, in that photograph, reaches its logical end: the planet rendered not as a fragile sphere but as a mote, and the whole of human history a matter of a single speck of dust caught in a sunbeam.

For most of the space age, this transformative experience was available only to the tiny cadre of professional astronauts — fewer than six hundred human beings in six decades. The new space age began, however tentatively and however unequally, to widen the circle. The private astronauts of the SpaceX missions — the crew of Inspiration4, the visitors to the space station, the flyers of Polaris Dawn — experienced the overview effect and spoke of it in the same transformed terms as the professionals before them. The suborbital passengers of New Shepard and Virgin Galactic, though they reached space only briefly, described their own versions of the shift in perspective. Here it is worth resisting the temptation to make the effect tidier than it is. It is often described as if seeing the Earth uniformly fills the beholder with serene, uplifting love — but the responses are in truth more varied and sometimes darker. When the actor William Shatner, at ninety, flew a few minutes above the line of space aboard Blue Origin's New Shepard in October 2021, becoming the oldest human to reach space, he came back not euphoric but shaken, and later wrote that what overwhelmed him was not the beauty of the Earth but the lifeless blackness beyond it — a sense of grief and death in the cold void that threw the fragile warmth of the planet into unbearable relief. Some astronauts report no dramatic transformation at all. The overview effect, in other words, is real and widely attested but not a single guaranteed switch that flips in every mind; it is a family of responses to the same overwhelming sight, and its variety is part of its honesty.

The very success of the private missions in widening access sharpened an old argument about what actually counts. Suborbital passengers on New Shepard and SpaceShipTwo crossed briefly above the boundary of space — the Kármán line, by convention a hundred kilometers up — floated weightless for a few minutes, glimpsed the curve of the Earth against the black sky, and came home changed, or said they did; skeptics wondered whether a flight lasting minutes could deliver the profound reorientation that months in orbit had worked on the professionals, or whether the very word "astronaut" had been cheapened by the sale of tickets. Researchers, meanwhile, began for the first time to study the overview effect systematically, surveying flyers before and after their journeys, precisely because a growing population of space travelers had finally made such study possible; where Frank White once had to seek out a few dozen astronauts to interview, the new space age was minting new candidates by the launch. Whether the effect would prove to be a durable and distinct psychological phenomenon or a heightened instance of the awe that any overwhelming vista can produce, the science remained, as this was written, young — outpaced, as so much in the new space age was, by the very events it was trying to understand.

The numbers give the shift its scale. For the whole of the twentieth century, reaching space had been the near-exclusive privilege of a few hundred government-trained professionals; by the mid-2020s, with suborbital hops and private orbital flights added to the government missions, the total number of human beings who had ever crossed into space was climbing past the seven hundred mark and rising faster than at any point in history, an ever-larger share of them private citizens rather than career astronauts. It remained a vanishingly small fraction of humanity — a rounding error against eight billion — and the newcomers were overwhelmingly wealthy. But the direction of travel had reversed a decades-long stagnation, and if the overview effect really does something lasting to those who experience it, then for the first time the population of the changed was growing not by ones and twos but, on some flights, by the capsule-load.

Even so, as spaceflight opened to more people, more human beings had the chance to see the Earth from above and to be changed by it, and there was, in this, a genuine and underappreciated significance. If the overview effect is real — if seeing the Earth from space really does instill a deeper sense of its unity and fragility, a stronger commitment to protecting it and to seeing humanity as one — then opening space to more people might have consequences beyond the commercial and the technological, might slowly seed a wider human population with the perspective that so moved the astronauts, might, in some small way, change how humanity sees itself and its home. This is, admittedly, a hopeful reading, and it must be balanced against the sobering fact that, for now, the people who get to experience the overview effect are overwhelmingly the wealthy and the privileged, the ones who can afford a seat on a private spaceflight or who are chosen for a government mission. The democratization of the overview effect, like the democratization of spaceflight generally, remained more aspiration than reality, and there was something uncomfortable in the spectacle of billionaires and their guests achieving a transformative communion with the Earth's fragility from a vantage the vast majority of humanity would never share. Whether the circle would widen enough to matter depended on the same uncertain question of falling costs that shadowed all the grand visions of the new space age. It might, or it might not. But the possibility was real, and it added a dimension to the story that the ledgers of launch prices and market share could not capture: the possibility that opening space to more people might change not just the economy or the technology but the human spirit.

There is a further irony worth dwelling on, one that sits at the heart of the whole enterprise. The most famous expression of the overview effect — the sense of the Earth's preciousness and fragility that astronauts bring back — sits in tension with the animating vision of SpaceX itself, which is to leave the Earth, to make humanity multi-planetary, to establish a second home on Mars. The astronaut who looks down and is overwhelmed by the beauty and fragility of the home planet, and the entrepreneur who dreams of settling another world lest humanity perish on this one, are responding to the same fundamental fact — the singularity and vulnerability of the Earth — but drawing opposite conclusions: one toward cherishing and protecting the home we have, the other toward building an escape from its potential ruin. This tension runs through the whole of the new space age, and it is not easily resolved. Is the future of humanity to be found in turning inward, in cherishing and healing the only home we have? Or in reaching outward, in spreading beyond the Earth so that the story of consciousness does not depend on a single fragile world? SpaceX's answer was the latter, but the overview effect whispers the former, and perhaps the truth is that both are needed — that a humanity wise enough to protect its home and bold enough to reach beyond it might be the one most likely to endure. The people who leave the Earth and see it whole carry that question back with them, and as more of them make the journey, the question they carry becomes, in some small way, humanity's own.

Chapter 29: Civilians in Orbit

Jared Isaacman was a billionaire who had made his fortune in payment processing, founding a company from his parents' basement at the age of sixteen — a business he built into the financial-technology firm Shift4 Payments, which by the time of his flights was handling a substantial share of the card transactions in American restaurants and hotels. Alongside his wealth he possessed an authentic passion for flying: he was an accomplished jet pilot who had flown high-performance military aircraft, set an around-the-world speed record in a light jet, and co-founded a company that owned one of the largest private fleets of fighter jets in the world, used to train military pilots. In 2021 he did something that no private individual had done before: he bought an entire SpaceX Crew Dragon flight, not to visit the space station but to fly a free orbital mission of his own design, and rather than filling the seats with fellow wealthy adventurers, he gave them away. The mission was called Inspiration4, and its crew, besides Isaacman, was chosen to represent a set of themes. Hayley Arceneaux, representing "Hope," was a physician assistant who had been treated for bone cancer as a child at St. Jude Children's Research Hospital, the institution the mission was raising money for, and who still carried a titanium rod in her left leg from the surgery that had saved it; she would become, at twenty-nine, the youngest American ever to reach orbit, and the first person to fly to space with a prosthesis. Sian Proctor, representing "Prosperity," was a geoscientist and educator who served as the mission's pilot, becoming the first Black woman to pilot a spacecraft. Chris Sembroski, representing "Generosity," was a data engineer and Air Force veteran who reached the crew almost by accident, through a charity raffle a friend had entered and then passed the seat to him. None of them were professional astronauts. They were, in the mission's framing, everyday people, and their flight was meant to demonstrate that space was opening to a broader humanity.

Inspiration4 launched on September 15, 2021, and spent three days in orbit — the first time a crewed mission had flown to orbit with no professional astronaut aboard at all, no government-trained commander, just four civilians in a spacecraft flying itself around the Earth. It is often assumed that such a tourist flight must have hugged the safe, familiar altitude of the space station, but in fact the opposite was true: the Crew Dragon climbed to roughly 585 kilometers, well above the station's orbit and higher than any human being had flown since the last Space Shuttle mission to service the Hubble Space Telescope in 2009. The crew conducted medical experiments on themselves, spoke by radio with the children at St. Jude, and gazed at the Earth through a specially installed glass dome — a cupola that replaced the docking mechanism Crew Dragon normally used to mate with the station — giving them a panoramic, uninterrupted view of the planet turning beneath them. Then they returned, splashing down safely after three days aloft. The mission raised some 250 million dollars for the children's hospital — Isaacman had pledged 100 million and challenged the public to match it, and when the public giving fell short of the 200-million-dollar goal, Elon Musk himself contributed the final 50 million — and it captured public imagination as a vision of a more accessible space age. It was, of course, made possible by a billionaire's fortune, and the "everyday people" had been carefully selected and were flying on one man's dime — the democratization was symbolic, not yet structural. But the symbolism mattered, and so did the particulars: a childhood cancer survivor with a metal rod in her leg, once a patient too sick to imagine such a thing, speaking from orbit to children in the very hospital that had healed her. For the first time, a spacecraft had carried an all-civilian crew to orbit and back, and it had done so safely, proving that private human spaceflight, free of any government mission, was genuinely feasible.

The three days aloft were not a joyride. The crew ran a battery of medical experiments on themselves — ultrasound scans, tests of balance and cognition, blood samples drawn and stored — turning their own bodies into a small research dataset on how ordinary, non-astronaut physiologies respond to spaceflight, exactly the kind of data a future of broader access would need. They also met the ordinary indignities of the frontier: the capsule's waste-management system gave trouble during the flight, a reminder that even a gleaming spacecraft is, at bottom, a cramped metal can in which four people must eat, sleep, and relieve themselves for days on end. And they carried the story to the public in a way no government mission had, a documentary crew having followed the selection and training from the beginning for a series that streamed on Netflix, so that millions watched a childhood cancer survivor, a raffle beneficiary, and a community-college geoscience professor become astronauts in something close to real time. The framing was unabashedly promotional, and yet the fact beneath it was not: people with no military or scientific pedigree had trained for a matter of months and flown safely to orbit and back, and that was genuinely new.

Isaacman was not finished. He conceived a more ambitious program, a series of missions called Polaris that would push the boundaries of private spaceflight further, developing techniques and demonstrating capabilities that went beyond anything private citizens had attempted. The first Polaris mission, Polaris Dawn, flew in September 2024, and it was a genuinely audacious undertaking. It carried a crew of four — Isaacman commanding; Scott "Kidd" Poteet, a retired Air Force lieutenant colonel and fighter pilot, as pilot; and two SpaceX engineers, Sarah Gillis and Anna Menon, as mission specialists. Gillis was a young woman who, in her day job, led the company's astronaut training program, teaching NASA's professionals and private crews alike how to fly the very spacecraft she now rode. The mission aimed to fly higher than any human had flown since the Apollo program, reaching an apogee of roughly 1,400 kilometers — 1,408, to be precise — taking the crew out into the inner reaches of the Van Allen radiation belts that surround the planet, the highest any human had traveled from Earth in more than half a century, since the last astronauts left the vicinity of the Moon in 1972, and the farthest anyone had ever gone while still, technically, in orbit around the Earth. From that altitude, the crew saw the planet not as the vast curved floor the station's astronauts see but as a distinct and distant sphere, a perspective almost no living person had witnessed. The high pass through the radiation belts was itself an experiment, a deliberate test of how the human body and the spacecraft's systems would fare in an environment that longer journeys to the Moon and Mars would have to cross.

The centerpiece of Polaris Dawn, however, was something even more daring: the first spacewalk ever conducted by private citizens, and the first spacewalk from a Crew Dragon. Spacewalks — leaving the pressurized safety of a spacecraft to venture into the vacuum, protected only by a suit — are among the most dangerous activities in spaceflight, and they had always been the exclusive domain of professional government astronauts, conducted from spacecraft and stations equipped with airlocks. Crew Dragon had no airlock; to perform the spacewalk, the entire capsule had to be depressurized, exposing all four crew members to the vacuum at once, with those remaining in their seats protected only by their suits — a fact that made the whole cabin, in effect, the airlock, and every person aboard a spacewalker whether they left the hatch or not. SpaceX had developed a new spacesuit for the purpose, evolved from the sleek pressure suit worn inside the capsule but reengineered for the exposure, with new joints and seals and a heads-up display in the helmet, and the mission would test it in the most demanding way possible. In orbit, the crew slowly purged nitrogen from their bodies over two days to avoid the bends, then depressurized the capsule, opened the hatch, and Isaacman and then Gillis each ventured partway out, holding onto a structure SpaceX had built at the hatch and dubbed the "Skywalker," testing the suit's mobility through a choreographed sequence of movements in the vacuum before returning inside and repressurizing. Each excursion lasted only minutes and neither astronaut fully left the vehicle, but the significance was undeniable. In a separate moment that captured the mission's blend of engineering and humanity, Gillis, a trained violinist, played "Rey's Theme" from the Star Wars films on a custom-made violin in orbit, the performance woven into a video linking space to young musicians on Earth as part of another fundraising effort for St. Jude. The spacewalk was brief and cautious, a first step rather than a full excursion, but it was historic: private citizens had walked in space, and a new spacesuit — a prototype of the kind that thousands might one day need, if the largest visions came true — had been proven in the vacuum.

The road to that hatch had been longer than planned. Polaris Dawn slipped well over a year past its original target, much of the delay swallowed by the spacesuit itself — for a suit meant not merely to keep a seated astronaut alive but to let a body bend, reach, and work in vacuum is one of the hardest things in all of spaceflight to build, in effect a personal spacecraft with joints, and SpaceX was designing one essentially from scratch with the explicit intention of eventually manufacturing them at scale. That ambition was the real point of the exercise. A brief spacewalk from a Crew Dragon had little practical use in itself; its value was as a rehearsal. If SpaceX meant one day to send not four but eventually hundreds or thousands of people beyond the Earth —

to the Moon, to Mars, into a raw vacuum where suits would be as ordinary as coats — it would need a suit it could mass-produce and mobility procedures it had actually tested on living bodies, and Polaris Dawn was the first true trial of both. The mission also gathered a trove of medical data on how the human body fares in the radiation of the Van Allen belts, information a company planning long journeys through exactly that environment could not otherwise buy. The Polaris program had been conceived as a series of three missions, and its final flight was intended, in the original plan, to be the first crewed voyage of Starship itself — a direct line drawn from a billionaire's private spacewalk to the very vehicle Musk hoped would one day carry settlers to Mars.

There was something telling, too, in who filled the other seats. Sarah Gillis and Anna Menon were not paying passengers but SpaceX employees — engineers who worked on the very programs they now flew, Gillis leading the company's astronaut training and Menon a mission-operations engineer who had helped write the procedures crews followed from the ground. That a private company could staff a record-setting spaceflight partly with its own rank and file, sending working engineers to the highest orbit humans had reached in half a century, said as much about how far the enterprise had come as any billionaire's ticket. For a handful of people, spaceflight had begun to shade from a national honor bestowed on a chosen few into something closer to a job assignment — a place one went, extraordinary still, but increasingly for work rather than for glory.

The Polaris missions pointed toward a future that Isaacman and SpaceX imagined together, in which private human spaceflight would grow more capable and more ambitious, developing the suits, the techniques, and the experience that would be needed for the larger goals ahead — including, ultimately, missions on the giant Starship then under development, which Polaris was explicitly meant to help prepare. The collaboration between Isaacman and SpaceX was more than a series of joyrides for the wealthy; it was a deliberate program to advance the state of private spaceflight, and it produced real firsts and real capabilities. Isaacman's own arc, meanwhile, took a turn that few could have predicted: his prominence as a private astronaut and his relationship with the space enterprise led, in time, to his nomination and appointment to lead NASA itself — a remarkable trajectory from private citizen buying his own flights to the administrator of the national space agency, a sign of how thoroughly the boundaries between the private and public space worlds had blurred in the new era. That an entrepreneur who had bought his way to space could end up running NASA said something about how much had changed, and about the growing entanglement of private wealth and public purpose in the new space age — an entanglement with both promise and peril, and one this book will have occasion to examine more critically.

The civilian missions demonstrated that the human-spaceflight capability SpaceX had built was not limited to ferrying government astronauts, and that a genuine market for private orbital flight existed, at least among the very wealthy and their beneficiaries. They were a glimpse of one possible future — space opening, however gradually and however unevenly, to a wider humanity. But even as SpaceX was making private human spaceflight a reality, the intended competitor in the Commercial Crew Program, Boeing's Starliner, was providing an object lesson in how hard human spaceflight remains, and in how far the old aerospace order had fallen behind the new. The contrast would reach its climax in a saga that saw astronauts launched on a troubled spacecraft and unable to return on it — a crew stranded, in the headlines' telling, until the reliable workhorse of the new space age came to bring them home.

Chapter 30: Starliner's Long Road

The story of Boeing's Starliner is, in a sense, the shadow that gives the SpaceX story its full definition. The two companies had been chosen together in September 2014 to build America's new crew vehicles, and the terms of the choice are worth remembering, because they sharpen everything that followed: Boeing received a contract worth 4.2 billion dollars, SpaceX one worth 2.6 billion, the established giant handed both the larger sum and the greater prestige, the upstart the smaller purse and the underdog's reputation. What happened over the following decade was a reversal so complete that it became a parable about the state of American aerospace — a tale of how the better-funded, more experienced contractor stumbled while the cheaper newcomer soared, and of what that reversal revealed about the two ways of building things that they represented.

Boeing was, by any historical measure, one of the greatest aerospace companies in the world. It had built bombers and airliners and spacecraft for generations; it had been a principal contractor on the Apollo program and the Space Shuttle; it had more experience putting human beings into space than almost any organization on Earth. When NASA chose Boeing to build Starliner, it was choosing safety, experience, the proven hand. And yet Starliner's development was troubled almost from the start, beset by the very problems that had come, in the 2010s and 2020s, to afflict Boeing across its businesses — the same corporate culture whose decline was blamed for the tragedies of its 737 MAX airliners, which crashed twice and killed 346 people, was, critics argued, visible in Starliner's struggles. The program ran over budget and behind schedule. Its first uncrewed test flight, in December 2019, went wrong in an especially revealing way: a software error in the spacecraft's mission clock left the capsule believing it was at a different point in its flight than it actually was, so that it fired its thrusters at the wrong time, burned through propellant, and never reached the orbit it needed to catch the station. Worse, during the troubleshooting, engineers on the ground discovered a second, unrelated software error — one that, had it not been caught and corrected by radioed command while the vehicle was still aloft, could have caused Starliner's service module to collide with the capsule during separation, a potentially destructive event. A subsequent review catalogued dozens of corrective actions and made plain that the mission could have ended in disaster. Boeing had to refly the uncrewed test at its own expense, which it managed successfully in May 2022, but the delays stretched on, and Starliner fell years behind Crew Dragon, which was by then flying astronauts routinely.

By the time Starliner finally carried its first crew, the scale of the divergence was almost hard to credit. Crew Dragon had first flown astronauts in the spring of 2020; in the four years that followed, it launched crew after crew — NASA rotations, all-private missions, the record-setting Polaris flights — carrying dozens of people to orbit and home again, while Starliner had yet to lift a single human being off the pad. The two companies had started together, on the same day, under the same program, chosen for the same task and judged, at the outset, with the established giant the safer bet; one had turned the job into a dependable service and a springboard to grander things, and the other, four years on, was still struggling to complete its first crewed test.

When Starliner finally launched with a crew aboard, on June 5, 2024 — years after Crew Dragon's first crewed flight — it did so carrying two veteran NASA astronauts, Butch Wilmore and Suni Williams, both former Navy test pilots with previous spaceflights behind them, on what was meant to be a roughly week-long test flight to the station and back, the final demonstration before Starliner could enter regular service. The launch succeeded and the capsule reached the station, but the journey there revealed problems: as Starliner made its approach, five of its twenty-eight reaction-control thrusters malfunctioned, forcing a tense delay while controllers coaxed several of them back to life, and the spacecraft had suffered a series of helium leaks in its propulsion system. These were not trivial glitches. The thrusters and the propulsion system were exactly what Starliner would need, in working order, to safely undock from the station, maneuver, and fire its engines to drop out of orbit and carry its crew home. With the thrusters having failed in flight — and, crucially, with the engineers unable to fully reproduce or explain the failures in ground testing — NASA and Boeing faced an agonizing question: was it safe to bring Wilmore and Williams home aboard Starliner, or was the risk too great?

The deliberations stretched over weeks and then months, as engineers on the ground ran a Starliner thruster through punishing test firings in the New Mexico desert, trying to recreate and understand a failure they could not confidently rule out for the critical return. The two astronauts, meanwhile, remained aboard the station, their planned week-long visit extending into an open-ended stay. It should be said, in fairness to them, that they bristled at the way their situation was described; both insisted they were neither stranded nor unhappy, that they were doing the work of the station and would fly Starliner home if asked. But the decision was not theirs. NASA, an agency shaped by the memory of Challenger and Columbia — disasters in which institutional and schedule pressures had contributed to accepting risks that killed crews — was determined not to repeat the mistake of flying a crew on a vehicle it did not fully trust. And in the end, on August 24, 2024, the agency made the cautious choice, the choice that the whole architecture of Commercial Crew, with its two independent providers, had been designed to make possible: it decided that Starliner would return to Earth empty, without its crew, and that Wilmore and Williams would come home instead on a Crew Dragon. In September, Starliner undocked and flew back autonomously, descending under parachutes to a soft landing at White Sands in New Mexico — proving, in a way, that it might well have carried the crew safely, but confirming that NASA had not been willing to bet their lives on "might."

There is a human texture to the long wait that the "stranded" headlines flattened. Wilmore and Williams were not idle captives; folded into the station's routine, they became working crew, and over the course of the extended stay Williams took over as commander of the orbiting laboratory, running the very outpost her troubled ride had delivered her to. Both were seasoned Navy test pilots — Williams among the most experienced spacewalkers in the corps, having once held the record for the most time spent spacewalking by a woman — and they met the situation with the professional equanimity of people who had spent their careers strapping into experimental machines, insisting publicly that they had food, work, and a way home, and that a test flight which surfaced problems was, in a sense, a test flight doing its job. That equanimity, however, could not keep the saga inside the technical realm. The image of two astronauts "stuck" in space month after month proved irresistible to the press and, in time, to politicians, and the question of when and how they would be brought home swelled into a public drama layered over the engineering one — a reminder that in the new era, with private companies and public agencies and outsized personalities all tangled together, even a cautious safety decision could harden into a headline and a grievance.

The two astronauts' stay aboard the station, originally planned for about a week, ultimately extended to roughly nine months. To bring them home, NASA reconfigured the next scheduled rotation flight, Crew-9, launching it in the autumn of 2024 with two of its four seats deliberately left empty so that Wilmore and Williams could fill them, and folding the pair into the station's normal crew rotation until that Dragon was due to return. In early 2025, it carried them safely back to Earth, ending one of the most closely watched sagas in recent spaceflight. The episode was widely, and somewhat unfairly, narrated as the astronauts having been "stranded" or "stuck" in space — in truth they were safe aboard a fully functioning station, integrated into its crew, never in immediate danger, and a lifeboat was docked and available the entire time — but the essential facts were stark enough. Astronauts had been launched on Boeing's spacecraft and could not be brought home on it; the vehicle built by the storied giant had failed to complete its most basic mission of safely carrying its crew round trip; and it had fallen to SpaceX's Crew Dragon, the workhorse of the new space age, to bring them back. The contrast that had been building for a decade was now impossible to ignore — and it came with a financial coda that underlined the point, for Starliner was a fixed-price contract, meaning Boeing, not the taxpayer, absorbed the overruns, and by this stage the company had booked well over a billion and a half dollars in losses on a program that had carried a grand total of two people to orbit and returned them on someone else's ship.

The financial wound pointed to a deeper strategic irony. NASA had structured Commercial Crew as a set of fixed-price contracts — a deliberate break from the cost-plus arrangements of the traditional aerospace world, under which a contractor's overruns had historically been passed along to the government as a matter of course. Fixed price was meant to discipline the contractors and protect the taxpayer, and in this it worked exactly as intended: when Starliner ran long and troubled, it was Boeing's shareholders, not the public, who swallowed the better part of two billion dollars in losses. But the same arrangement that SpaceX had turned to its advantage — using the fixed milestone payments, plus its own capital and its own reusable rockets, to build a vehicle it could fly cheaply and often — became a trap for a company built around the older model, one long accustomed to being paid for effort rather than results. By the mid-2020s there was open speculation that Boeing might abandon the program altogether rather than keep pouring money into a fixed-price commitment that only bled it further, and the spectacle of a storied prime contractor quietly contemplating retreat from human spaceflight, while a company two decades its junior flew astronaut after astronaut, was perhaps the sharpest single image of how thoroughly the ground beneath American aerospace had shifted.

For NASA, the episode was both a vindication and a warning. The two-provider strategy had worked exactly as designed — when one vehicle faltered, the other stood ready, and no American was left without a way home — and that redundancy was the very reason the agency had insisted, over years of budget pressure and congressional skepticism, on funding two companies rather than betting everything on one. Yet the near-total reliance on Crew Dragon that resulted was itself a kind of fragility, an echo of the single-point dependence the redundancy had been meant to banish. An agency that had set out to cultivate a competitive marketplace for human spaceflight found itself, by the mid-2020s, leaning on a single dominant supplier — cheaper and more reliable than anyone had dared hope, but a near-monopoly all the same, and a reminder that the new order carried risks as distinctive as the ones it had replaced. The triumph of the commercial model and the concentration of power in one company's hands were, disconcertingly, the same fact seen from two sides.

The lesson of Starliner's long road is not that Boeing's engineers were incompetent — they were not — nor that building crewed spacecraft is easy for anyone — it is not. Human spaceflight is brutally difficult, and SpaceX had its own failures, its own explosions, its own near-misses on the road to Crew Dragon; the very capsule that first proved the vehicle had been destroyed on a test stand only a year before Demo-2. The lesson is subtler and has to do with organizational culture and the ways of building things. SpaceX had approached the problem with the iterative, vertically integrated, test-and-fix method that had become its signature, surfacing its failures early, in testing, on the ground, and correcting them, and it had built a vehicle it could trust and fly again and again. Boeing had approached it in the traditional aerospace mode — more subcontractors, more paperwork, more reliance on analysis and review in place of hardware pushed to destruction — and had been undone by the very problems, the schedule slips and cost overruns and quality and software failures, that had come to characterize the decline of the legacy primes. It is often assumed that the story ended with Starliner scrapped, but that was not so; NASA continued to insist it wanted two independent crew providers, and Boeing kept working toward certification, precisely because the redundancy was worth having. Still, the Commercial Crew Program had been intended as a competition, and in a sense it delivered one, but the competition had a clear and instructive winner. The new way of building rockets and spacecraft, the way SpaceX had pioneered, had outperformed the old, and it had done so on the most demanding task in the business, the one where lives were at stake. That verdict — rendered not in theory but in the concrete fact of which spacecraft could safely carry astronauts home — was perhaps the most complete vindication SpaceX ever received. And it came even as the company was pouring its energy and its ambition into a new enterprise that had nothing to do with the space station and everything to do with Musk's original, animating dream: a constellation of satellites that would blanket the Earth, generate the fortune to fund the journey to Mars, and quietly become one of the most consequential and controversial pieces of infrastructure in the world.

Chapter 31: The End of the Station

The International Space Station is the largest and most expensive structure humanity has ever built in space — a sprawling laboratory the size of a football field, its main truss stretching more than a hundred meters from end to end, assembled in orbit over more than a decade at a cost of well over a hundred and fifty billion dollars, crewed continuously since November 2000 by rotating teams of astronauts from the United States, Russia, Europe, Japan, Canada, and their partners. For a quarter of a century it has been the center of human activity in orbit, the destination for Crew Dragon and the Soyuz, the workplace for the private astronauts and the professional crews, the symbol of international cooperation in space that endured even as relations on Earth deteriorated — through the war in Ukraine, the sanctions, the collapse of nearly every other channel between Washington and Moscow, the two segments of the station kept turning together, dependent on each other's power and propulsion, because they had been built that way and could not easily be pulled apart. But the station is old, its oldest modules dating to the 1990s, its systems aging, its maintenance ever more demanding — one Russian module has leaked air for years, the crews sealing hatches and hunting hairline cracks — and its end is now in sight. Around the end of the decade, the International Space Station is scheduled to be retired, deorbited, brought down in a controlled descent to burn up over the ocean, ending an era of human spaceflight.

To grasp what will be lost, it helps to hold the record in mind. Human beings have lived aboard the International Space Station without interruption since November 2000 — a continuous human presence in space now stretching past a quarter of a century, the longest unbroken habitation off the Earth in history, sustained through wars and a pandemic and the rise and fall of the governments below. The station was assembled the hard way, module by module, across more than forty flights of shuttles and rockets over more than a decade, astronauts and cosmonauts bolting and wiring together in orbit a structure that grew until it enclosed a pressurized volume comparable to the cabin of a jumbo jet and spread solar wings wide enough to shadow a football field. It is, by a wide margin, the most expensive single object ever built by human hands, and it is also, on any clear night, among the most democratic of spectacles: bright enough to be seen with the naked eye as a swift, steady point of light sliding across the sky, visible in its turn to nearly everyone on Earth, a shared human artifact that a child in a backyard can watch pass overhead. When it is gone, that light will go with it.

It is often imagined that a retiring station simply falls from the sky, or, alternatively, that it might be nudged up to a safe "graveyard" orbit and left to circle harmlessly forever. Neither is true, and the reason is instructive. The station is far too massive — some four hundred and twenty tonnes — to boost to a higher orbit with any propulsion humanity can affordably send up; and it is far too massive to allow it to come down on its own, uncontrolled, for a structure that size would not fully burn up, and its surviving fragments could rain across a populated swath of the Earth. It must instead be brought down deliberately, its descent aimed with precision at the emptiest place on the planet: Point Nemo, the oceanic pole of inaccessibility in the remote South Pacific, some 2,700 kilometers from the nearest land, a stretch of ocean so far from anything that the nearest human beings are often the astronauts overhead. Point Nemo is already a spacecraft cemetery, the resting place of the Russian station Mir and hundreds of other deorbited craft, and it is there that the International Space Station is meant to end. To perform that final maneuver, NASA turned, characteristically, to SpaceX: in June 2024 the agency awarded the company a contract worth up to 843 million dollars to build a United States Deorbit Vehicle, a heavily modified Cargo Dragon fitted with an enlarged trunk packed with propellant and power and studded with dozens of Draco thrusters — forty-six of them, against the ordinary Dragon's sixteen — whose sole purpose is to dock with the station near the end and fire, over a series of burns, to drive the whole enormous structure down into the atmosphere on a controlled path to the sea. The total cost of the effort was estimated at around a billion and a half dollars: the price, in the end, of taking the great station apart.

The need for so specialized and costly a vehicle is itself a measure of the problem. For routine reboosts — the periodic nudges that keep the station from spiraling down through the tenuous upper atmosphere — Russian Progress cargo ships had long sufficed, firing their modest engines to lift the outpost a little higher. But a controlled deorbit is a different order of task: to drive four hundred tonnes down along a precise path, on a schedule, so that whatever survives the fire falls into a targeted patch of empty ocean and nowhere near a city, requires far more propellant and thrust than any existing cargo craft can bring, and leaves no room for the kind of last-minute failure that no crewed rescue could ever fix. Hence a purpose-built machine, flown once, to do a single irreversible thing. The politics add their own weight. Russia, whose segment provides much of the station's propulsion and reboost capability, signaled that it would step back from the partnership toward the end of the decade, even as the United States and its other partners agreed to extend operations to around 2030; and once the deorbit sequence begins there is no reversing it, no boosting the station back up if the commercial successors prove not to be ready. The end, when it comes, will be deliberate, planned years in advance, and final.

There is precedent for the spectacle, and it is a sobering one. When Russia deorbited the Mir station in 2001 — a structure only a fraction of the International Space Station's size — its flaming fragments streaked across the sky over the South Pacific near Fiji, some surviving pieces reportedly as large as small cars as they plunged into the sea. The International Space Station is several times more massive, and its controlled destruction will be correspondingly greater, a deliberate, engineered meteor shower over the emptiness of Point Nemo marking the end of the outpost that taught humanity how to live in orbit. That the same company now flying the crews, launching much of the cargo, and building the intended successors should also be the one entrusted to bring the old station down is the final measure of how completely SpaceX had woven itself into every thread of the enterprise — present, by the end, at both the birth of the new order and the funeral of the old.

The retirement of the station poses a question that the new space age must answer: what comes next? For the United States and its partners, the station has been the anchor of human presence in low Earth orbit, the place where astronauts live and work, where research is conducted, where the experience of long-duration spaceflight is accumulated. To retire it without a successor would be to abandon low Earth orbit, to give up the continuous human presence that has been maintained for a quarter of a century, and — not least — to leave the field to China, whose Tiangong station, completed in 2022 and permanently crewed, would, once the International Space Station is gone, become the only continuously inhabited outpost in orbit. Tiangong, roughly a fifth the mass of the International Space Station and home to three taikonauts at a time, was smaller and simpler than the aging giant it might outlast — but it was wholly China's own, built without foreign partners after China had been excluded from the International Space Station by American law, and its solitary survival, however temporary, would carry a symbolism no policymaker in Washington could ignore. The prospect of China holding that distinction alone, even temporarily, was unwelcome to American policymakers, and it added urgency to the question of what would replace the aging station.

The answer, characteristically for the new space age, was to turn to the commercial sector. Rather than build and own a new government space station in the traditional way — an enormously expensive proposition — NASA moved to encourage private companies to build commercial space stations, which the agency would then use as a customer, buying access for its astronauts and its research, just as it bought cargo and crew transport. Several ventures took up the challenge. Axiom Space, which had been flying private astronauts to the International Space Station, planned to build its own station, initially by attaching modules to the existing outpost and later detaching them to form a free-flying station of its own. A consortium led by Blue Origin and Sierra Space proposed a station called Orbital Reef, pitched as a "mixed-use business park" in orbit. A partnership of Voyager Space and the European giant Airbus advanced a design called Starlab, a single large module meant to launch in one piece. And a newer entrant, the company Vast, moved fastest of all, aiming to launch a small first station, Haven-1, on a Falcon 9 and claim the title of the first commercial station in orbit — the kind of aggressive, compressed timeline that the new space economy had made thinkable. The vision that united them was of a commercial economy in low Earth orbit, a set of privately owned and operated stations serving government astronauts, private researchers, manufacturers, and even tourists, replacing the single government-run outpost with a plurality of commercial ones.

Whether this vision would be realized on the timeline required was, like so much in the new space age, uncertain. Building a space station is hard and expensive, and the commercial stations were, as this book was written, still in development, facing the familiar challenges of funding, engineering, and schedule; more than one had already slipped its dates or restructured its plans. There was a real risk of a gap — a period after the International Space Station's retirement and before the commercial stations were ready, during which the United States and its partners might have no crewed outpost in orbit, and China might stand alone. Bridging that gap, ensuring that the commercial stations were ready in time, became a significant concern, and it depended on the commercial ventures delivering on their ambitious plans, which was far from guaranteed. The end of the International Space Station thus loomed as both an opportunity — to transition to a vibrant commercial economy in orbit — and a risk — to lose the continuous human presence that had been so hard-won, and to cede ground to a rival.

SpaceX's role in this transition was, as everywhere, central and multifaceted. Its Crew Dragon and, in the future, its Starship would provide the transport to and from whatever stations replaced the International Space Station. Its launch capabilities would carry the commercial stations to orbit. Its deorbit vehicle would end the old station's life. And SpaceX itself harbored ambitions in orbit beyond mere transport, with Starship envisioned not only as a launch vehicle but potentially as a habitat, a pressurized volume vastly larger than any station module, that could serve as an orbital outpost or a base on the Moon or Mars. The figures make the prospect vivid: the entire pressurized volume of the International Space Station, painstakingly assembled from module after module over more than a decade, is roughly 900 cubic meters, about the interior of a large passenger jet; the pressurized volume of a single Starship is on the same order or greater, meaning that one vehicle, launched whole in a single flight, might match the living space that took thirty-odd assembly missions to build. That a single mass-produced rocket could deliver to orbit, in one launch, what the great station represented was the kind of discontinuity that made the new space age feel genuinely new. Whether it would be realized depended, once again, on Starship's success, but it pointed toward a future in which the very meaning of a space station might change, from a fragile, expensive assembly of small modules to a spacious, cheap habitat delivered whole to orbit.

The end of the International Space Station marks, in a sense, the end of the first era of human spaceflight — the era of government programs, national agencies, and a single cooperative outpost — and the beginning of a new one, commercial, plural, and uncertain. The station stood for a quarter of a century as the symbol of what international cooperation could achieve in space, a place where the nations of the Earth worked together above the divisions of the ground. What replaces it — commercial stations, Chinese outposts, Starship habitats, or some mixture not yet foreseen — will reflect the character of the new space age: more commercial, more competitive, more fragmented, and more uncertain than the era it succeeds. The retirement of the great station, scheduled for the end of the decade and to be carried out, fittingly, by a SpaceX vehicle, is thus not merely the deorbiting of an aging structure but a passage from one age of human spaceflight to another, and how that passage is managed — whether the continuous human presence in orbit is maintained or lost, whether the commercial vision is realized or falters, whether cooperation or competition defines the new era — is one of the open questions of the transformed space age, and one in which SpaceX, as in so much else, sits at the center.

Chapter 32: A Sky Full of Satellites

The idea behind Starlink was, at its core, a bet on falling launch costs — a bet that SpaceX was uniquely positioned to make because it was the company driving those costs down. For decades, entrepreneurs had dreamed of providing internet access from space by means of satellites, and for decades those dreams had mostly ended in bankruptcy. The classic problem was geography and physics. Traditional communications satellites sat in geostationary orbit, some thirty-five thousand seven hundred and eighty-six kilometers above the equator, where a satellite orbits at the same rate the Earth turns and thus appears to hover over a fixed spot on the ground. From that great height, a single satellite could cover a vast area — three well-placed geostationary satellites can see nearly the whole planet — but the distance imposed a fatal penalty. Here it is worth doing the arithmetic, because the number is the whole story. A signal traveling up to a geostationary satellite and back down covers more than seventy thousand kilometers, and even light, the fastest thing there is, crawling along at three hundred thousand kilometers per second, needs roughly a quarter of a second to make each leg; add the return path and the switching and processing at each end, and the round-trip delay — the latency — comes to something like half a second or more. Half a second does not sound like much until you try to hold a conversation across it, or click a link, or fire a shot in an online game, and everything answers you a beat late. Geostationary satellites could deliver internet, in other words, but a slow, laggy internet, forever out of step with the interactive applications that define the modern network, and they could serve only so many users before their capacity was exhausted.

The alternative was to put the satellites much closer to the Earth, in low orbit, only a few hundred kilometers up, where the same calculation runs the other way. A signal to a satellite five hundred and fifty kilometers overhead and back travels barely a thousand kilometers round trip and returns in a handful of milliseconds — fast enough to rival, and in some cases beat, the fiber and copper of the ground. But low orbit imposed its own brutal arithmetic. A satellite only a few hundred kilometers up moves rapidly across the sky and can see only a small patch of the Earth's surface at any moment; to provide continuous coverage of the whole planet, you would need not a handful of satellites but thousands, an entire constellation blanketing the sky, with enough satellites overhead at every point on Earth at every moment that a user could always reach one. And building, launching, and operating thousands of satellites had always been prohibitively expensive — the launch costs alone, at the old prices, would have run to tens of billions of dollars, dooming the economics before the first customer signed up. This was not a hypothetical failure mode; it was history. In the 1990s a company called Iridium, spun out of Motorola, actually built and launched a low-orbit constellation — sixty-six satellites, a technical triumph — to deliver satellite telephone service, and in 1999 it filed for bankruptcy after defaulting on some 1.5 billion dollars of debt, one of the largest failures of its day. A more ambitious venture called Teledesic, backed by Bill Gates and the cellular pioneer Craig McCaw, proposed hundreds of satellites for an "Internet in the Sky" and folded without launching an operational system at all.

It is often said that Iridium proved the whole idea of a satellite constellation unworkable, but that is precisely the wrong lesson, and the distinction matters for understanding why Starlink could succeed where its ancestor failed. Iridium's technology worked; its satellites flew and its phones connected. What killed it was economics and timing. Its handsets were bricklike and cost several thousand dollars, its per-minute charges were steep, and just as it switched on, ordinary terrestrial cellular networks were spreading across the very markets it had counted on, offering cheaper service to anyone within reach of a tower. Iridium had bet that the world would still need expensive satellite phones, and the world, cabled and towered faster than anyone expected, largely did not. There is a small, telling detail buried in its story: Motorola had originally planned seventy-seven satellites and named the venture Iridium after the chemical element with seventy-seven electrons; when the design was trimmed to sixty-six, the name that would have been chemically accurate was Dysprosium — from the Greek for "hard to get at" — and, perhaps wisely, they kept Iridium. The constellation, salvaged out of bankruptcy for a fraction of its cost, flies still, serving ships and soldiers and remote sensors. But its collapse was the cautionary tale that hung over every constellation proposal for two decades: the constellation was too expensive to build, and the market was not yet ready.

The market's readiness had, by SpaceX's day, utterly transformed. When Iridium and Teledesic made their bets, the internet was still a novelty and a landline telephone did most of the world's talking; by the 2010s, connectivity had become something close to a necessity of modern life, the medium of work, school, banking, medicine, and simple human contact — and vast stretches of the planet still lacked it. Roughly a third of humanity remained offline, not for want of wanting it but because the economics of stringing fiber or raising cell towers to remote villages, scattered farms, ships at sea, and far-flung islands had never closed. That was precisely the gap a constellation could fill: the places terrestrial networks found unprofitable to reach were exactly the places a satellite passing overhead could serve as easily as any other. Where the pioneers of the 1990s had been early to a market that did not yet exist, SpaceX arrived at one that was starving — and that difference, as much as any rocket, separated a cautionary tale from a business.

What changed the calculus was the reusable rocket. If launch costs could be driven down by an order of magnitude — and SpaceX's reusable Falcon 9 was doing exactly that — then the economics of a low-orbit mega-constellation shifted from impossible to merely daunting. Suddenly it became conceivable to launch thousands of satellites at a cost the business could bear, especially if you owned the rocket company doing the launching and could fly the satellites on your own vehicles at cost. SpaceX was in the unique position of being both the launch provider and the constellation operator, able to launch its own satellites on its own reusable rockets, capturing the savings and controlling its own destiny. No competitor had that advantage. A rival constellation would have to buy launches on the open market, at prices that included someone else's profit — and, more pointedly still, its fiercest rival might have to buy some of those launches from SpaceX itself. SpaceX could launch itself, and did: the first batch of sixty Starlink satellites went up on a Falcon 9 in May 2019, packed flat and stacked like a deck of cards, deploying into a 550-kilometer orbit, and from there the launches came in a relentless cadence, sixty and then more at a time, until the constellation numbered in the thousands within a few years. The reusable rocket that had been built to lower the cost of reaching space now enabled a business that would have been impossible without it, and that business, in turn, would generate the revenue to fund the next and greatest step, the journey to Mars. It was a virtuous cycle, at least in conception: cheap launch enables the constellation, the constellation funds the Mars rocket, the Mars rocket fulfills the mission.

The strategic logic went deeper still. SpaceX's rockets, however cheap, served a launch market that was inherently limited — there are only so many satellites the world's governments and companies need to launch each year, and that market, even expanded by falling costs, was worth some billions of dollars annually, enough to sustain a launch company but not enough, Musk judged, to fund the staggering expense of developing a Mars-colonization vehicle and actually settling another planet. To generate that kind of money, SpaceX needed to sell not launches but a service directly to millions of customers, and internet access was the obvious candidate: a global market worth hundreds of billions of dollars a year, with vast numbers of people underserved by terrestrial networks — rural populations, remote communities, ships at sea, aircraft, anyone beyond the reach of cables and cell towers. If SpaceX could capture even a modest share of the global internet market by beaming connectivity from space, it would generate revenues dwarfing anything the launch business could provide, and it would do so from a captive infrastructure it controlled end to end. Starlink was, in this sense, the answer to the question of how to pay for Mars — the business that would turn the cost advantage of reusable rockets into the fortune needed to fund humanity's expansion beyond Earth.

The bet, it turned out, would pay off on a scale that dwarfed its failed predecessors. Where Iridium had launched sixty-six satellites and nearly died of the effort, SpaceX would loft thousands, flying its own satellites on its own rockets in batches every week or two until, within a few years of that first launch of sixty, the constellation numbered in the thousands and served millions of subscribers across dozens of countries — ranchers in the American West, villages in the Amazon, research camps in Antarctica, airliners at cruising altitude. Its consequence quickly outran commerce. When Russia invaded Ukraine in 2022 and shattered much of the country's communications, thousands of Starlink terminals poured in and became, almost overnight, a backbone of Ukrainian military and civilian connectivity — a vivid demonstration that a private internet constellation had become strategic infrastructure, and, in the disputes that followed over who controlled it and how it could be used, a demonstration too of the unsettling power such infrastructure placed in the hands of a single company and a single man. All of that lay ahead of the first sixty satellites; but it was implicit in the idea from the start, in the recognition that whoever could actually build and afford a low-orbit constellation would come to hold something no one had held before.

That last fact — the sudden concentration of a new kind of power — had a visible counterpart in the sky, and it, too, arrived with the very first satellites. Almost as soon as the first Starlinks reached orbit in 2019, astronomers raised the alarm: the spacecraft were far brighter than anyone had anticipated, and their glinting trails streaked across long-exposure telescope images, threatening to pollute the very sky through which humanity studies the cosmos. SpaceX responded with darkening coatings and deployable sunshade visors that dimmed the later satellites, a partial remedy that never fully satisfied the observatories, and the episode became an early skirmish in a larger question the constellation forced open: what happens when a single company can alter, for every person on Earth at once, something as ancient and commonly held as the appearance of the night sky. The worry was not only optical. Radio astronomers fretted that the constellation's downlink transmissions might swamp the faint cosmic signals their instruments were built to catch, and international bodies that had never before had to regulate the brightness or the emissions of the heavens found themselves improvising rules for a problem that had not existed a decade earlier — one more instance of the technology arriving faster than the norms to govern it.

The plan was breathtaking in scale. SpaceX proposed to build and launch a constellation of thousands, and eventually tens of thousands, of small satellites in low Earth orbit, arranged in "shells" at various altitudes and inclinations to provide continuous global coverage. The satellites would communicate with user terminals on the ground — flat, pizza-box-sized phased-array antennas, with no moving parts, that steer their beams electronically to track a satellite racing across the sky, which customers would install at their homes or businesses, or mount on vehicles, ships, and planes — and with ground stations connected to the internet, and, in the later generations, with one another, by lasers firing between satellites so that data could cross oceans in orbit without ever touching the ground. As a satellite passed overhead, a user's terminal would connect to it, and as that satellite moved on and another rose over the horizon, the terminal would hand off seamlessly to the next, maintaining a continuous connection from a constantly shifting fleet of spacecraft. It was an audacious technical and logistical undertaking, requiring not just the satellites and the rockets but a whole ecosystem of ground stations, user terminals, and orchestrating software. And it required manufacturing satellites not in the handfuls or dozens that had always characterized the space industry but in the thousands — an industrial challenge as formidable, in its way, as the rockets themselves. The old space industry had built satellites the way craftsmen build cathedrals, one bespoke, hand-assembled masterpiece at a time, each costing hundreds of millions of dollars and taking years; to build a constellation of thousands, SpaceX would have to reinvent that entire tradition, turning a craft industry into something closer to an automobile assembly line, producing spacecraft by the day. That transformation, as much as the rockets or the business model, is what made Starlink possible — and it is why, within a few years, the objects most likely to catch your eye in a clear night sky, a silent train of bright dots sliding across the stars, would be not the constellations the ancients named but the one SpaceX built. It is where we turn next.

Chapter 33: Building Satellites Like Cars

The traditional communications satellite was a masterpiece of bespoke engineering and a monument to expense. It was large, often the size of a bus, weighing several tons, packed with specialized components, built by hand over years by teams of engineers, tested exhaustively, and priced in the hundreds of millions of dollars. It was built this way because it had to work, alone, for fifteen or twenty years in the unforgiving environment of geostationary orbit, some thirty-six thousand kilometers above the equator, with no possibility of repair; a single failure could lose a fortune, so every satellite was engineered and tested to an extreme standard of reliability, and built in tiny numbers. The whole industry ran at that cadence. A large Western manufacturer might turn out a dozen or so of these exquisite machines in a good year, each one effectively a prototype, each one negotiated and insured and fussed over like a cathedral. This model was utterly incompatible with Starlink. A constellation of thousands of satellites could not be built one bespoke spacecraft at a time; it would take forever and cost more than any business could bear. To make Starlink work, SpaceX had to do to satellites what it had done to rockets: reinvent how they were made, driving down the cost and driving up the volume by orders of magnitude.

The philosophy SpaceX brought to satellite manufacturing inverted the traditional one. Instead of building a few exquisite, ultra-reliable, long-lived satellites, it would build enormous numbers of cheaper, simpler, shorter-lived ones, accepting that individual satellites would fail or wear out and simply replacing them, treating the constellation as a whole as the reliable system rather than any single spacecraft. This was a profound shift in thinking, borrowed more from the world of consumer electronics and mass manufacturing than from the world of traditional aerospace. A Starlink satellite was designed to be produced on something resembling an assembly line, in a factory, in large quantities, using where possible standard, mass-produced components rather than the exotic space-grade parts that cost fortunes. It was designed to be flat and compact, so that many of them could be stacked together and packed into a single Falcon 9 launch — dozens at a time, flat-packed like plates, deployed in orbit not by the delicate spring-loaded dispensers of tradition but by a brutally simple maneuver in which the rocket's upper stage rolled and the whole stack was released to drift apart under its own gentle momentum. And it was designed with a limited lifespan of about five years, after which it would be deliberately deorbited, using the last of its propellant to lower itself into the atmosphere where it would burn up harmlessly, and be replaced by a newer, better version — a constellation in constant renewal, always incorporating the latest improvements, rather than a fixed set of satellites frozen at the technology of their launch date.

It is worth pausing here to correct a misconception that clung to Starlink from the beginning. It was often said, in alarmed press coverage, that the constellation would rain thousands of dead satellites down on people's heads, or leave a permanent junkyard overhead. In fact the design cut precisely the other way. Because the satellites fly low — the main shell sits at roughly 550 kilometers — a failed satellite does not linger for centuries as debris the way objects in higher orbits do; atmospheric drag drags it down within a few years even if its engine is dead, and a healthy one is steered down on command. The satellites are built, deliberately, to vaporize completely on reentry, leaving nothing to reach the ground. The short life that critics mistook for wastefulness was in truth a feature: it kept the orbital neighborhood self-cleaning and let the whole system be refreshed with better hardware every few years, the way a phone company upgrades its towers.

This manufacturing achievement is, in some ways, as impressive as the rockets, and it is far less appreciated. SpaceX built the capacity to produce satellites at a rate the industry had never seen — not a few dozen a year, as had been typical, but thousands. At a satellite plant in Redmond, Washington, north of Seattle, the company industrialized spacecraft the way Detroit once industrialized the automobile, and the numbers it eventually reached would have been science fiction to the old guard. In the documents SpaceX filed ahead of its public offering in 2026, the company disclosed that the Redmond line was turning out satellites at an average rate of about seventy a week over the winter and spring of 2025 into 2026 — on the order of ten spacecraft a day, something like three and a half thousand a year, from a single factory. An earlier boast by one Starlink executive that the company could build six satellites a day, which had sounded outlandish when he said it, had quietly been surpassed. To achieve this required rethinking every aspect of satellite design and production: standardizing components, simplifying designs, automating assembly, and applying the lessons of high-volume manufacturing to a product that had always been made in artisanal quantities. Even the customer hardware moved onto a mass footing; the flat "Dishy" terminals and their kits poured out of a separate plant in Bastrop, Texas, by the tens of thousands a day, a scale of consumer-electronics production that no satellite company had ever attempted. The contrast with the industry's old habits was almost vertiginous: the same civilization that once treated a single communications satellite as a national event, unveiled to the press and insured like a skyscraper, now built spacecraft and their ground terminals on rolling lines the way earlier decades had built televisions and automobiles, in numbers so large that no one bothered to name the individual units. What had been handcraft became throughput, and the psychological shift — from cherishing each satellite to treating the fleet as a renewable, upgradeable resource — was as radical as any of the hardware.

The company iterated rapidly on the satellite design, producing successive versions that were more capable, more powerful, and cheaper to build, each generation an improvement on the last, in the same restless, iterative spirit that drove the rockets. The earliest operational satellites weighed a couple of hundred kilograms and did one job — relay a signal down to a nearby ground station. Later versions grew heavier and vastly more capable; the so-called V2 Mini, riding a bus about twice the size of its predecessor and adding more powerful phased-array antennas and E-band backhaul, was reckoned to deliver roughly four times the capacity of the earlier model. Later generations added the capability the early satellites conspicuously lacked: laser links between spacecraft, optical terminals that let the constellation route data through space from satellite to satellite, each link carrying data at up to two hundred gigabits per second, without ever touching a ground station. The result was a mesh of light strung across the sky. By early 2024 SpaceX engineers disclosed that the network's inter-satellite lasers — some nine thousand of them at the time — were moving on the order of forty-two million gigabytes of data every day, a figure that made the laser mesh, quietly and almost unnoticed, one of the largest data backbones ever built by anyone, anywhere. The satellites grew larger and more powerful over time, to the point that the biggest versions could only be launched by the giant Starship then under development, tying the growth of the constellation to the progress of the Mars rocket.

To grasp what this industrialization meant, it helps to hold the two worlds side by side. A traditional geostationary communications satellite might cost between one hundred and fifty and four hundred million dollars, take three to five years to design and build, and be produced essentially one at a time, each unit a negotiated contract fussed over by insurers who would demand a premium of tens of millions more before it ever left the ground. A Starlink satellite, by contrast, was reckoned to cost on the order of a few hundred thousand dollars — less, in real terms, than the insurance alone on one of the old giants — and rolled off the line in a matter of days. The reliability philosophy inverted just as sharply. The old satellite was tested to exhaustion precisely because it was irreplaceable; a single flaw discovered in orbit was a catastrophe with no remedy. SpaceX could afford a looser, statistical standard: if a small fraction of each batch failed, it hardly mattered, because a replacement was already being built and a fresh launch was only weeks away. The constellation, not the spacecraft, was the thing that had to be reliable — and a constellation of thousands, constantly replenished, could be extraordinarily reliable even while any individual member was cheap, imperfect, and disposable. This was the same conceptual leap that had once separated the hand-built automobile from the mass-produced one, or the artisan's clock from the machine-stamped watch, transplanted into orbit for the first time.

The launches, too, were unprecedented in their cadence. To build a constellation of thousands and then tens of thousands of satellites, and to keep replacing the ones that aged out, SpaceX had to launch at a rate no one had ever sustained. Starlink launches became the backbone of SpaceX's flight schedule, with Falcon 9 rockets carrying batches of satellites to orbit again and again, dozens of times a year and then more than a hundred times a year, a relentless drumbeat of launches that both built the constellation and, not incidentally, justified and refined the reusable-rocket infrastructure. Here the arithmetic of the enterprise becomes vivid. Consider a single Falcon 9 lofting a stack of a couple dozen satellites: the booster that carries them has flown before, perhaps a dozen or twenty times, and will land on a drone ship and fly again; the satellites cost a fraction of a traditional spacecraft; and SpaceX is both the launch provider and the customer, so the transaction happens at internal cost. No competitor buying rides on someone else's rocket, one payload at a time, at commercial prices, could come close to matching the economics. The synergy was total: Starlink gave SpaceX a constant, high-volume customer for its own rockets — itself — which kept the launch cadence high, the boosters flying, the costs low, and the manufacturing and operations sharp, while the rockets gave Starlink the cheap, frequent access to orbit that made the constellation economically viable. Each business made the other stronger. By flying its own satellites on its own rockets at cost, SpaceX enjoyed an advantage no competitor could match, and it turned that advantage into the fastest and largest build-out of orbital infrastructure in history.

The results, measured in raw numbers, are staggering, and they accumulated with a speed that outran most observers' ability to keep track. The first batch of operational Starlink satellites launched in May 2019 — sixty of them at once, flat-packed into the Falcon 9 fairing like a deck of enormous cards, a startling sight as they spread out across the sky in a train of bright dots that alarmed skywatchers who had never seen anything like it. Reports of a mysterious "string of pearls" gliding silently overhead flooded into observatories and police switchboards alike in the nights that followed; astronomers, who had not been consulted, watched with mounting dread, understanding at once what a few thousand such objects would do to their long-exposure images. From there the constellation grew relentlessly, batch after batch, launch after launch, until within a few years Starlink comprised thousands of satellites, more than any other operator had ever flown, and then more than all other operators combined. By the mid-2020s Starlink had become, by a wide margin, the largest satellite constellation in history, and it kept growing. By 2026 the constellation numbered well over ten thousand satellites in orbit — a figure that, according to analysts, represented roughly two-thirds of all the active satellites in space, meaning that a single private company operated the majority of humanity's functioning spacecraft. This concentration was without precedent, a transformation of the near-Earth environment accomplished in a handful of years by one firm. What had been a sky sparsely populated by a few thousand satellites belonging to many nations and companies became a sky dominated by Starlink.

There is a lasting significance in all of this that is easy to miss amid the numbers, and it is best captured by a comparison. For the whole of the space age, from Sputnik onward, humanity had launched somewhere on the order of ten thousand satellites in total, across seven decades and every nation that ever reached orbit. SpaceX built a single company's constellation of comparable size in under seven years, from one factory, on reused rockets, and did it while the price of the customer hardware fell from three thousand dollars at the start of the beta to under fifteen hundred to build, and the satellites themselves became cheap enough to throw away and replace. That is the quiet revolution: not any single clever satellite, but the industrialization of the satellite itself, the moment when spacecraft stopped being cathedrals and started being cars. You meet the consequence of it every time a ship at sea, a plane at altitude, or a farmhouse at the end of a dirt road connects to the internet through a dish that found its own satellites overhead — hardware descended, in a direct manufacturing line, from that South Texas and Washington State experiment in making satellites the way Henry Ford made automobiles. The scale of it raised questions — about astronomy, about orbital congestion, about the wisdom of one company controlling so much — to which we will return. But first, the business: who was buying this space-based internet, and how a constellation of satellites became, in a few short years, a service used by millions and a strategic asset that reshaped the map of global connectivity.

Chapter 34: How Starlink Works

To a customer, Starlink is simple: a flat dish, a subscription, and fast internet where there was none before. Beneath that simplicity lies one of the most complex engineering systems ever built — a constantly moving constellation of thousands of satellites, a network of ground stations, a fleet of user terminals, and the software to orchestrate it all into a seamless connection. Understanding how it works illuminates both the achievement and the reasons no one had done it before.

Start with the problem the system solves, and with the misconception it had to overcome. For decades, satellite internet meant geostationary satellites — a handful of enormous spacecraft parked over the equator at thirty-six thousand kilometers, each one hanging motionless above a fixed spot on the ground. The advantage was that a dish could point at one and never move again; the fatal disadvantage was distance. A signal climbing to geostationary orbit and back travels the better part of a hundred and fifty thousand kilometers on the round trip, and light, for all its speed, needs roughly half a second to make it, an eternity that made web pages crawl and rendered video calls and online games all but impossible. It was often assumed that satellite internet was simply, inherently, slow — a technology of last resort. Starlink's founding insight was that this was a consequence of altitude, not of physics: put the satellites low, only a few hundred kilometers up, and the round trip collapses to a few thousandths of a second, comparable to a good terrestrial connection. The price of going low, however, is motion. A satellite in low Earth orbit does not hang over one place; it races across the sky at some twenty-seven thousand kilometers an hour, high enough overhead to be useful for only a few minutes before it drops below the horizon and another must take its place. To provide continuous coverage, then, requires not one satellite but a great many, arranged so that at every moment, from every point on the served part of the Earth, at least one satellite is overhead and reachable.

Starlink's satellites are organized into "shells" — groups at particular altitudes and orbital inclinations, the bulk of them circling at around 550 kilometers on paths tilted about fifty-three degrees to the equator — arranged so that their coverage overlaps and combines to blanket the populated latitudes of the Earth. A useful way to picture it: at any given instant, a customer's dish has perhaps a handful of satellites climbing across the patch of sky above it, each of them visible for only three or four minutes from horizon to horizon. As the satellites sweep across the sky, the terminal connects to whichever is best positioned, and as that satellite moves on and another rises, the terminal hands off from one to the next, maintaining a continuous connection from an ever-changing fleet of spacecraft passing overhead — a constant, invisible relay race conducted hundreds of kilometers up, with a new runner taking the baton every few minutes, thousands of times a day, without the user noticing a single handoff. That choreography, repeated across millions of terminals at once, is the thing that had never been made to work before.

A single worked comparison makes the payoff concrete. When you click a link on an old geostationary connection, the signal must climb to a satellite parked thirty-six thousand kilometers up and come back down — and the reply must make the same journey — so that even at the speed of light the round trip alone consumes something like six hundred milliseconds, better than half a second of dead air before anything happens on your screen. On Starlink, with the satellite only five hundred and fifty kilometers overhead, that same round trip collapses to a few tens of milliseconds, low enough that a video call does not stutter and an online game remains playable. The difference is not a matter of clever engineering so much as of geometry: Starlink is fast for the same reason a neighbor's shout reaches you sooner than one from across a valley. But the geometry that buys low latency imposes the punishing requirement of numbers, because a satellite so close to the ground can see only a small patch of Earth and stays in view only briefly, which is precisely why the system needs thousands of them where the old approach needed three.

The user terminal — the dish that customers install — is itself a marvel, and its affordability was as important to Starlink's viability as the satellites. To track the fast-moving satellites and hand off between them, the terminal cannot use a simple fixed antenna; it needs to steer its beam electronically, aiming at one satellite and then rapidly redirecting to the next, without physically moving. This is accomplished with a "phased array," a flat panel containing many small antenna elements whose signals are combined and precisely timed — phased — so that the beam they form points in a chosen direction and can be swung to a new one in a fraction of a second, purely by electronics, with no moving parts at all. Phased-array antennas had traditionally been extraordinarily expensive, the stuff of warship radars and missile-defense systems, costing tens or hundreds of thousands of dollars apiece; the notion of putting one on a suburban roof for the price of a laptop would have been laughed out of any engineering meeting. One of SpaceX's most underrated achievements was to drive the cost of a capable phased-array terminal down from around three thousand dollars at the start of the public beta in late 2020 to under fifteen hundred dollars to build within a few years — and then to sell it to consumers for less than that, swallowing the difference in order to grow the network. The company was, in effect, subsidizing a small supercomputer of an antenna in every customer's yard, betting that the monthly subscriptions would pay it back. The terminal, once installed and pointed at the open sky, aligns itself automatically, finds the satellites, and maintains the connection with no effort from the user, who sees only fast, reliable internet, and who has no idea that the flat white panel is quietly re-aiming an invisible beam several times a minute.

That self-sufficiency was a deliberate and hard-won design goal, and it is worth appreciating what it replaced. Installing a traditional satellite dish had always been a job for a trained technician with a signal meter, painstakingly aligning the dish to a fraction of a degree toward a fixed point in the sky. Starlink shipped its terminal in a flat box to customers who had never aimed an antenna in their lives, with instructions that amounted to little more than "plug it in and point it at open sky." The terminal did the rest, sweeping its electronic beam across the heavens to locate the passing satellites, learning over its first minutes which patches of sky were blocked by trees or rooflines, and settling into service on its own — a piece of radar-grade technology deliberately dumbed down at the interface so that anyone could use it. It was this combination, a supercomputer's worth of antenna wrapped in the simplicity of a kitchen appliance, that turned a military capability into a consumer product, and it was as essential to Starlink's reach as anything happening in orbit.

The satellites, for their part, must connect the users to the wider internet, and here the system evolved in an important way. In the early configuration, each satellite acted as a relay between the user terminals below it and a ground station within its view that was connected to the internet — the signal went from the user's dish up to a satellite, down to a ground station, and thence to the internet, and back. This "bent pipe" worked, but it carried a hard geographic constraint: a satellite could only serve a user if it could also see a ground station, which meant the system fell silent over the middle of oceans, across the poles, and in vast remote regions far from any ground infrastructure. To overcome this, later Starlink satellites added laser links — optical connections between satellites, invisible infrared beams reaching across hundreds and even thousands of kilometers of vacuum, allowing spacecraft to pass data directly from one to another through space, at the speed of light, without needing to touch a ground station. With laser links, the constellation became a mesh network in the sky, able to route a user's data from satellite to satellite across the globe until it reached one within view of a ground station, extending coverage to the middle of oceans, the poles, and anywhere else beyond the reach of ground infrastructure. By 2024, SpaceX engineers revealed, the network carried something like nine thousand of these laser links, moving on the order of forty-two million gigabytes of data every day at a combined throughput measured in the terabits per second — a global backbone of light, floating overhead, that few members of the public even knew existed. There is a further, counterintuitive twist buried in the physics: light travels through the vacuum of space roughly half again as fast as it travels through the glass of a fiber cable, so over very long distances a signal bouncing between laser-linked satellites can, in principle, reach its destination faster than one racing through undersea cable — a fact that made the laser mesh interesting not only to remote villages but to financial traders for whom milliseconds are money.

Orchestrating all of this — thousands of satellites in constant motion, millions of user terminals, ground stations, laser links, handoffs, routing — required software of formidable complexity, and here SpaceX's software-centric culture was again an asset. The constellation had to be managed as a coherent system: satellites maneuvered to maintain their orbits and avoid collisions, capacity allocated to users, connections handed off seamlessly, traffic routed efficiently through the mesh, all autonomously and in real time, across a system spanning the globe and the sky. The scale of the traffic-management problem alone is worth dwelling on. With more satellites in orbit than every other operator combined, SpaceX found itself responsible for the lion's share of the collision-avoidance maneuvers performed anywhere in low Earth orbit; its satellites received warnings of potential close approaches and, guided by software rather than by controllers on the ground, nudged themselves out of harm's way autonomously, day and night. The software that ran Starlink was as much a part of the achievement as the hardware, and it reflected the same iterative, rapidly improving approach that characterized everything SpaceX did, updated and refined continuously as the constellation grew and the service expanded. And the whole system was designed to keep getting better: as newer, more capable satellites replaced older ones, as the constellation grew denser, as the software improved, the service grew faster and more capable, a network in constant renewal rather than a fixed installation frozen at the technology of its deployment. A customer who bought a Starlink dish in 2021 and never touched it again would have found, years later, that their connection had quietly grown faster and more reliable while they slept, upgraded by satellites overhead that had not even been built when they signed up. It is a strange inversion of the usual relationship between a customer and a piece of hardware: the dish on the roof aged, as all hardware does, yet the service it delivered improved year over year, because the part that did the real work was not the panel the customer owned but the churning, self-renewing machine in the sky that no customer ever touched. In that sense Starlink behaved less like a product one buys and more like a public utility that keeps rebuilding itself — except that this utility floated hundreds of kilometers overhead and was rebuilt, in its entirety, every few years.

The result of all this complexity, from the user's point of view, was something that felt like magic: fast internet, anywhere under the open sky, from a dish that set itself up and a network the user never saw. But the magic rested on a staggering foundation — thousands of mass-produced satellites, launched by the hundreds on reusable rockets, connected by lasers, tracked by phased-array terminals, orchestrated by software, all working together to relay a signal from a dish in a remote field to the global internet and back, through a constellation sweeping overhead at thousands of kilometers an hour. No one had built such a thing before, because the pieces — cheap mass-produced satellites, cheap reusable launch, cheap phased-array terminals, the software to run it all — had not existed, and SpaceX had had to create them all, together, in a single integrated system. That integration, the building of every piece of a fantastically complex whole, was the achievement, and it was why Starlink, for all the competitors who would try to follow, remained years ahead: not because any single piece was uniquely brilliant, but because SpaceX had assembled the entire system, from the rockets to the routing, into a working whole. You meet the fruit of that integration in the most ordinary places now — a cruise ship in mid-Pacific streaming a live sports match, an Antarctic research base on a video call, a wildfire crew coordinating over a connection thrown up in a place where the towers had burned down — each of them the far end of an invisible relay race, run every minute of every day, hundreds of kilometers overhead, by a machine the size of a small nation that almost no one ever sees. That the whole apparatus should feel, to the person tapping at a laptop, like nothing at all — like ordinary internet, unremarkable and taken for granted — is perhaps the surest measure of how completely the engineering succeeded.

Chapter 35: Ten Million Customers

The people who signed up for Starlink in its early days were, overwhelmingly, those whom the terrestrial internet had failed. They lived at the end of long dirt roads, on farms and ranches far from any cable or fiber line, in mountain valleys and forests and deserts where the nearest broadband connection was a fantasy and the only options were painfully slow satellite services or nothing at all. For these customers — rural households, remote businesses, isolated communities — Starlink was transformative in a way that is hard to overstate for anyone who has always had fast internet. A family that had struggled for years with a connection too slow to stream video or support remote work or online school could install a Starlink dish, point it at the sky, and suddenly have fast, responsive broadband, indistinguishable in daily use from a city connection. The relief was often visceral. In rural forums and small-town newspapers, the same story repeated itself in a thousand variations: the farmer who could finally file crop reports without driving to town, the teenager who could at last attend online classes, the remote worker who no longer had to choose between a career and the land they loved. The service was not cheap — the hardware cost hundreds of dollars and the monthly fee ran higher than typical urban broadband — but for people with no good alternative, it was worth it, and the demand was immediate and intense. Starlink had found a market that was underserved, hungry, and willing to pay, and it grew fast.

The growth curve, once the constellation reached critical mass and the service became widely available, was extraordinary, and it is worth setting down the numbers, because they tell the story better than any adjective. Starlink passed its first million subscribers around the end of 2022, only a few years after the first operational satellites launched. It roughly doubled to some two and a third million by the end of 2023, then to well over four and a half million by the end of 2024, then past nine million by the close of 2025 — the pace accelerating as the constellation grew, coverage expanded, and word spread. In February 2026, Starlink crossed ten million active subscribers around the world — a customer base built essentially from nothing in about six years, a rate of growth that few businesses of any kind have ever matched, and remarkable for a service that required customers to buy specialized hardware and pay a premium price. And ten million was only the count of subscriptions. Behind each one there might be a household of five, a business with fifty employees, a fishing boat, a mid-ocean container ship, or an airliner full of passengers all sharing a single dish, so that the number of people actually touching Starlink at any given moment ran into the tens of millions and beyond. From a standing start, Starlink had become one of the largest internet service providers in the world by geographic reach, and a major one by subscriber numbers, serving customers on every continent — Antarctica included, where a research station's dish gave scientists at the bottom of the world a connection better than many suburbs — and in scores of countries.

It is often said that Starlink is a niche product for wealthy Westerners at the end of their driveways, an expensive toy for people who could mostly get by without it. That was true of the earliest adopters, but it became steadily less true as the constellation matured, and by the mid-2020s it badly understated what the service had become. In much of sub-Saharan Africa, where laying fiber across vast distances had never penciled out and cellular coverage thinned to nothing beyond the cities, Starlink arrived as the first genuinely fast internet many communities had ever known, and demand in some countries outran the company's ability to supply terminals. Schools that had rationed a single slow connection among hundreds of students, clinics that had relied on couriers to move medical records, small traders locked out of the digital economy — for these, the dish was not a luxury but a leap across a gap that had defined their disadvantage for a generation. The digital divide, that phrase policymakers had worn smooth over decades of conferences, turned out to be crossable in some places not by a national fiber program but by a flat white panel pointed at the sky. That inversion carried its own complications — a foreign company now sat astride the connectivity of communities that had spent decades waiting for their own governments to act — but the immediate human fact was undeniable, and it was arriving in places that the slow machinery of national infrastructure had, in truth, given up on.

The market broadened well beyond rural households as the service matured. Starlink developed offerings for a range of customers whose defining feature was mobility or remoteness: maritime services for ships, from commercial vessels and cruise liners to fishing boats and private yachts, which had always struggled with slow, expensive connectivity at sea; aviation services to give airline passengers fast internet in flight, with major carriers signing up to install Starlink across their fleets and, in a telling detail, several of them offering it to passengers for free, having concluded that reliable Wi-Fi at thirty-five thousand feet was now table stakes; services for recreational vehicles and mobile users, so that travelers could have connectivity anywhere; and enterprise and government services for organizations operating in remote areas, from mining and energy operations to disaster-response teams. Each of these markets had been poorly served by existing options, and each represented a substantial source of revenue. The maritime and aviation businesses were especially lucrative, since a shipping line or an airline would pay far more per connection than a rural household, and they lifted the company's average revenue per user well above what the consumer base alone would suggest.

The spread between those markets was startling once the company's finances came into view. A residential household might generate on the order of two thousand dollars a year; a ship at sea, paying for connectivity it could get nowhere else, might generate something closer to thirty-odd thousand; and a single equipped airliner, serving hundreds of passengers across a fleet, could be worth on the order of hundreds of thousands of dollars a year. This is why a curious statistic emerged as Starlink grew: even as the subscriber count exploded, the average revenue per user drifted downward — from roughly a hundred dollars a month in the constellation's early years toward something nearer sixty-five by early 2026. It would be easy to read that decline as a warning sign, and some analysts did. In fact it reflected a deliberate strategy. Having saturated the well-off rural markets of North America and Europe, Starlink pushed aggressively into Africa, Southeast Asia, and Latin America, cutting prices sharply to reach customers who could never pay American rates, trading revenue per head for sheer reach and volume — the classic move of a utility that has decided its future lies in serving the many at a thinner margin rather than the few at a fat one.

And Starlink kept pushing into new domains. One of the most significant expansions was "direct-to-cell" — a capability in which specially equipped Starlink satellites could communicate directly with ordinary, unmodified mobile phones on the ground, acting as cell towers in the sky, providing basic connectivity to standard handsets in areas beyond the reach of terrestrial networks. This was a genuinely hard trick: a phone in a user's pocket is designed to talk to a cell tower a few kilometers away, not to a spacecraft racing overhead two hundred kilometers up, and closing that link required both new satellites carrying large antennas and a great deal of clever signal processing. In partnership with mobile carriers — most prominently T-Mobile in the United States, whose "T-Satellite" service moved from beta into commercial launch in July 2025 — SpaceX deployed more than six hundred and fifty direct-to-cell satellites, enough that the company could plausibly describe the result as the largest mobile network on the planet by sheer coverage area, blanketing hundreds of thousands of square miles of territory where no ground tower had ever stood. The promise was to eliminate the "dead zones" where phones simply do not work, and to keep people connected in remote wilderness, at sea, and in disasters that knock out ground infrastructure — a lost hiker able to text for rescue from a canyon with no bars, a community able to reach the outside world after a hurricane flattened its towers. It was a striking extension of the constellation's reach, turning Starlink from a service that required a special dish into one that could touch the phone already in a user's pocket.

Nowhere did the value of that reach show more vividly than in disasters, and the pattern established itself early. In January 2022, the undersea volcano Hunga Tonga erupted with a violence heard thousands of kilometers away, and its shockwave severed the single fiber-optic cable that tied the Pacific kingdom of Tonga to the rest of the world, plunging an entire nation into digital darkness at the very moment it most needed to call for help. SpaceX donated dozens of terminals and sent a team to neighboring Fiji for months to stand up a link, restoring a lifeline that traditional repairs would have taken weeks or longer to mend. The same script played out again and again. When Hurricane Helene tore through the mountains of the American Southeast in the autumn of 2024, drowning towns and toppling the cell towers that rescue crews depended on, Starlink kits were flown and trucked into the disaster zone by the hundred, and the company waived fees for affected residents through the end of the year. There was, to be sure, a hard-nosed logic beneath the generosity — every emergency was also a demonstration, and skeptics noted that the free months tended to convert into paying subscriptions — but for a family with no other way to reach the outside world, the distinction was academic. A technology conceived to serve remote farmhouses had become, almost by accident, a piece of the world's emergency infrastructure, the thing reached for when the ordinary wires went down.

The financial implications of all this were enormous, and although SpaceX was a private company that had long guarded its exact figures, the veil lifted considerably as it prepared to go public in 2026. Starlink had grown from a speculative venture that skeptics doubted would ever make money into the largest single source of SpaceX's revenue: the connectivity business generated something on the order of eleven billion dollars in 2025, up roughly fifty percent on the year before, and — the detail that most confounded the doubters — it was the one part of SpaceX that turned a healthy operating profit, having crossed into positive cash flow in 2024. It is often said that Starlink was a money pit that could never justify the staggering cost of building and constantly replenishing a constellation of thousands of satellites; in fact, by the mid-2020s it had become the profit center that subsidized everything else, the very engine Musk had always claimed it would be. The service had achieved the scale and the recurring revenue that he had envisioned when he first conceived it as the business that would fund Mars: a global utility, serving millions, throwing off the kind of steady cash flow that a launch business alone — a lumpy trade of a few dozen contracts a year — could never provide. Precisely how the economics would evolve as the constellation grew and competition emerged remained matters of estimate and debate, and the reader should treat any single revenue figure as an approximation. But the direction was unmistakable. Starlink had become the financial engine of SpaceX, the business that transformed the company from a launch provider dependent on a limited market into a diversified space enterprise with a vast and growing consumer business, and it was this that underpinned the company's soaring valuation and, in June 2026, its historic public offering.

There is a larger significance to Starlink's success that goes beyond SpaceX's balance sheet, and it cuts in more than one direction. On one hand, Starlink extended fast internet to millions of people the terrestrial networks had left behind, connecting the unconnected, bridging a digital divide that had persisted for decades, bringing the possibilities of the online world — education, commerce, communication, information — to remote and underserved populations across the globe. That is a genuine good, and it should not be dismissed. You meet it in places the map of connectivity had always shown as blank: a schoolhouse in the Amazon basin, a clinic in the Sahel, a village in the Pacific that had never had a working telephone. On the other hand, Starlink concentrated an extraordinary and unprecedented amount of critical global infrastructure in the hands of a single private company controlled by a single individual — a company that could, in principle, turn the service on or off, prioritize or throttle it, extend or withhold it, according to its own commercial and political judgment. This concentration of power over a piece of infrastructure that millions had come to depend on, and that had strategic and even military significance, would become one of the defining controversies of the new space age, dramatized above all by Starlink's role in a war. The very ubiquity that made Starlink a blessing — the fact that ten million subscriptions and tens of millions of people had come to lean on it — was also what made its concentration so fraught, for the more indispensable a utility becomes, the more its private ownership matters. To that tension, to Starlink's role in a war, and to the sharp questions it raised about who should control the wires of the world, we now turn.

Chapter 36: The Price War

Starlink did not conquer the market for satellite internet without a fight, and the contest it entered — over prices, over spectrum, over subsidies, over the rules of the game — was as much a commercial and regulatory battle as a technological one. Behind the simple experience of a customer installing a dish and getting fast internet lay a fierce competition for a market worth hundreds of billions of dollars, and a tangle of regulatory fights over the finite resources — orbital slots, radio spectrum, government subsidies — that satellite internet required. Understanding these battles illuminates the business behind the constellation and the obstacles even a dominant player faced.

The first battlefield was price and value. Starlink was not cheap: the hardware cost customers hundreds of dollars, and the monthly subscription ran higher than typical urban broadband. Here it is worth correcting a common assumption — that the high sticker price meant Starlink was gouging its customers. In fact the opposite was closer to the truth. The company was selling its terminals for less than they cost to manufacture, absorbing a loss on every dish in order to lower the barrier to signing up; a terminal that cost well over a thousand dollars to build might be sold to a rural household for a few hundred, on the calculation that the monthly fees would repay the subsidy over the life of the subscription. It was the razor-and-blades logic of the mobile-phone industry transplanted into space. For the customers Starlink served best — those in rural and remote areas with no good alternative — the price was worth it, and demand was strong. But as Starlink sought to grow beyond its core market of the underserved, it faced competition from terrestrial broadband, from cellular networks, and from rival satellite services, and it had to balance the price it charged against the market it could capture. Starlink experimented ceaselessly with pricing, offering different tiers and plans for different customers and markets — cheaper "deprioritized" service in congested regions, premium plans for those who wanted guaranteed speed, discounted or even free hardware where it needed to seed a new country — adjusting prices to balance growth against profitability, and working to drive down the cost of its hardware to widen its accessible market. The economics were demanding: building and constantly replenishing a constellation of thousands of satellites was enormously expensive, and Starlink had to generate enough revenue, from enough subscribers paying enough, to cover those costs and turn a profit — a milestone that skeptics had doubted it would reach, but that it attained as its subscriber base grew into the millions.

The tactical texture of this pricing war is worth seeing up close, because it looked less like a single price and more like a shifting mosaic. In markets where it wanted to accelerate adoption, Starlink slashed the price of the dish in promotions — sometimes to a fraction of the sticker price, occasionally offering the hardware at a steep discount or bundling it with prepaid service — swallowing still more of the loss to win the customer. In regions where its capacity was strained by too many subscribers, it did the opposite, introducing congestion surcharges or steering new sign-ups toward pricier plans to throttle demand, and it invented a ladder of tiers — a cheaper, "deprioritized" residential service that yielded to paying traffic at busy times; a premium tier with guaranteed performance for those who would pay; roaming and mobile plans for travelers; and steeply differentiated rates for ships and aircraft. It was, in short, the pricing playbook of a mature utility crossed with that of a consumer-tech startup still buying growth, deployed simultaneously across scores of countries at different stages of maturity. The single question underlying every move was the same: how to convert a fixed and enormous cost — a constellation that had to be built and rebuilt whether ten customers used it or ten million did — into a recurring revenue large enough to justify the whole endeavor.

The second battlefield was spectrum and regulation, and it was here that some of the fiercest fighting took place, largely invisible to the public. Satellite internet depends on radio spectrum — the specific bands of airwaves used to communicate between satellites and the ground — and spectrum is a finite, fiercely contested resource, allocated and regulated by national governments and, internationally, by the International Telecommunication Union in Geneva, which parcels out orbital slots and frequency rights on a first-come, first-served basis that put a premium on moving early. Starlink had to secure the spectrum rights it needed, in country after country, navigating the regulatory processes and the competing claims of other users, and it faced determined opposition from rivals and incumbents. The satellite operator Viasat, among others, argued that Starlink's swarm interfered with existing systems and even challenged its environmental clearances; a bruising domestic fight erupted over the twelve-gigahertz band, in which Dish Network and its allies sought to repurpose frequencies that Starlink relied upon for ground-based 5G use, a dispute Starlink fought with unusual public ferocity, warning regulators and even mobilizing its own customers to write in, because a loss would have degraded service for every dish in the country. Similarly, Starlink needed regulatory approval to operate in each country it served, and the terms of that approval — what it could offer, at what price, under what conditions, with what degree of local control — varied enormously and were the subject of negotiation and, at times, of conflict. Some nations welcomed Starlink as a boon to connectivity; others were wary of a foreign company controlling critical communications infrastructure, or sought to protect domestic providers, or demanded local gateways, local data storage, and local partners as the price of entry. The regulatory landscape was a patchwork of national rules and negotiations, and navigating it was a major part of Starlink's business, a battle fought in regulatory filings and diplomatic negotiations as much as in the marketplace.

The unevenness of that map was striking. By the mid-2020s Starlink was licensed and operating in well over a hundred countries, yet great blank spaces remained where it was banned outright or held up indefinitely. Russia and China forbade it, unwilling to permit an American-controlled network beyond the reach of their censors and security services; Iran outlawed it even as dissidents smuggled terminals across the border. India, a vast and coveted market, kept the company waiting for years amid wrangling over security rules and how satellite spectrum should be assigned. In some countries the service ran ahead of the paperwork, terminals arriving through gray markets and operating before formal approval, which occasionally provoked crackdowns and confiscations. Behind these national decisions lay a genuine sovereignty anxiety that money alone could not resolve: to let Starlink in was to route a growing share of a nation's communications through a constellation controlled by a foreign company and, ultimately, by one foreign individual — a bargain some governments would strike for the sake of connectivity and others would refuse on principle. The commercial battle and the geopolitical one, in other words, were never fully separable, a truth that would only sharpen as satellite internet became an instrument of statecraft.

Beneath the country-by-country licensing lay a scarcer and more technical contest still, over radio spectrum and the rules that govern how it may be used. A satellite constellation cannot simply broadcast on whatever frequencies it likes; it must fit within internationally coordinated bands and obey limits — expressed in forbidding regulatory language about power flux density — designed to keep one operator's signals from swamping another's. Rivals learned to fight Starlink on this ground with real effect. In the American proceeding over the twelve-gigahertz band, Dish Network and its allies argued that letting them build a terrestrial 5G network on those frequencies would not harm satellite users, while SpaceX marshaled studies purporting to show it would cripple Starlink service for millions, and even urged its own customers to flood the regulators with comments — a rare instance of a satellite operator turning its subscriber base into a lobbying army. Elsewhere, incumbents such as Viasat pressed the argument that Starlink's sheer numbers raised interference and even environmental concerns that regulators had never before had to weigh. These were not sideshows. Spectrum is the invisible foundation on which the whole edifice rests, and a ruling in an obscure docket could do more to shape Starlink's prospects in a country than any amount of engineering.

The third battlefield was government subsidies, particularly for rural broadband, and it produced the single most stinging defeat of Starlink's early business history. Many governments, seeking to extend internet access to underserved rural and remote populations, offered subsidies to providers who would serve those areas, and Starlink, which excelled precisely at serving the remote and underserved, was a natural candidate for such support. In the United States, SpaceX had been provisionally awarded roughly eight hundred and eighty-six million dollars over ten years through the Federal Communications Commission's Rural Digital Opportunity Fund — money to connect some six hundred and forty thousand homes and businesses across thirty-five states. Then, in August 2022, the FCC reversed course and rejected the award, and it upheld that rejection on appeal the following year. The commission's stated reasoning was that Starlink had not shown it could reliably deliver the program's required benchmark of a hundred megabits per second down and twenty up to every location it had committed to; critics inside and outside the agency argued the technology was still developing and public money should not be staked on an unproven, expensive, privately controlled system. Supporters of the company, and Starlink itself, cried foul, calling the decision political and pointing out that a real, working service was being denied support while slower incumbents kept theirs. The episode became a lasting sore point, cited for years afterward as evidence either of prudent stewardship of public funds or of regulatory hostility to a disruptive newcomer, depending on the teller. In some other cases and countries Starlink won significant subsidy awards; in still others it was denied or lost them amid disputes over eligibility, performance, and policy. The subsidy battles reflected the larger questions about Starlink's role: was it the best way to connect the unconnected, deserving of public support, or an expensive, privately controlled option that public money should not favor over alternatives? The answers varied, and the subsidy contests were a recurring feature of Starlink's expansion.

Underlying all these battles was the fundamental competitive dynamic that shaped the satellite-internet market: Starlink's enormous head start and structural advantages, and the efforts of rivals and regulators to challenge or contain them. Starlink's advantage — its early and massive constellation, its ability to launch its own satellites cheaply on its own reusable rockets, its integrated system — was formidable, and competitors like Amazon's Project Kuiper faced the daunting task of catching up from years behind while paying commercial prices for launches Starlink flew at cost. The scale of that handicap became almost comic. Amazon, having promised regulators it would deploy thousands of Kuiper satellites, launched its first operational batch only in the spring of 2025 — years after Starlink had begun serving customers — and faced a federal deadline requiring it to orbit roughly half its constellation, some sixteen hundred satellites, by the summer of 2026, a pace its patchwork of contracted rockets could barely sustain. In a detail that captured the whole dynamic, Amazon found itself buying launches from SpaceX, paying its arch-rival to help loft the satellites meant to compete with Starlink, because no other provider had enough rockets to go around. That single fact said more about the structure of the market than any market-share chart: the incumbent did not merely lead, it owned the road its challengers had to travel. But the market was large and growing, the strategic importance of satellite internet drew in national and regional efforts to build alternatives, and the regulatory and subsidy landscapes offered points of leverage where rivals and governments could contest Starlink's dominance. The price war, in its full sense — the competition over prices, spectrum, subsidies, and market access — was thus a genuine and ongoing contest, in which Starlink's dominance was real but not unchallenged, and in which the outcome, like so much in the new space age, remained to be determined by the interplay of technology, commerce, regulation, and geopolitics.

The larger lesson of the price war is that dominance in the new space age, however grounded in genuine technological and commercial advantage, was never simply a matter of building better technology. It required navigating the finite resources and contested rules — spectrum, orbital slots, regulatory approval, public subsidy — that governed the domain, and it drew the opposition of rivals, incumbents, and governments with their own interests and their own leverage. You meet the residue of these battles every time a Starlink dish works seamlessly in one country and is banned outright in the one next door, or when a rural broadband map is redrawn by a subsidy decision made in a hearing room a thousand miles away. Starlink had transformed satellite internet and built a commanding lead, but it had done so through battles fought in regulatory filings, subsidy contests, and spectrum disputes as much as in engineering, and it faced continuing competition and contestation as rivals and nations sought to challenge its dominance and to build alternatives. The constellation in the sky was, in the end, embedded in a web of terrestrial commerce, regulation, and politics, and its fortunes depended on those earthbound contests as much as on the satellites overhead. It was a lesson every would-be disruptor of a regulated industry eventually learns: the hardest part is rarely the machine, and almost always the thicket of rules, rivals, and rents that grows up around anything worth having.

Chapter 37: The Wires of War

When Russia invaded Ukraine in February 2022, one of the first things the invaders attacked was the country's communications infrastructure, and one of the first things Ukraine's defenders reached for was Starlink. The appeal, when it came, was startling in its informality. On the twenty-sixth of February, two days into the full-scale invasion, Ukraine's vice prime minister and digital minister, Mykhailo Fedorov, did not pick up a secure diplomatic line; he opened Twitter and addressed Elon Musk directly, publicly, asking him to provide Starlink terminals to a country whose networks were under attack. Roughly ten hours later Musk replied in kind: "Starlink service is now active in Ukraine. More terminals en route." The satellites, as it happened, were already overhead — the real obstacle had been bureaucratic, the landing rights and regulatory permissions that SpaceX had been seeking before the war, which the minister's public plea effectively swept aside — and within weeks thousands of terminals were flowing into the country, many of the fees waived. By that May, on the order of a hundred and fifty thousand Ukrainians were using Starlink every day. What followed was a demonstration, unplanned and unprecedented, of how a private satellite constellation could become a decisive factor in a major war — and of the profound and troubling questions that arise when critical military communications depend on infrastructure owned by a private company and controlled by a single man.

The depth of the reliance that followed is hard to overstate. Within a year, tens of thousands of terminals were scattered across the country, and Starlink had become, in the words of Ukrainian officials, essential infrastructure — not a convenience layered on top of the war effort but a load-bearing part of it. Reconnaissance drones streamed their video back over it; artillery crews received their targeting coordinates through it; commanders in dugouts kept a picture of a chaotic front through it; and soldiers, in the quiet between actions, used it to call home. A Ukrainian drone operator interviewed during the fighting described the terminal in his trench as being as vital as ammunition, and the comparison was not hyperbole: an army that had built much of its tactical edge on cheap drones and rapid, distributed communication had, in effect, wired that edge through a commercial internet service beamed from orbit. It was a vivid illustration of a broader truth the war kept teaching — that the line between civilian and military technology, always blurry, had all but dissolved, and that a product designed to stream video to a farmhouse could, with no change to the hardware, become one of the nervous systems of a modern army.

Starlink proved enormously valuable to Ukraine. It gave Ukrainian forces and civilians a resilient means of communication that was extremely difficult for Russia to destroy or jam, because it relied not on vulnerable ground infrastructure but on a constellation of thousands of satellites overhead. Ukrainian military units used Starlink to coordinate operations, to control drones, to direct artillery, to maintain command and communications across the front, to keep functioning when terrestrial networks were destroyed. Hospitals, government offices, and ordinary citizens used it to stay connected amid the chaos of war; teachers held classes over it, families reached relatives through it, and news left besieged cities on it. Starlink became, by many accounts, an essential piece of Ukraine's ability to resist, a technological advantage that helped a smaller country hold off a much larger invader. It was, in a sense, exactly the kind of resilient, distributed, hard-to-destroy communications network that militaries had long dreamed of, delivered not by a government program but by a commercial internet service, repurposed on the fly for war. That very success made it a target: Russia poured resources into electronic warfare aimed at jamming the terminals, and the two sides settled into a running technical duel — Russian jammers against SpaceX software updates pushed out to counter them — a contest that continued for years and that itself demonstrated how central this commercial system had become to the fighting.

The speed of that duel became a small legend of the war. When Russian forces began jamming the terminals in the opening weeks, SpaceX engineers were said to have pushed out a software fix within days that blunted the attack — a nimbleness that astonished military officials accustomed to procurement cycles measured in years, and that a senior Pentagon electronic-warfare official reportedly praised as running circles around anything a traditional defense contractor could manage. The episode carried a lesson that ran deeper than the war itself: a commercial network built to shrug off ordinary outages, updated continuously the way a consumer app is updated, turned out to be more resilient under deliberate attack than bespoke military systems that were frozen at the moment of their delivery. It was an accidental proof of concept for a whole philosophy of warfare, in which cheap, mass-produced, software-defined commercial hardware might outperform the exquisite, expensive, slow-moving equipment of the defense establishment — and it did not escape the notice of planners in Washington, Beijing, or Moscow.

But the very features that made Starlink valuable — that it was a single, integrated system, controlled by one company and, ultimately, by one person — also made it a source of deep unease, and the unease crystallized into controversy over incidents in which Musk's personal decisions shaped the course of military operations. The most notorious concerned an attempted Ukrainian attack, in the late summer and autumn of 2022, on the Russian Black Sea Fleet at its anchorage in Sevastopol, in occupied Crimea. Ukraine had built explosive-laden naval drones — uncrewed boats — that navigated using Starlink, and it wanted the service enabled all the way up to the Crimean coast so the drones could reach the fleet. Musk declined. His account, which he gave publicly, was that "there was an emergency request from government authorities to activate Starlink all the way to Sevastopol," with "the obvious intent being to sink most of the Russian fleet at anchor," and that had he agreed, "SpaceX would be explicitly complicit in a major act of war and conflict escalation." He feared, he said, being drawn into an offensive strike that might trigger a Russian response of unimaginable severity.

The episode became one of the most misreported stories of the war, and it is worth setting the record straight, because the correction matters. It was widely reported — and asserted in Walter Isaacson's 2023 biography of Musk — that Musk had secretly ordered his engineers to switch off Starlink coverage that was already supporting the attack in progress, dramatically pulling the plug on Ukrainian drones as they closed on the Russian ships. That version was wrong, and Isaacson publicly corrected it: the coverage near Crimea had never been switched on in the first place. SpaceX had geofenced the area — the service simply did not extend that far — and when the Ukrainians discovered this and asked Musk to activate it, he refused. The distinction between turning a service off in the middle of a battle and declining to turn it on is not a trivial one, and yet even in its corrected form the episode laid bare an extraordinary reality: a private businessman, through his control of a commercial satellite network, was in a position to influence the conduct of a war between nations, making decisions of profound geopolitical consequence according to his own judgment, without the accountability that governs governments.

This was a genuinely novel and unsettling situation, and it provoked serious debate among policymakers, military officials, and observers about the wisdom and the dangers of depending on privately owned infrastructure for critical national-security functions. On one side, Starlink had done enormous good, helping Ukraine defend itself and providing a capability no government network could match; and it was not unreasonable for a private company to be cautious about becoming a direct participant in offensive military operations, with all the risks of escalation that entailed. Musk was, after all, being asked to decide, in effect, whether an American company should help sink a nuclear power's fleet — a decision that in any prior era would have belonged to presidents and generals. On the other side, the idea that a single unelected individual, however brilliant or well-intentioned, should hold such power over the communications of a nation at war struck many as intolerable, a concentration of strategic power in private hands that no democracy should accept. The episode accelerated efforts to establish clearer arrangements. Chief among them was Starshield, a distinct service that SpaceX developed specifically for government and military customers, built on Starlink technology but operated under government contracts and control rather than at the company's discretion — an attempt to put the military use of the constellation on a firmer, more accountable footing, so that a nation buying the capability would own the decision to use it rather than depending on the mood of a chief executive. Starshield grew quickly into a serious line of business in its own right, with the American national-security establishment contracting for satellites and services tailored to government needs — encrypted, hardened, and, crucially, under the customer's control rather than the company's — and it signaled how thoroughly the Pentagon had concluded that the future of military space lay less in building its own bespoke systems than in buying, and adapting, what a commercial company had already built at scale. In a few short years the relationship had inverted the old order in which governments led and contractors followed: on this frontier it was the contractor that had built the capability first, and the government that was scrambling to harness it on terms it could trust. The Pentagon itself moved to contract directly for Starlink service in Ukraine, precisely so that the arrangement would rest on a formal agreement rather than on Musk's goodwill. The push for such formality had a concrete trigger. In the autumn of 2022, as the costs of providing and defending the service mounted, SpaceX reportedly warned the Pentagon that it could not go on funding Starlink in Ukraine indefinitely out of its own pocket — a message that leaked and briefly cast Musk as threatening to pull the plug, before he reversed course and declared that the company would keep paying regardless. The flap was short-lived, but it dramatized the underlying fragility of the arrangement: a capability on which a nation's defense had come to depend was riding on the finances and the temper of a single company, with no contract to guarantee it. Within months the Defense Department had moved to put the Ukrainian service on a formal, funded footing, converting an improvised act of corporate goodwill into something closer to accountable national infrastructure, at least in principle. But the underlying issue did not disappear. As Starlink and services like it became more deeply woven into the infrastructure of nations and militaries, the question of who ultimately controlled them, and to whom they answered, would only grow more pressing.

The Ukraine experience was a preview of a broader transformation in the relationship between private space companies and national power. Starlink and Starshield made SpaceX not just a contractor to governments but a strategic actor in its own right, a provider of capabilities that nations had come to depend on for their security. The company's constellation was increasingly integrated into the plans and operations of the American military and its allies, valued precisely because it offered something governments could not easily build themselves: a vast, resilient, constantly renewed network of thousands of satellites, delivering communications that were hard to disrupt and available almost anywhere. This gave SpaceX enormous strategic importance and, with it, enormous leverage — the kind of position that a government must take seriously, because the loss or the withholding of the capability would be deeply damaging. It also raised the stakes of the concerns about concentration and control that Starlink had begun to provoke. A private company operating the majority of the world's satellites and providing critical military communications to major powers was something new under the sun, and the world had not yet worked out how to think about it, let alone how to govern it.

The war in Ukraine thus revealed both the promise and the peril of the new space age in its starkest form. The promise: a commercial technology, born of the drive to lower launch costs and fund a journey to Mars, could deliver capabilities of immense value, connecting the unconnected and helping a nation defend its freedom. The peril: those capabilities concentrated unprecedented power in private hands, beyond the reach of the democratic accountability that governs public power, dependent on the judgment and the goodwill of individuals and corporations whose interests might not always align with the public's. You meet the shadow of this tension every time a general has to wonder whether a system his forces depend on will be there tomorrow, and whether the answer rests with a ministry or with a man. It is a question the old order, in which such power belonged only to states, never had to ask, and one the new order has not yet learned how to answer. This tension — between the extraordinary benefits of what SpaceX had built and the dangers of the concentration it represented — runs through the whole of the new space age, and it is nowhere more visible than in the story of Starlink, the constellation that connected millions and, in doing so, quietly rewrote the relationship between private enterprise and national power. But Starlink raised other, more literal concerns as well — about the sky itself, about the astronomers whose view of the heavens it obstructed, and about the growing risk of a low Earth orbit so crowded with satellites that it might one day become dangerous to use. To those concerns about the crowded heavens we now turn.

Chapter 38: The Satellite and the State

Starlink's role in the war in Ukraine had revealed, in the starkest terms, that a satellite constellation was no longer merely a commercial internet service but a strategic asset, and that revelation set off a scramble among the nations of the world to reckon with a new reality: that whoever controlled the constellations in low Earth orbit held a form of power that states could not afford to cede to a single company or a single country. The result was a wave of national and regional efforts to build alternatives to Starlink, a contest for orbital communications infrastructure that became one of the defining competitions of the new space age, driven not by commerce alone but by the imperatives of sovereignty and security.

The concern was straightforward. Starlink, controlled by an American company and, ultimately, by a single American individual, had become critical infrastructure for communications, including military communications, and its role in Ukraine had shown that its owner's decisions could shape the course of conflicts. For any nation that was not the United States — and even, in some respects, for the United States itself — this represented a dependence and a vulnerability. A country that relied on Starlink for connectivity, or whose military came to depend on it, was placing a critical capability in the hands of a foreign company beyond its control, subject to the decisions, the commercial interests, and the political entanglements of its owner. In a world of rising geopolitical tension, this was intolerable to many governments, and it drove a determination to build sovereign alternatives — constellations under national or regional control, immune to the decisions of a foreign billionaire. There was, moreover, a subtler race running beneath the strategic one, and it turned on a scarcity few members of the public appreciated. The radio frequencies and orbital paths that a mega-constellation needs are allocated internationally on a broadly first-come, first-served basis, and every satellite launched stakes a claim. A nation that waited risked finding the best slots and the most useful spectrum already occupied by others, its access to a shared commons foreclosed not by treaty but by the simple fact of arrival. The scramble was thus, in part, a land rush in the sky, in which planting satellites early was a way of planting a flag.

The mechanics of that land rush deserve a closer look, because they explain much of the frantic pace. The International Telecommunication Union, the Geneva-based body that coordinates the world's use of radio frequencies and orbital positions, hands out rights on a broadly first-come basis, with deadlines requiring an operator to actually deploy a meaningful fraction of a filed constellation within a set number of years or forfeit the claim. The predictable result was a flurry of enormous filings — constellations of tens of thousands of satellites registered on paper, sometimes by states or shell companies with no evident means of building them, in a bid to reserve spectrum and slots before rivals could. Analysts took to calling these "paper constellations," and they clogged the international registers with claims that might never fly. But the tactic had a logic: even an unbuilt filing could block or complicate a competitor's plans, and the frequencies best suited to satellite broadband were finite. Starlink's own early and aggressive filings had helped it secure the spectrum on which its lead was built; latecomers, national and commercial alike, now raced to stake their own claims before the most valuable bands were spoken for. The scramble in orbit was mirrored, in other words, by a quieter scramble through the filing cabinets of a UN agency, where the future geography of the sky was being provisionally drawn.

China moved most decisively. Wary of dependence on American infrastructure and determined to project its own influence, China embarked on the construction of not one but two state-backed mega-constellations. One, known as Guowang — the name means, roughly, "national network" — was a state venture approved for some thirteen thousand satellites, its first examples reaching orbit in December 2024. The other, variously called Qianfan, "Thousand Sails," or the "G60" constellation, was backed by the Shanghai municipal government and planned to swell to fifteen thousand satellites or more, with launches beginning in August 2024 and an ambition to loft hundreds within its first year. Together the Chinese programs envisioned tens of thousands of spacecraft, intended both to serve China's own communications and strategic needs and to offer an alternative to Starlink for other nations — particularly those in the developing world, along the routes of the Belt and Road, where a Chinese-run network might come bundled with Chinese influence. Western analysts noted the dual-use character of these systems plainly: a civilian broadband network and a military communications backbone are, in orbit, very nearly the same thing. The Chinese constellations were, in part, an extension of the geopolitical competition between the United States and China into low Earth orbit, a bid to ensure that China would not be dependent on American space infrastructure and to give China a tool of influence and connectivity to offer to others. The pace was deliberately brisk: China's commercial launch industry, long a laggard, was spurred into a crash effort to develop the reusable rockets and high launch cadence that such constellations demanded, so that the race to fill orbit doubled as a race to build the industrial base that filling it required — the same virtuous loop between launch and payload that had powered SpaceX, now consciously imitated at national scale.

Europe, too, pursued its own sovereign constellation, driven by the same desire for strategic autonomy, for a communications capability under European control rather than dependent on American or Chinese systems. Its answer was a program with the unlovely bureaucratic name of IRIS² — Infrastructure for Resilience, Interconnectivity and Security by Satellite — a multi-orbit European Union system of a few hundred satellites, underwritten by billions of euros of public and private money and structured so that European governments would retain sovereign control of secure communications. It was, characteristically, smaller and slower than its American and Chinese counterparts, and it drew criticism at home for its cost and its pace; but its very existence was a declaration that Europe intended not to be left dependent on the constellations of others. And other nations and blocs weighed their own options, whether building constellations, partnering with existing providers on terms that preserved some control, or seeking guarantees against the withholding of service.

The deeper implication of all this was a possible fracturing of what had been imagined, in the internet's optimistic youth, as a single global network. On the ground, the "splinternet" — the walling-off of national internets behind censorship and control — was already a familiar worry; in orbit, the sovereign-constellation race threatened to extend that logic upward, so that a citizen's connection might soon depend on which power's satellites happened to be overhead and willing to serve. A developing nation courted by both Washington and Beijing might find itself choosing not merely a broadband provider but a patron, its digital life routed through an American constellation or a Chinese one, with all the dependencies and leverage that implied. Europe's IRIS², for all that it was mocked at home as slow and costly, was in this sense less a business than an insurance policy: a guarantee that at least secure European government communications would not, in some future crisis, be hostage to a decision made in California or in the offices of a state-owned Chinese enterprise. The constellations overhead, in other words, were beginning to redraw the map of alliances and dependencies below, quietly turning the question of who provided your internet into a question of whose sphere you inhabited.

Alongside these national efforts ran the commercial competition, as companies sought to challenge Starlink's dominance of the emerging market for satellite internet. The most significant commercial rival was Amazon's Project Kuiper, backed by Jeff Bezos's fortune and Amazon's vast infrastructure, which aimed to deploy thousands of satellites and compete directly with Starlink for consumers, enterprises, and governments. Kuiper faced the daunting challenge of building its constellation years behind Starlink, which already had a commanding head start and the self-reinforcing advantage of launching its own satellites on its own cheap reusable rockets — an advantage Kuiper, which had to buy its rides on other companies' rockets, could not match. The awkwardness of that position was captured by a single fact that said much about the state of the industry: Amazon, having booked launches with United Launch Alliance, Arianespace, and Bezos's own still-maturing Blue Origin, was eventually obliged to buy launches from SpaceX as well — paying its fiercest rival to help put its competing satellites into orbit, because no one else had enough rockets. Amazon launched its first operational Kuiper satellites only in the spring of 2025, and even then faced a federal deadline requiring it to orbit roughly half its planned constellation, some sixteen hundred satellites, by mid-2026 — a schedule that looked daunting given the shortage of available launches. The gap between Starlink's thousands already flying and a rival's first few dozen was a measure not of ambition but of the compounding advantage that reusable, self-owned rockets conferred. Other constellations, some inherited from earlier ventures, some newly conceived, added to the competitive landscape, all vying for a share of the market that Starlink had pioneered and dominated. The competition was fierce, and its outcome uncertain, but the very proliferation of rivals — commercial and national — testified to the strategic and commercial importance that satellite constellations had assumed, and to the determination of states and companies alike not to leave the orbital high ground to a single dominant player.

The scramble for sovereign constellations had implications that went beyond communications, touching on the deeper question of how the orbital environment would be governed and shared. Low Earth orbit is a finite resource, and as multiple mega-constellations — Starlink, the two Chinese networks, Kuiper, the European system, and others — filled it with tens of thousands of satellites, the problems of congestion, collision risk, and debris that we have examined grew more acute. The competition for orbital infrastructure thus threatened to crowd the skies not with one company's satellites but with many, each nation and company racing to establish its presence, with little coordination and weak governance to manage the shared risks. The tragedy of the commons loomed in its classic form: each actor, pursuing its own rational interest in establishing a constellation, contributed to a collective crowding of orbit that endangered the environment on which all depended, and no authority stood above them with the power to impose restraint. And the geopolitical dimension added a further danger, for a low Earth orbit divided into rival national constellations, in a time of tension, was a low Earth orbit in which conflict — the jamming, the interference, even the destruction of satellites — became a more plausible and more catastrophic prospect.

There is a useful historical parallel that clarifies what was at stake. A generation earlier, the United States had built the Global Positioning System, a constellation of navigation satellites originally intended for the military, and had eventually opened it to the world; in time GPS became so woven into the fabric of daily life — guiding ships and aircraft, timestamping financial transactions, synchronizing power grids and mobile networks — that whole economies quietly depended on a single nation's space asset. Rival powers, unwilling to rest their critical infrastructure on American goodwill, built their own: Russia's GLONASS, Europe's Galileo, China's BeiDou. The mega-constellation race rhymed with that earlier contest, but at a far larger scale and with far higher stakes, for a broadband and communications network is a more intimate dependency than a positioning signal, and the constellations now numbered in the thousands rather than the dozens. What GPS had been to navigation, Starlink threatened to become to connectivity itself — indispensable, and therefore, to any state that did not control it, a vulnerability that had to be answered. Seen in that light, the scramble for sovereign constellations was not paranoia but the predictable next move in a game whose rules the space age had been teaching for fifty years: whatever becomes essential in orbit, every great power will eventually insist on owning a version of.

The transformation of the satellite from a commercial convenience into a strategic asset, and the resulting scramble among states to build sovereign constellations, marked a significant maturation of the new space age — the point at which the orbital high ground became, unmistakably, a domain of great-power competition and national security, not merely of commerce. You meet the consequence of it in the quiet accumulation overhead: where once a handful of nations kept a few hundred satellites among them, now rival powers each intend to fly tens of thousands, turning the near-Earth sky into contested territory as surely as any sea lane or airspace below. SpaceX, through Starlink, had shown the world what a mega-constellation could do, and in doing so had triggered a global race to replicate and rival it, a race driven by the deepest imperatives of sovereignty and security. The company that had set out to fund a journey to Mars by selling internet from space had, almost incidentally, reshaped the geopolitics of the orbital environment, made constellations a matter of national power, and set off a competition that would fill the skies with the satellites of rival states and companies. It was one more way in which the consequences of SpaceX's innovations ran far beyond anything the company had intended, transforming not just an industry but the strategic landscape of the world, and raising questions about the governance of the orbital commons that the institutions of the Earth were only beginning to confront.

Chapter 39: The Crowded Heavens

The astronomers noticed first. On the night after the first batch of sixty Starlink satellites launched in May 2019, observers around the world saw something startling: a train of bright points of light moving across the sky in a line, the freshly deployed satellites catching the sunlight as they crossed the darkness, strung out like a string of glowing beads. To the public it was a curiosity, even a wonder — reports poured in of a mysterious procession gliding silently overhead, and more than a few callers wondered aloud whether they had seen a fleet of UFOs. To professional astronomers it was an alarm. For the satellites, especially when newly launched and still low, were bright — bright enough to be seen with the naked eye, bright enough to leave streaks across the long-exposure images that astronomers use to study the faint and distant objects of the universe. And this was only sixty satellites. SpaceX planned thousands, then tens of thousands, and other companies planned their own constellations. Astronomers looked at the trajectory and saw a future in which the sky, humanity's oldest window on the cosmos, would be crisscrossed by an ever-growing swarm of artificial satellites, streaking through telescope images, interfering with observations, and, in the accumulation, degrading the darkness itself.

The concern was not sentimental but scientific and practical. Modern astronomy depends on capturing faint light from distant objects over long exposures, and a bright satellite crossing the field of view during an exposure leaves a streak that can ruin the image or corrupt the data. As the number of satellites grew, the frequency of these intrusions would grow with it, until, astronomers feared, a significant fraction of images — especially from the wide-field survey telescopes designed to scan large swaths of sky — would be affected. The prime example, and the one that focused minds most sharply, was the great new observatory rising on a Chilean mountaintop, the Vera C. Rubin Observatory, whose Legacy Survey of Space and Time was designed to photograph the entire visible sky every few nights with a camera the size of a small car. Precisely because it drank in such enormous swaths of sky at once, Rubin was maximally exposed to satellite trails, and its scientists calculated that a meaningful share of its twilight exposures could carry a streak. That mattered for everything from the study of distant galaxies and the mapping of dark matter to, most ironically of all, the search for near-Earth asteroids — the very hunt for hazardous rocks that might one day threaten the planet, now complicated by the glare of our own machines. The problem was worst around twilight, when satellites low on the horizon remained lit by the sun after the ground had darkened, precisely the hours when certain critical observations must be made. Radio astronomy faced its own version of the threat, as the satellites' transmissions risked drowning out the faint radio whispers that radio telescopes strain to detect from across the universe. The astronomical community, which had not been consulted before the constellations began going up, found itself confronting a transformation of its observing environment imposed by commercial interests, and it pushed back, publishing studies, raising alarms, and pressing SpaceX and other operators to mitigate the harm; the International Astronomical Union went so far as to establish a dedicated center to defend the dark and quiet sky.

The scale of the projected intrusion, once astronomers ran the numbers, was sobering. Studies modeling a future sky populated by tens of thousands of satellites — not only Starlink but the rival constellations then being planned — suggested that from certain latitudes, on summer nights when the sun never dips far below the horizon, hundreds of satellites might be sunlit and visible at once, and that a meaningful fraction of every wide-field exposure would carry at least one trail. For a survey like Rubin's, which would take some tens of thousands of images a year and whose entire scientific value rested on comparing them to spot anything that moved or changed, even a small per-image contamination rate compounded into an enormous cleanup problem. And there was a subtler loss beneath the practical one. Astronomy is among the oldest of human activities, older than agriculture or writing, and for the whole of that history the night sky had been a shared inheritance, the same overhead for a Babylonian priest and a modern child. The prospect that this inheritance might be permanently altered — not by some natural catastrophe but by the business decisions of a handful of companies — struck many astronomers as a kind of quiet theft, the enclosure of a commons that had belonged, until now, to everyone and to no one. Nor was the loss confined to professionals: the amateur astronomers and night-sky photographers who had always been the public's ambassadors to the cosmos found their long exposures increasingly scarred by trails, and the growing ranks of "dark sky" advocates, who had spent years fighting terrestrial light pollution to preserve a few remaining pockets of true darkness, now faced a source of interference they could not switch off by lobbying a town council. The night, once the one environment no zoning board or company could touch, had acquired an owner of sorts.

To its credit, SpaceX responded, and its engagement with astronomers was more genuine than critics expected. It experimented with concrete fixes and let scientists measure the results. An early attempt, a satellite nicknamed DarkSat, wore an experimental darkening coating on one side; astronomers who studied it found it dimmer than an ordinary Starlink but still around magnitude six and a half — better, yet not enough, and the dark coating brought its own thermal headaches. The next attempt, VisorSat, added a deployable sunshade to block sunlight from striking the most reflective surfaces, and it fared somewhat better, measured at roughly magnitude six. SpaceX kept iterating on later generations with dielectric mirror films and adjustments to how the satellites were oriented, especially during the bright period just after launch. These efforts had real success, dimming the satellites appreciably, though they never made them vanish. The target the astronomers set — a recommendation that satellites at these altitudes be kept fainter than magnitude seven, dim enough to be invisible to the naked eye and, crucially, dim enough that Rubin's software could subtract their trails from an image rather than lose the frame — proved a threshold that the constellation approached but did not reliably meet. Still, the mitigations represented a genuine attempt to address the concern rather than dismiss it, and they stood in contrast to some other operators, launching their own mega-constellations, who were not always as responsive. The fundamental tension remained: the commercial and social value of the constellations, delivering internet to millions, pushed toward ever more satellites, while the value of a dark and unobstructed sky, for science and for the human heritage of the night, pushed the other way, and there was no authority with the power to strike the balance. The sky is a global commons, belonging to everyone and governed by no one, and the decision to fill it with satellites was being made, in effect, by a handful of private companies pursuing their own business plans, with the rest of humanity — astronomers, other nations, future generations — presented with a fait accompli.

Beyond the visual and scientific concerns lay a more physically dangerous one: the risk of orbital congestion and debris. Low Earth orbit is a finite resource, and as it fills with satellites, the risk grows that two of them will collide, or that a satellite will be struck by a piece of debris — a defunct spacecraft, a spent rocket stage, a fragment from a previous collision, any of them traveling at some ten times the speed of a rifle bullet, fast enough that a fleck of paint can crack a window and a lost bolt can wreck a spacecraft. Scientists have long warned of a nightmare scenario, sometimes called the Kessler syndrome after Donald Kessler, the NASA scientist who described it in 1978, in which collisions cascade, debris begets debris, and low Earth orbit becomes so cluttered with high-velocity fragments that it turns dangerous or even unusable — a shell of shrapnel around the planet that could destroy any satellite that ventured into it and could, in the worst case, cut humanity off from the routine use of space for generations. The mega-constellations, adding tens of thousands of satellites to orbit, dramatically increased the number of objects that must be tracked and maneuvered to avoid collisions, and the burden of "space traffic management" — keeping all these objects from hitting each other — grew heavier by the year.

That the danger was not merely theoretical had already been demonstrated. In February 2009, high over Siberia, a defunct Russian military satellite named Cosmos 2251 slammed into an active American communications satellite, Iridium 33, at a closing speed of some forty-two thousand kilometers an hour — the first accidental hypervelocity collision between two intact satellites in history. The impact obliterated both and flung some two thousand trackable fragments into orbit, many of which are still up there, still being cataloged and dodged, years later. It was a small, sharp preview of the cascade that Kessler had warned about: two objects became two thousand in an instant, each of the new pieces now a hazard to everything else. When operators and regulators weighed the mega-constellations, it was collisions like that one — multiplied by the tens of thousands of new satellites now going up — that haunted the calculations, for the arithmetic of a runaway cascade is unforgiving: once the debris population crosses a certain density, collisions can breed faster than atmospheric drag can clean them up, and the process, once begun, might not be reversible on any human timescale.

Here a widespread misconception deserves correcting, because it inverts the real risk. Starlink is often blamed as the chief menace to orbit, on the crude logic that it has by far the most satellites. But the gravest debris threats came not from active, maneuverable satellites in low orbits but from the dead junk of the old space age and from deliberate acts of destruction. SpaceX argued, with genuine justification, that it was a responsible operator: its Starlink satellites flew low, at altitudes from which they would naturally deorbit and burn up within a few years if they failed, rather than lingering as debris for centuries as objects in higher orbits do; they were equipped with propulsion to maneuver and avoid collisions autonomously, performing hundreds of thousands of avoidance maneuvers as the constellation grew; and the company shared tracking data and coordinated to prevent conflicts. The truly reckless behavior lay elsewhere — above all in the anti-satellite weapons tests in which nations deliberately blew up their own spacecraft to demonstrate they could. China's destruction of a defunct weather satellite in 2007 and Russia's destruction of one of its old satellites in 2021 each scattered thousands of trackable fragments across orbit in an instant, debris that will threaten spacecraft — including Starlink's — for decades, and that on occasion forced the crew of the International Space Station to shelter in their return capsules. Set against that, Starlink's self-cleaning low orbits were a model of restraint. And yet the caveat is real: the sheer number of satellites SpaceX was launching, and the number that other operators — rival constellations from other companies and nations — planned to add, meant that the total population of objects in low orbit was rising to levels never before seen, and the long-term risks of that crowding were genuinely uncertain. No one truly knew how safe or how dangerous a low Earth orbit filled with tens of thousands of satellites would be, and the experiment was being run in real time, with the stakes being the continued usability of the orbital environment on which so much of modern life had come to depend.

The crowded heavens, then, represent one of the clearest examples of the new space age's double edge. The same capabilities that connected millions and advanced human presence in space also crowded the sky, obstructed the view of the cosmos, and raised the specter of an orbital environment despoiled by congestion and debris — costs borne not just by SpaceX or its customers but by everyone, by astronomers and by future generations and by humanity's relationship with the night sky itself. You meet the loss in a small and personal way on any clear night far from city lights: look up for long enough and you will see them, faint moving stars that are not stars, sliding steadily across the constellations, a reminder that the oldest and most universal human view — the untouched dark of the heavens — is now, subtly and permanently, a human artifact. And these costs were being incurred through decisions made largely by private companies, driven by commercial logic, with only weak and fragmented governance to weigh the broader interest. Whether the balance struck would prove wise or foolish, whether the benefits would justify the costs, whether humanity would manage the crowding responsibly or blunder into disaster — these questions remained open, part of the vast unfinished experiment that the new space age had become. They are questions we will return to when we consider the deeper problems of concentration and governance that SpaceX's dominance raised. But for now we leave the constellation, humming in its orbits, connecting the world and crowding the sky, and turn to the project that Starlink was always meant to fund, the reason for all of it in Musk's telling: the towering, fully reusable, stainless-steel rocket rising on the Texas coast, designed to carry human beings to Mars — Starship.

Chapter 40: Big Falcon Rocket

Everything SpaceX had done — the Falcon 1 that nearly bankrupted it, the Falcon 9 that conquered the launch market, the Dragon that carried cargo and crew, the Starlink constellation that filled the sky and the coffers — was, in Elon Musk's telling, prologue. None of it was the point. The point, from the day the company was founded in 2002, was Mars: the establishment of a self-sustaining human city on another planet, so that the light of consciousness would no longer flicker on a single, vulnerable world. The rockets and the satellites were means to that end — ways to build the skills, the organization, and above all the money required for the staggering undertaking of settling Mars. And settling Mars would require a rocket unlike anything ever built: enormous, fully reusable, capable of carrying a hundred tons or more to orbit and then, refueled in space, onward to the surface of another planet, and cheap enough to fly not a few times a year but thousands of times, launching the fleets of ships that a Martian city would demand. That rocket, which went through several names and several designs before settling into its final form, was Starship.

The vision behind Starship was almost absurdly grand, and Musk laid it out publicly over a series of presentations across the 2010s, each iteration larger and more concrete than the last. He spoke of a fully reusable transportation system that would make humanity multi-planetary — a rocket and spacecraft that could be launched, landed, refueled, and flown again like an airliner, driving the cost of sending a person or a ton of cargo to Mars down from the unthinkable to the merely enormous, and then lower still. He imagined fleets of these ships departing for Mars roughly every twenty-six months, at the recurring alignment astronomers call the synodic period, when Earth and Mars swing close enough that the journey demands the least fuel; each ship would carry colonists and cargo, building up over decades a city of a million people capable of surviving and growing on its own, independent of resupply from Earth. It was a vision drawn straight from the science fiction of his youth, from the dream of humanity spreading through the solar system, and he pursued it with a seriousness that his critics found grandiose or delusional and his admirers found inspiring. He was fond of framing it not as optimism but as insurance — a hedge, as he put it, against the possibility that some calamity, natural or self-inflicted, might one day extinguish a species confined to a single planet. Whatever one made of the destination, the vehicle required to reach it drove the engineering, and the engineering was real even where the destination remained a dream.

The early conceptions of the Mars rocket were staggering in scale, and they arrived with a showman's flourish. At an astronautics congress in Guadalajara in 2016, Musk unveiled something he called the Interplanetary Transport System, a vehicle of colossal dimensions, far larger than the Saturn V that had carried astronauts to the Moon, capable of lifting hundreds of tons and carrying a hundred people or more to Mars at a time. The concept was so large and so expensive that even Musk eventually acknowledged it was not financially feasible as first imagined, and SpaceX revised it downward — though "downward" is relative, since even the scaled-back version remained the largest and most powerful rocket ever conceived. Through successive iterations the design evolved and was renamed. The following year, at a congress in Adelaide, the Interplanetary Transport System had become the Big Falcon Rocket, or BFR — a name whose cruder informal expansion the internet cheerfully supplied, and which SpaceX engineers themselves adopted with a grin. In time the BFR became, at last, Starship, the name applied both to the whole two-stage system and to its upper stage, the spacecraft that would actually carry cargo and people to their destinations. The lower stage, the giant booster that would provide the initial thrust to lift the whole assembly off the Earth, was named Super Heavy. Together, Super Heavy and Starship formed a rocket that, when stacked, stood something like a hundred and twenty meters tall — taller than the Saturn V, and indeed the tallest and most powerful launch vehicle ever built — and was designed to produce more thrust at liftoff than any rocket in history.

The decision to cluster thirty-three engines at the base of the booster was, to many observers, the most alarming choice in the whole design, and for a reason rooted in a painful precedent. Half a century earlier, the Soviet Union had tried to reach the Moon with a giant rocket called the N1, whose first stage was ringed by thirty engines; all four attempts to fly it ended in failure, at least one of them in one of the largest non-nuclear explosions in history, and the program was quietly abandoned. The lesson the industry drew was that many-engine rockets were a fool's errand — too many parts, too many things to fail, too much plumbing to coordinate. Musk's answer was that the Soviets had failed not because clustering was wrong but because they had lacked the computing power to manage it; a modern rocket, with sensors and flight software watching each engine hundreds of times a second, could detect a failing engine, shut it down, and rebalance the thrust of the others in an instant, so that losing one or even several engines need not doom the flight. Falcon 9, with its nine engines, had already demonstrated the principle in practice, once completing a mission after an engine failed in flight. Redundancy, in this reading, was not the weakness of the many-engine design but its hidden strength — a wager that thirty- three cheap, mass-produced engines watched by clever software would prove more robust than a handful of irreplaceable giants.

The scale of Starship is difficult to convey without standing beneath it, but some numbers help. The Super Heavy booster was designed to be powered by thirty-three engines, clustered at its base — a return, in a sense, to the many-engine philosophy of Falcon 9, but taken to an extreme, with all thirty-three firing together to produce a thrust in the neighborhood of sixteen or seventeen million pounds, roughly double that of the Saturn V. The engine itself, called Raptor, was a technical achievement in its own right: a full-flow staged-combustion engine burning liquid methane and liquid oxygen, a notoriously difficult design that no operational rocket had ever flown before, chosen partly because methane can, in principle, be manufactured on Mars from the planet's atmosphere and subsurface ice, closing the loop that a return journey would require. The fully stacked vehicle, fueled, would weigh several thousand tons on the pad — the mass of a fully loaded naval destroyer, balanced atop a column of flame. It was designed to lift on the order of a hundred tons or more to low Earth orbit in its reusable configuration — a payload capacity that dwarfed anything else flying and that would, if realized at the intended low cost, transform not just the Mars project but the entire economics of space, making it possible to launch things of a scale and at a price never before imaginable. And, crucially, every part of it was meant to be fully and rapidly reusable: both the Super Heavy booster and the Starship upper stage would return from their missions, be recovered, and fly again, with no part of the vehicle thrown away. This was the great leap beyond Falcon 9, which recovered only its first stage; Starship was designed for total reusability, the holy grail that would finally make spaceflight as economical as Musk had always insisted it could be.

One element of the architecture, easy to overlook, was in truth its linchpin, and it repays a concrete accounting: refueling in orbit. A fully fueled Starship could reach low Earth orbit with a large payload, but it would arrive there with its tanks nearly empty, far too depleted to press on to the Moon or Mars. The solution SpaceX settled on was audacious in its simplicity — launch the crewed or cargo ship to orbit, then send up a succession of tanker Starships, each carrying only propellant, to top off its tanks before it departed. The catch was the number of trips. Filling a single Mars-bound ship might require on the order of a dozen or more tanker launches in relatively quick succession, which is to say that the whole scheme was only feasible if rockets could be launched not occasionally but routinely, cheaply, and by the fleet — precisely the capability that full and rapid reusability was meant to unlock. The pieces thus locked together like a proof: Mars required refueling, refueling required many launches, many launches required reusability, and reusability required the stainless-steel, engine-dense, twice-recovered machine that Starship was designed to be. Take away any one piece and the whole edifice fell, which is why Musk treated cheap reuse not as a business nicety but as the precondition for everything else.

Achieving full reusability at this scale required solving problems that Falcon 9 had never had to face, especially for the upper stage. Recovering a first-stage booster, as Falcon 9 did, is hard but tractable: the booster separates while still relatively low and slow, and can flip around and fly back. Recovering an upper stage — the part that actually reaches orbital velocity — is far harder, because it must survive reentry from orbital speed, some twenty-eight thousand kilometers an hour, enduring the same ferocious heat that had always required the fragile, expensive heat shields of capsules and the Space Shuttle, whose brittle tiles had to be inspected and often replaced one by one after every flight, at ruinous cost. For Starship to be fully reusable, its upper stage had to come screaming back into the atmosphere from orbit, survive temperatures that could melt steel, and then land gently enough to fly again — and do all this cheaply and repeatedly. This was the central engineering challenge of Starship, and it drove some of the most distinctive and initially baffling choices in the whole design, above all the decision, made late in 2018, to build the entire vehicle not of aluminum or carbon fiber, the materials of modern aerospace, but of stainless steel — the stuff of kitchen sinks and cutlery, a material the rocket industry had abandoned decades earlier.

That choice struck many observers as bizarre when Musk announced it, and it is worth dwelling on, because it turned out to reflect a deep and characteristically first-principles reasoning about heat, cost, and strength. Musk called it "the best design decision" on the whole project, and delighted in how "counterintuitive" it seemed. The intuition it violated was that rockets must be built as light as possible, and steel is heavier than the exotic carbon-fiber composites the aerospace world prized. But Musk's argument ran deeper. Stainless steel was cheap — a few dollars a kilogram against hundreds for aerospace-grade carbon fiber — and it could be worked by welders in an open field rather than in a spotless clean room, which suited a company that meant to build rockets like water towers and iterate on them fast. More important still was heat. Carbon fiber weakens dangerously as it warms and cannot take much reentry heating without an elaborate shield; stainless steel, by contrast, keeps most of its strength at cryogenic temperatures and can tolerate far higher heat before failing, which meant a steel skin could survive reentry with a much lighter, simpler thermal protection system than a composite ship would need. Musk even mused publicly about a radical scheme in which the hot side of the ship would "sweat" — bleeding water or fuel through tiny pores in a double-walled steel skin to carry heat away, a transpiration-cooled hull like nothing ever flown — an idea SpaceX ultimately set aside in favor of ceramic tiles, but which captured how thoroughly the steel decision reopened the whole question of how to keep a spacecraft from burning up. Once the trade-offs of the whole vehicle were tallied, the counterintuitive material won: a heavier but tougher, cheaper, more heat-resistant ship could, in the end, carry more payload for less money than a lighter, costlier, more fragile one.

Before we follow the design into steel, it is worth registering how much of Starship remained, and remains, aspirational. The vision of a million-person Martian city, of fleets of ships departing every two years, of humanity becoming multi-planetary within a lifetime — this was, and is, a dream, unproven and perhaps unprovable, dependent on solving not only the engineering of the rocket but a thousand other problems, from keeping humans alive on the long journey and the hostile surface to the vast questions of who would go, who would pay, and why. Skeptics argued, not unreasonably, that the Mars vision was a fantasy, that the real value of Starship lay in its more prosaic uses — launching enormous payloads cheaply, deploying the next generation of Starlink satellites, serving as the lunar lander that NASA had contracted it to build for the return of astronauts to the Moon — and that the talk of Martian cities was salesmanship, a motivating myth rather than a serious plan. There is force in that skepticism, and this book does not pretend the Mars city is imminent or even certain to be possible. But the dream, whatever its ultimate fate, drove the creation of a genuinely revolutionary machine, and the machine was real even where the destination was not. To understand how SpaceX set out to build the largest rocket in history, we must go to a scrubby stretch of the South Texas coast, where a company that had learned to iterate on hardware in public began to build, test, and blow up a series of stainless-steel prototypes in full view of the world.

Chapter 41: Why Stainless Steel

When Elon Musk announced, toward the end of 2018, that SpaceX would build its Mars rocket out of stainless steel, the reaction across the aerospace world ranged from puzzlement to disbelief. Modern rockets and spacecraft were built of aluminum alloys and, increasingly, of advanced carbon-fiber composites — light, strong, sophisticated materials that the entire industry had spent decades mastering. SpaceX itself had been developing Starship's predecessors with carbon fiber, investing heavily in the tooling and expertise to build enormous composite structures; the company had even trucked out a colossal composite fuel tank and, at a 2016 presentation in Guadalajara, shown it off as a centerpiece of the Mars vision. Stainless steel, by contrast, was heavy, old-fashioned, the material of household appliances and industrial tanks, seemingly abandoned by rocketry back in the early days when the pursuit of lightness had driven designers to aluminum. To propose building the most advanced rocket ever conceived out of the same steel as a kitchen sink seemed, to many, like a step backward, even a folly. But the choice, once Musk explained the reasoning behind it — in a pair of candid interviews, one with Ars Technica in November 2018 and one with Popular Mechanics that January — revealed itself as a characteristic piece of first-principles thinking, and it may have been one of the most important decisions in Starship's development.

There is a myth worth correcting at the outset, one that Musk himself was fond of puncturing. It is often said that rocketry left steel behind in the 1950s because steel was too heavy for the job. In fact steel never quite left. The Atlas — America's first intercontinental ballistic missile, and the rocket that carried John Glenn into orbit in 1962 — was built from thin-walled stainless steel, its skin so gossamer that the empty tanks would crumple under their own weight unless kept inflated with pressure, like a metal balloon. The Centaur upper stage, still flying into the twenty-first century, used the same balloon-tank trick. Steel, in other words, had a distinguished if half-forgotten pedigree in spaceflight, and Musk, who read voraciously in the history of his own field, was reaching back past the composite era to retrieve an old idea and push it to a new extreme. That he did so, and that the industry recoiled, says as much about how orthodoxies calcify as it does about the metal itself.

The reasoning turned on the brutal problem of heat. Starship's upper stage had to survive reentry from orbital velocity, plunging back into the atmosphere at more than twenty-five times the speed of sound, its surface heated by friction with the air to temperatures that would soften or melt most metals. The conventional solution, used by the Space Shuttle and by every capsule, was to cover the vehicle in a heat shield — fragile, expensive tiles or ablative material that protected the underlying structure from the heat. But heat shields were exactly the kind of costly, delicate, maintenance-intensive technology that had made the Shuttle so expensive to reuse; the orbiter carried roughly twenty-four thousand tiles, many of them individually shaped and individually numbered, each requiring inspection and, often, replacement after every flight — a labor that consumed months and armies of technicians and that helped kill the Shuttle's economics. For Starship to achieve cheap, rapid reuse, it needed to survive reentry without rebuilding its heat protection after every flight, and that meant either a far more robust shield or a structure that could take the heat itself. Stainless steel offered a path to both.

Steel's great virtue, in this application, is that it retains its strength at high temperatures far better than aluminum. Aluminum begins to soften and lose structural integrity at a few hundred degrees, which is why an aluminum vehicle must be shielded from reentry heat entirely; steel remains strong at temperatures hundreds of degrees higher. Musk put a number on it in the Popular Mechanics conversation: steel, he said, could take roughly fifteen hundred to sixteen hundred degrees Fahrenheit — some eight hundred and twenty to eight hundred and seventy degrees Celsius — where the aluminum-lithium and carbon-fiber structures could not. That extra margin meant a steel structure could tolerate the heat of reentry with much less protection, and could survive localized overheating — a tile lost here, a hot streak there — that would destroy an aluminum vehicle outright. It allowed SpaceX to use a simpler, lighter heat-protection scheme: ceramic tiles on only the windward side of the ship, the belly that faced the worst of the plasma, with the bare steel of the leeward side left to radiate its heat into space. Musk went further still in those early days, floating an idea straight out of science fiction: a "sweating" rocket, its double-walled steel skin bleeding tiny quantities of methane or water through microperforations to carry heat away by transpiration. That particular scheme was quietly abandoned in favor of the tiles, but the fact that it was even contemplated shows how radically the switch to steel had reopened the design space.

Steel also had advantages that mattered enormously for a vehicle meant to be built in large numbers at low cost, and here the numbers were startling. Aerospace-grade carbon fiber, Musk explained, ran to something like a hundred and thirty-five dollars a kilogram — and because roughly a third of the expensive material ended up scrapped as offcuts and rejects, the effective cost climbed toward two hundred dollars a kilogram. Stainless steel, by contrast, cost about three dollars a kilogram. The disparity was so large it scarcely needed elaboration: for the price of the composite skin of a single ship, SpaceX could buy the steel for dozens. Steel was also forgiving to work with, weldable with techniques and equipment far simpler and more available than the specialized layup, curing, and autoclaving that carbon fiber demanded, which meant SpaceX could build Starship prototypes quickly and cheaply, in the open air of a South Texas field, without the vast climate- controlled clean rooms that composite construction required. And steel performed at the opposite extreme of temperature too, growing not more brittle but stronger and more ductile in the deep cold of cryogenic propellants — a property that let SpaceX use a cold-worked, cryogenically treated stainless, its strength coaxed upward by the very chill of the liquid oxygen and methane it would hold. Asked where such exotic material came from, Musk was almost dismissive: "It's just 301 stainless. Let me put it this way: 304 stainless is what they make pots out of. There's plenty of it."

The cost and manufacturability arguments were, in the end, as important as the thermal ones, and they reflected the deeper philosophy that had always driven SpaceX. To build a Mars rocket in the numbers Musk imagined — fleets of ships, launching thousands of times — the vehicle had to be cheap to produce, and stainless steel, whatever its penalty in weight, was cheap to produce. Musk argued, with his characteristic willingness to accept a penalty in one dimension for a gain in another, that the extra mass of steel was more than offset by its lower cost, its higher temperature tolerance, and its ease of manufacture. It is worth being precise about the trade-off, because it is the crux of the whole decision. Steel is roughly three times as dense as aluminum, so on a naive accounting a steel tank should weigh far more. But at cryogenic temperature the cold-worked 301 alloy is dramatically stronger, so the walls can be made proportionately thinner; and because the steel needs so much less thermal protection, the vehicle sheds the mass of a heavy shield. Run the sums, Musk contended, and the weight penalty of a fully optimized steel ship was smaller than intuition suggested — small enough that the overwhelming gains in cost, heat tolerance, and buildability swamped it. It was a trade the traditional aerospace world, obsessed with minimizing weight above all else, would almost never have made, and that is precisely why it was revealing: SpaceX was optimizing not for the elegance of a single vehicle but for the economics of an entire transportation system, and by that measure steel won.

There was, too, a subtler thermal virtue in leaving the steel bare and bright, one that the mirror finish of the finished ships makes literal. A polished metal surface radiates heat poorly and reflects it well, and both properties help a spacecraft. In the vacuum and glare of space, a reflective skin bounces away much of the sunlight that would otherwise soak into the tanks and boil the cryogenic propellants, a real concern for a vehicle that may have to loiter in orbit for the long refueling campaigns the Moon and Mars missions demand; during the fury of reentry, meanwhile, the steel on the lee side, away from the plasma, can shed heat by radiating it into the sky. Engineers speak of a material's emissivity and reflectivity as if they were dry abstractions, but on Starship they became design decisions you could see: the choice to leave the hull unpainted, gleaming, was not merely for the retro-futurist romance of it but because a coat of paint would have added mass and hurt the thermal balance for no gain. The rocket was silver because silver, here, was cheaper, lighter, and cooler all at once — an unusually happy convergence of the practical and the beautiful.

The decision was also, characteristically, made at speed and against the grain of the company's own recent investment. SpaceX had spent real money and real years on the composite architecture; to throw it over for steel was to write off sunk costs that a more sentimental organization would have defended to the last. Musk did it in a matter of months, and the first tangible product of the new direction appeared with almost comic haste. In the winter of 2018 and 2019, a squat, water-tower-shaped prototype called Starhopper took shape in the open Texas scrub, welded together not by aerospace craftsmen in a hangar but, in part, by contractors more accustomed to building water towers and stainless tanks — men who knew how to roll and weld steel cylinders and needed only to be pointed at a rocket. Its nose cone, famously, was blown off and dented in a gust of wind early on and simply left off for its first tests. There was something both absurd and profound in the sight: the most advanced rocket program on Earth beginning its Mars campaign with a lumpen steel cylinder assembled by the same trade that built the country's grain silos, precisely because that was cheap, fast, and good enough to learn from.

The choice had one further, unexpected consequence, and it was aesthetic. The finished Starship, its bare stainless-steel skin gleaming in the Texas sun, looked less like a modern rocket than like something out of the golden-age science fiction of the 1950s — a silver spaceship straight off a pulp-magazine cover, or off the pages of a Tintin adventure, the very image the old dreamers had drawn before real rockets turned out to be white-painted aluminum. Musk relished the resonance and said so; the retro-futurist look was, he cheerfully admitted, a happy accident, but a fitting one. And it is the form in which most people now meet the decision. The gleaming silver towers that rise from the Texas coast, that catch the low sun and throw it back in a hard metallic glare, that photograph like something imagined rather than built — their skin is the direct, visible expression of a wager Musk made against the entire orthodoxy of his industry, that the humble metal of the kitchen sink, cheap and tough and old, was the right stuff for the road to Mars. Whether that wager pays off is still, as the Starship program grinds through its flights, an open question; but few individual engineering calls in the modern space age have been so contrarian, so widely doubted at the outset, and so quietly vindicated in the building.

The other heart of Starship was its engine, the Raptor,

and it too represented a leap beyond anything SpaceX had done before. Where the Merlin engines of Falcon 9 burned kerosene, Raptor burned methane — specifically, liquid methane and liquid oxygen, a combination sometimes called methalox. The choice was driven, once again, by the Mars mission: methane can, in principle, be manufactured on the surface of Mars, from the carbon dioxide of the thin Martian atmosphere and water ice in the soil, through the well-understood chemistry of the Sabatier reaction, which means a methane-burning rocket could be refueled on Mars for the journey home rather than having to haul all its return propellant from Earth — an essential capability for any sustainable presence on another world. Beyond that, methane burns cleaner than kerosene, leaving less soot to foul the engine and complicate reuse, and it strikes an attractive balance of performance and density. But Raptor was radical not only in its propellant but in its internal cycle, the arrangement by which it fed that propellant to its combustion chamber — a cycle called full-flow staged combustion, one of the most efficient and most difficult ever attempted, and a thing no engine had ever managed to fly. With the material chosen and the engine taking shape on the test stands, SpaceX turned to the place where Starship would be built and flown: a remote patch of the South Texas coast that the company would transform into a factory, a launch site, and eventually a city of its own.

Chapter 42: The Raptor

At the base of Starship, thirty-three of them clustered together in a great circle of nozzles, are the engines that make the whole enterprise possible — the Raptors, among the most advanced rocket engines ever built, and one of the hardest technical achievements in SpaceX's history. If the stainless steel was Starship's skin and the grand architecture its skeleton, the Raptor was its heart, and its development, spanning years of difficulty and iteration, is a story of engineering ambition pushed to the very edge of the possible — and, in one important respect, past an edge that had turned back everyone who reached it before.

To appreciate the Raptor, one must understand what a rocket engine must do and how engines are graded. An engine burns propellant and expels the hot gas at high speed to produce thrust, and its efficiency — how much thrust it gets from a given amount of propellant — is measured by a quantity called specific impulse, which depends largely on the temperature and pressure of combustion and the design of the nozzle. Higher pressure and temperature mean more performance, but they also mean more stress on the engine, more heat to manage, more ways for things to go catastrophically wrong. Chamber pressure is the number engineers watch most jealously, because it governs how much power can be wrung from an engine of a given size, and because it is so punishing to achieve: every additional bar of pressure is another increment of force trying to burst the chamber and tear apart the pumps. When Musk remarked in early 2019 that a Raptor test had reached a chamber pressure of about two hundred and sixty-eight bar — edging past the vaunted Russian RD-180, long the pressure benchmark of the industry — it was a small, technical boast that carried a large meaning: SpaceX had built, in a few years, an engine operating at pressures the great state design bureaus had spent decades approaching.

Engines are also graded by their cycle, the internal arrangement by which propellant is fed into the combustion chamber. The simplest cycles waste some propellant overboard or accept lower performance; the most sophisticated extract the maximum energy but are fantastically difficult to build. The Raptor used the most demanding cycle of all: full-flow staged combustion. In such an engine, both the fuel and the oxidizer are first burned in separate preburners — one running fuel-rich, one oxidizer-rich — producing hot, high-pressure gas that drives the turbopumps, the whirling machinery that forces propellant into the main chamber at tremendous rates. That hot gas is then fed, all of it, into the main combustion chamber, where final combustion occurs. Nothing is dumped; every atom of propellant passes through the main chamber, and because the turbines can run cooler for a given power — the heat spread across two gas streams rather than concentrated in one — the engine can push to higher pressures and longer life than any lesser cycle. The efficiency is superb. But the cycle is brutally hard to realize. It requires running two preburners and two turbopumps at extreme temperatures and pressures, managing the flows of hot fuel-rich and oxidizer-rich gas, and — hardest of all — keeping the whole violent process stable, resisting the combustion instabilities that can, in milliseconds, hammer an engine to fragments.

Here is the fact that ought to sit at the center of any honest account of the Raptor: full-flow staged combustion had been dreamed of for more than half a century, and no one had ever made it fly. The Soviets got there first in ambition. Between 1962 and 1970, the engine bureau NPO Energomash developed the RD-270, a single-chamber full-flow monster intended for a super-heavy rocket, the UR-700, and ran it through some twenty-seven hot-fire tests. The cycle worked at the level of components; the machinery pumped and burned as the theory promised. But combustion instability in the main chamber kept destroying hardware, and when the UR-700 was cancelled the RD-270 was shelved, a proof that the idea was real and a warning of how savagely it fought back. The Americans tried later and more modestly. In the mid-2000s, Aerojet Rocketdyne built an Integrated Powerhead Demonstrator, which ran the preburners and turbomachinery of a full-flow engine — the "powerhead" — but never became a flight engine; funding lapsed, the requirement evaporated, and the program faded by the end of the decade. So the ledger, when SpaceX began, read plainly: two of the most formidable rocket-engine establishments on Earth had each taken up full-flow staged combustion and each, in the end, set it down. It had defeated every previous effort.

SpaceX set out not merely to build such an engine but to build one that could be mass-produced cheaply, reused many times, and clustered by the dozens in a single booster — a compounding of the difficulty that made the undertaking almost audacious to the point of folly. And it burned an unusual propellant besides: liquid methane and liquid oxygen, methalox, rather than the kerosene of the Merlin or the hydrogen of some other high- performance engines. The choice of methane was driven by the Mars mission, since methane can in principle be manufactured on Mars from the atmosphere and subsurface ice, allowing a methane-burning vehicle to refuel for the return journey; but methane also offered practical advantages, burning cleaner than kerosene, leaving less residue to foul the engine and complicate reuse, and striking a good balance of performance and density. Methalox was, in a sense, the ideal propellant for a reusable Mars rocket, and the Raptor was designed around it. But burning methane in a full-flow staged-combustion engine at the extreme pressures SpaceX demanded compounded the difficulty yet again, and the Raptor's development became a long, hard slog of testing, failure, and refinement — engines exploding on test stands, problems with combustion stability and cooling and the turbopumps, each difficulty overcome through the iterative grind that characterized all of SpaceX's engineering. The milestone, when it came, was almost anticlimactic in its setting: on July 25, 2019, a single Raptor lifted the squat Starhopper prototype a few feet off a Texas field, and in that brief, ungainly hop the full-flow staged-combustion engine — the cycle the Soviets and the Americans had abandoned — flew for the first time in history.

Over the years that followed, the Raptor evolved through successive versions, each more powerful, more reliable, and — crucially — simpler and cheaper to build than the last, and the numbers tell the story of relentless improvement. The first flight-configuration Raptor produced on the order of a hundred and eighty-five tons of thrust at sea level. The Raptor 2 that followed pushed past two hundred and thirty tons, even as SpaceX stripped out plumbing, integrated components, and drove down the parts count. The Raptor 3, the version maturing toward the end of the program's early period, aimed higher still — around two hundred and eighty tons of thrust at sea level, at chamber pressures that in testing reached three hundred and fifty bar, a figure that would have seemed fantastical to the engineers of the RD-270. Yet the most striking thing about the Raptor 3 was not its power but its plainness. Earlier Raptors bristled with external pipes, wires, and heat shielding, a visible tangle of complexity; the Raptor 3 was designed to shed almost all of it, integrating the cooling and shielding into the engine itself so that the finished unit looked startlingly clean, a smooth bell where its ancestors had been a thicket. This was deliberate, and it was the harder achievement. Musk had long insisted that "the best part is no part, the best process is no process," and the progression from Raptor 1 to Raptor 3 was that maxim made metal — each version doing more with fewer pieces, at a fraction of the cost, on a shorter build time.

That obsession with cost and simplicity was not an aesthetic indulgence but a strict requirement of the mission, and the arithmetic makes it vivid. Falcon 9 flew with nine Merlins; a single Starship stack demanded thirty-three Raptors in the booster and six in the ship, thirty-nine engines per vehicle — and the Mars vision called not for a handful of vehicles but for fleets of them, flying again and again, which meant Raptors would have to be turned out not by the dozen but by the thousand. Mass-producing a full-flow staged-combustion engine cheaply was, if anything, a harder problem than building one at all; it is one thing to hand-craft a temperamental masterpiece and quite another to make it on a line, reliably, at a price that closes the business case. SpaceX's progress in driving the Raptor's cost down while pushing its performance up — building, at its Texas and California plants, toward a cadence of an engine roughly every day — was one of the quieter but more remarkable engineering achievements of the whole program, less photogenic than a booster catch but arguably more fundamental to whether the enterprise could ever pay for itself.

It is worth dwelling for a moment on the sheer violence contained within one of these engines, because the statistics have a way of numbing what they describe. Inside a Raptor, the turbopumps spin fast enough to empty a domestic swimming pool in seconds, driving propellant into a chamber held at pressures several hundred times that of the atmosphere at sea level, where methane and oxygen burn at temperatures approaching the surface of the Sun; the whole assembly, no larger than can be craned by a few workers, generates in the neighborhood of a hundred thousand horsepower in its pumps alone. Multiply that by thirty-three, pack the engines so tightly at the base of the booster that their bells nearly touch, and light them within a fraction of a second of one another, and you begin to grasp why clustering was long considered the graveyard of ambitious rocket design. The Soviet N1, with a mere thirty engines in its first stage, had failed on all four of its flights, undone in part by the impossibility of managing so many engines in concert with the technology of the 1960s. SpaceX proposed to do it with three more engines, of a far more demanding cycle, and to bring the booster home and fly it again. That the company eventually made thirty-three Raptors light and run together, flight after flight, is an achievement whose difficulty is easy to underestimate precisely because SpaceX made it look, in time, almost routine.

The Raptor's difficulties were also, at times, Starship's difficulties, for an engine so advanced and so numerous was a natural source of trouble. Clustering thirty-three engines in a single booster meant that the failure of even a few could compromise a flight, as happened on the very first integrated Starship launch in April 2023, when several Raptors failed or shut down during ascent, robbing the vehicle of thrust and control and contributing to its loss. The engines' complexity and their operation at the edge of the possible made them a recurring challenge, and much of Starship's development was, in effect, the maturing of the Raptor — making it reliable enough, and reproducible enough, to power the enormous vehicle consistently. As the design advanced through its versions, the engines grew more dependable, and the wholesale engine-outs that had plagued the earliest flights diminished; by the middle flights, boosters were completing their ascents even after losing an engine or two, absorbing failures that would once have doomed them, a mark of both hardier engines and smarter control. Yet trouble never entirely left. As late as the summer of 2025 an entire ship, Ship 36, was lost in a tremendous explosion on the test stand before it ever flew, its destruction traced to a burst pressure vessel in the nose rather than the engines themselves — a reminder that the whole machine, not merely its Raptors, lived at the ragged edge.

The Raptor was, in the end, both an extraordinary achievement and a persistent source of difficulty, and it is too easy to remember only the first. It was the first full-flow staged-combustion engine ever to fly, a cycle that had beaten the Soviet Union and the United States in turn; it was mass-produced and reusable, built by the thousand at a cost that would have been unthinkable a decade earlier; it was the beating heart of the largest and most powerful rocket ever assembled. And it was, at the same time, a temperamental marvel whose very ambition made it hard to perfect, whose failures wrote themselves across the sky in the early flights, whose maturing consumed years of a company famous for speed. It embodied, as much as any single component could, the character of the whole Starship enterprise: reaching for the edge of the possible, achieving what had never been done, and struggling, flight by flight and version by version, to make the extraordinary reliable. When you see the ring of thirty-three nozzles at the base of a Super Heavy — a fan of engines packed so tightly they nearly touch, each one a machine the Soviets could not tame — you are looking at the hardest thing SpaceX ever chose to build, and at the reason the rest of the vehicle above it could exist at all.

Chapter 43: Starbase

At the southernmost tip of Texas, where the Rio Grande empties into the Gulf and the United States meets Mexico, there is a flat, windswept stretch of coastal scrubland called Boca Chica. For most of its history it was a quiet, nearly empty place — a scattering of houses, a state park, a wildlife refuge home to sea turtles and migratory birds, a beach at the end of a lonely road. SpaceX had acquired land there in the early 2010s, initially intending to build a launch site for Falcon rockets, breaking ground in 2014 with a ceremonial scoop of soil and promises of a modest commercial spaceport. But as the Starship program took shape, Boca Chica became something far more ambitious than anyone had first imagined: the birthplace of the Mars rocket, a combined factory and spaceport where SpaceX would build, test, and launch its stainless-steel behemoths, and which the company would eventually christen Starbase — transforming a sleepy corner of the Texas coast into one of the most closely watched pieces of ground on Earth.

What happened at Boca Chica in the late 2010s and early 2020s was unlike anything in the history of rocket development, and much of it happened in the open, visible to anyone who cared to watch. Rather than developing Starship behind closed doors and unveiling a finished vehicle, SpaceX built its prototypes outdoors, in a sprawling, chaotic construction site, and iterated on them in full public view —

welding together enormous stainless-steel cylinders, testing them to destruction, learning from the failures, and building the next version. The company's approach to Starship was the purest expression of its iterative, hardware-rich philosophy: build fast, test hard, fail often, learn, and build again, accepting that many prototypes would be destroyed in the process because the information gained was worth more than the hardware lost. This was possible only because steel was cheap and Boca Chica was remote; SpaceX could afford to blow up prototype after prototype, and it had the space to do it. The openness was not merely tolerated but, in a sense, exploited. A community of enthusiasts, journalists, and space fans took up residence in the area or watched via livestreams, and a small industry of observation grew up around the site — outlets like NASASpaceflight stationing reporters in the town, and a company called LabPadre mounting cameras with names like "Nerdle Cam" that trained on the build site around the clock, so that a global audience could watch, live and unedited, every weld, every crane lift, every cryogenic test, and every explosion. The development of Starship became a kind of continuous public spectacle, an open-air laboratory whose triumphs and disasters played out before the world in real time.

The first steps were humble and, to the uninitiated, almost comical. An early test article — a squat prototype that looked more like a water tower than a spaceship, and was assembled, fittingly, with the help of contractors who built industrial tanks — was nicknamed Starhopper and used to test the Raptor engine and the basics of controlled flight. Its story is a small parable of the whole endeavor: an early gust of wind toppled its nose cone and crumpled it, so SpaceX simply left the nose off and flew the stubby remainder as it was. On August 27, 2019, Starhopper performed the largest of its short "hops," rising about a hundred and fifty meters into the air under the power of a single Raptor, translating sideways, hovering, and settling back down onto a nearby pad — the crude beginnings of the same controlled-landing capability that Falcon 9 had mastered, now applied to the new vehicle and the new engine. Starhopper was ungainly and strange, and its hops were small, but they proved the Raptor could fly and the basic control worked, and they were the first flights of the Starship program. From there, SpaceX built a succession of taller, more complete prototypes, designated with the prefix "SN" for "serial number" — SN1, SN2, and onward — each a full-diameter, nine-meter Starship built section by section in the open, and each pushing the development further.

Many of them failed, sometimes spectacularly, and the failures were treated not as embarrassments to be hidden but as data to be harvested. SN1 crumpled and blew apart during a cryogenic pressure test in early 2020, its collapse captured on the ever-watching cameras and replayed around the world. Others ruptured or leaped off their stands during testing as SpaceX deliberately pushed them past their limits to find where the limits were — a philosophy Musk summarized with a phrase that became something of a motto around the site, that a test which does not end in failure has not really told you where the edge is. Before a single one of these prototypes flew high, several had been sacrificed on the ground, each teaching a lesson about welds, about tank pressures, about the strange behaviors of cold steel under stress, that no amount of paper analysis could have supplied. It was, by the standards of traditional aerospace, an almost shocking rate of destruction, and by the standards of SpaceX it was simply the cost of learning quickly.

One episode from this early stretch deserves to be remembered, because it captures both the showmanship and the humbling reality of the enterprise. In late September 2019 — on the eleventh anniversary, as it happened, of the Falcon 1's first success — Musk staged a grand evening presentation at Boca Chica, using as his backdrop a towering, fully assembled prototype called Starship Mk1, its polished skin lit dramatically against the night. It was pure spectacle, a Mars rocket unveiled beneath the Texas stars, and the imagery flew around the world. Less than two months later, in November 2019, that same Mk1 was filled with cryogenic fluid to test its tanks to the limit, and its top bulkhead blew clean off in a great white plume, launching into the air and thudding to the ground as vapor boiled across the site. SpaceX's response was almost cheerfully unbothered: the point of the test, the company noted, had been to push the systems to their maximum, so the outcome was "not completely unexpected," and Mk1, Musk allowed, had really only ever been a "manufacturing pathfinder" whose flight design was already obsolete. The show rocket had failed on the stand, and the program simply moved on to the next one. That was the rhythm of Boca Chica in those years — a triumphant unveiling and a burst tank separated by weeks, the grand and the messy braided together, the whole thing narrated live to an audience that had never before been let so close to the making of a rocket.

The most dramatic phase of the early testing came in late 2020 and the first half of 2021, when SpaceX began flying complete Starship prototypes on high-altitude test flights, launching them some ten kilometers into the sky and attempting to bring them back to a landing — a maneuver unlike any rocket landing before it. The Starship upper stage was designed to reenter and descend not tail-first, like a Falcon 9 booster, but belly-first, falling flat through the atmosphere like a skydiver, using four large flaps to control its plunge and slow itself, before flipping upright at the last moment, reigniting its engines, and landing vertically — a complex, balletic maneuver that SpaceX called the "belly flop" and the "landing flip." The high-altitude flights of prototypes SN8, SN9, SN10, and SN11, across a bruising five months, attempted this maneuver, and the results were a study in the iterative method at its most vivid and most punishing. SN8, on December 9, 2020, flew beautifully — rose to altitude, cut its engines one by one as planned, tipped over into the horizontal belly-flop descent, and then, at the last, flipped upright for landing, only to come in too fast: low pressure in its methane header tank starved the engines at the critical moment, and the ship slammed into the pad in a tremendous fireball. Musk, characteristically, declared it a success anyway, tweeting that SpaceX had gotten all the data it needed and congratulating the team; the fuel-header pressure problem, he noted almost cheerfully, was the culprit, and it would be fixed. SN9, that February, flipped and could not right itself in time, and crashed. SN10, in March, actually landed — settled onto its legs, upright and whole, the first Starship to touch down from altitude — and then, some eight minutes later, having apparently survived, exploded on the pad, destroyed by damage and a propellant leak from the hard touchdown, a cruel coda that turned an apparent triumph into another fireball. SN11 launched into thick fog that March and was lost in an explosion partway through its flight, invisible on the cameras, its failure announced by falling debris. Four prototypes, four fiery ends, each teaching something the next would carry.

And then, on May 5, 2021, SN15 flew — rose to altitude, throttled down, performed the belly flop, executed the landing flip, reignited its engines, and settled onto the landing pad, intact, upright, and whole. A small fire at its base, quickly doused, was the only blemish. For the first time, a full Starship prototype had flown high and landed successfully, mastering the extraordinary belly-flop-and-flip maneuver that had destroyed its predecessors. SN15 carried improvements SpaceX had folded in from all that had come before — better avionics, a reworked propellant architecture in the aft skirt, a new configuration of Raptors — and it worked. It was a milestone that, like the first Falcon 9 landing before it, marked the transition of a seemingly impossible feat into demonstrated reality. SN15 stood on its pad after landing, a gleaming steel spaceship that had flown and returned, and it proved that the fundamental flight and landing profile of the Starship upper stage could be made to work. The company had, through a rapid, brutal, public campaign of build-test-fail-repeat, cracked one of the hardest parts of the problem — and had done it in less than half a year from the first high flight, a pace that left the traditional industry, accustomed to measuring such campaigns in years, faintly incredulous.

It is worth pausing on how easily this stretch of history gets misremembered, because the misremembering flatters SpaceX in the wrong way. It is often said that the belly-flop landings were the great early triumph of Starship, the proof that the vehicle worked. In fact those high hops, impressive as they were, involved only the upper stage, flying alone to modest altitudes and carrying no booster, no thirty-three-engine Super Heavy, no reentry from orbital velocity. They demonstrated a landing maneuver, and a genuinely novel one — but they left untouched the vastly harder problems that stood between SN15 and the actual mission. The real goal, the thing the whole program was building toward, was to stack the Starship upper stage atop the giant Super Heavy booster, with its thirty-three engines, and launch the fully assembled vehicle — the largest and most powerful rocket ever built — on a flight toward orbit, and then bring both stages home from the fury of reentry. That would require not just a working Starship but a working Super Heavy, a launch pad and tower capable of handling the colossal vehicle, and the resolution of countless problems of scale that a ten-kilometer hop never had to confront.

The watching, meanwhile, had become a phenomenon in its own right, and it is part of how most people came to know the place. On launch mornings the causeway to South Padre Island, a few miles up the coast, filled with cars and camera tripods; motels sold out; families drove from across Texas to stand on the sand and feel the ground shake. The nearest town, Brownsville, one of the poorest cities in the United States, found itself suddenly tied to the fortunes of a Mars rocket, its diners and hotels swelling on test days, its economy quietly reoriented around the launches down the road. What had been a lonely dead-end beach became a destination, a pilgrimage site for the space-curious, its rhythms set by road closures and countdowns. That, as much as the factory and the towers, was the measure of how completely

Boca Chica had changed: a place people had once sought out for its emptiness was now sought out for the loudest machines on Earth.

By then the site itself had become as much a story as the rockets. The little cluster of buildings had swelled into a genuine factory, its high bays and "tents" turning out ship and booster sections around the clock; the first great launch and catch tower, nicknamed Mechazilla, was rising beside the pad, its arms a promise of the audacious reuse to come; a workforce of thousands had descended on a corner of Texas that a few years earlier had counted its permanent residents in the dozens. The company had begun, informally, to call the whole complex Starbase, a name that would in time become official in a way few company sites ever have. What had started as a modest plan to launch Falcon rockets from a quiet beach had become the industrial heart of the Mars project, a place where the future was being welded together in the open air and blown up on livestream when it failed. And it was all prologue. Mastering the hop had taken five months and four wrecked prototypes; mastering the full stack would take another two years, and when the first launch of the complete Starship finally came, in April 2023, it would be at once a triumph and a disaster — clearing the pad in a roar of thirty-three engines and a storm of shattered concrete, and then tearing itself apart in the sky. That was the beginning of the hardest and most public phase of the whole SpaceX story, and it would unfold, like everything at Boca Chica, before the watching eyes of the world.

Chapter 44: The Town at the End of Texas

Boca Chica was, before SpaceX came, one of the emptiest and most peaceful places on the Texas coast — a flat expanse of tidal flats, dunes, and scrub at the very southern tip of the state, where the Rio Grande meets the Gulf of Mexico, home to a small, scattered community, a state park, a beach at the end of a lonely road, and a national wildlife refuge that shelters sea turtles, ocelots, and migratory birds. It was a place people came to for its emptiness and its wildness, a last stretch of undeveloped coast. Then SpaceX chose it as the birthplace of the Mars rocket, and the emptiness ended. The transformation of Boca Chica into Starbase — a factory, a spaceport, and eventually an incorporated city — is a story of extraordinary ambition colliding with a fragile place and a small community, and it deserves to be told honestly, because it is part of the cost of what SpaceX built.

The transformation was rapid and total. Where there had been quiet flats and a few dozen homes, SpaceX erected a sprawling industrial complex — assembly buildings, launch pads, tank farms, test stands, the great launch towers with their catching arms — and filled it with thousands of workers, building and testing and launching the largest rockets ever made. The noise was tremendous: the roar of engine tests, the thunder of launches, the industrial din of round-the-clock construction. The launches and tests required closing the public beach and State Highway 4, the single road that ran to it, sometimes for days at a stretch, restricting access to a place that had been open to all — and the closures fell, by law and by the calendar, on the very weekdays and holidays when Cameron County residents were most likely to want the beach. The failures scattered debris across the surrounding land: the cratering of the first flight's launch mount, which threw pulverized concrete for hundreds of meters and rained dust on the town of Port Isabel across the bay; the explosions of prototypes; the fragments that fell from failed flights into the flats and the water. The environmental toll was real, and it was documented not only by activists but by the federal government's own biologists. The land abutted the Lower Rio Grande Valley National Wildlife Refuge and lay along a coast that is one of the last strongholds of the Kemp's ridley, the most endangered sea turtle on Earth, which nests on these very beaches. Conservationists and the U.S. Fish and Wildlife Service recorded fires sparked in refuge habitat, debris strewn across nesting grounds, disturbance to shorebirds, and the general degradation of a fragile ecosystem under the pressure of an industrial rocket operation. Regulators required environmental reviews, and those reviews — and the tension between SpaceX's drive to build and fly quickly and the obligation to protect the environment — became a recurring source of friction, delay, and litigation, with environmental groups repeatedly hauling the Federal Aviation Administration into court over the adequacy of its assessments.

The small community that had lived at Boca Chica found itself directly in the path of the transformation. The settlement, a modest grid of homes called Boca Chica Village, had a few dozen residents, many of them retirees drawn by the quiet and the price, who had bought into the emptiness precisely because nothing ever happened there. Beginning around 2019, SpaceX moved to buy them out, and the process left a residue of hard feeling. The company's offers, some residents complained, were framed as generous but pegged to appraisals they considered low, and arrived with the implicit message that a rocket-testing site was no place for a home and that staying would mean living beside detonations and evacuation orders. A handful resisted, publicized their grievances, and became, briefly, symbols of the small person standing against a giant; most, in the end, sold and left. What had been a quiet residential community became a company town in the most literal sense, its land and homes owned by SpaceX, its streets renamed — one of them, pointedly, "Memes Street" — its character entirely remade in the company's image.

And in 2025, the transformation was formalized in a way that startled even close observers: the area voted to incorporate as a city. The election, held on May 3, 2025, in Cameron County, was as lopsided as the demographics made inevitable. Of the roughly two hundred and eighty-three people eligible to vote — nearly all of them SpaceX employees or their families, living on land the company controlled — two hundred and twelve cast ballots in favor of incorporation and just six against, far past the hundred and forty-three votes the measure needed. The new municipality took the name the company had long used informally: Starbase. Its first mayor, running unopposed, was Bobby Peden, a SpaceX vice president; its two city commissioners, also unopposed, were SpaceX employees as well. A company that had already been described, half in jest, as running a company town now literally elected the town's government from its own payroll. The incorporation of Starbase was a striking development — a private corporation's industrial site becoming an official city, with the power to levy taxes, write ordinances, and administer the land around its own factory — and it raised pointed questions about the blurring of corporate and civic power. Nor were those questions merely abstract. In the same season, the Texas legislature advanced a bill handing the new city the authority to close Boca Chica Beach and its access road on weekdays, a power that had belonged to the county — so that the decision to shut the public off from a public beach, on the days it suited a launch schedule, would now rest with a government the launching company had elected. To SpaceX and its supporters, Starbase was a proud symbol, the frontier town of the Mars age, a place where the future was being built and where the people who built it had chosen, freely and overwhelmingly, to govern themselves. To critics, it was something more troubling — a company town in an era that had thought company towns a relic, its civic institutions indistinguishable from a corporation's org chart.

There is a long American shadow behind that unease, and it is worth naming, because the company town is not a new American thing but a recurring one. In the 1880s the railroad-car magnate George Pullman built a model town south of Chicago and named it after himself, owning the houses, the shops, the church, and the water, and governing his workers' lives with a paternal thoroughness that curdled, in the depression of 1894, into one of the bitterest strikes in the nation's history; an Illinois court eventually ordered the company to divest the town, ruling that owning a municipality was "incompatible with the theory and spirit of our institutions." That ruling is more than a century old, and yet the questions it wrestled with — whether a corporation should hold the civic power of a government over the people who depend on it for their wages — returned, almost verbatim, to a Texas beach in 2025. History does not repeat, but it rhymes, and Starbase rhymed with Pullman more than its boosters cared to admit.

The maturing of Starbase could be read, too, in the people SpaceX brought to run it, and one hire in particular signaled how serious and how permanent the place had become. In May 2023, the company installed as its Starbase general manager Kathryn Lueders — Kathy Lueders — who had just retired from NASA, where she had led the entire human-spaceflight enterprise and, before that, run the Commercial Crew program that returned American astronaut launches to American soil. For the woman who had overseen NASA's astronaut launches to come and manage a private rocket town on the Texas border was a striking crossing of the old world and the new, and it underscored a truth about Starbase that the company-town framing can obscure: this was not a fringe experiment but the central industrial project of the most consequential space company on Earth, staffed at the top by the very people who had run the government's own program. The frontier town had become an establishment, even as it remained a company's private ground.

The friction with the outside world, meanwhile, played out increasingly in courtrooms and agency filings. After the debris-strewn first flight in 2023, a coalition of conservation and Indigenous groups sued the Federal Aviation Administration, arguing that it had licensed Starship launches without the rigorous environmental study the law required, and the litigation ground on in the background of the flight campaign, a slow legal counterpoint to the fast cadence of the rockets. Regulators required environmental monitoring, turtle and bird surveys, and mitigation; SpaceX complied and also chafed, with Musk periodically venting his frustration at the pace of the paperwork. The pattern that emerged was neither the clean villainy of the company's harshest critics nor the frictionless progress of its boosters, but something messier and more characteristically American — a powerful enterprise pushing hard against a thin web of environmental and civic rules, the rules bending and slowing but not breaking, and a fragile coast absorbing the difference. It was governance by attrition, and which side you thought was winning depended largely on whether you had come to Boca Chica for the turtles or for the rockets.

The tensions at Boca Chica captured, in a single place, the larger costs and controversies of the SpaceX enterprise. The environmental cost — the damage to a sensitive ecosystem, the disturbance to protected wildlife, the debris and the noise and the beach closures — was a genuine harm, borne by the land and its creatures and by the people who valued the coast for its wildness. The community cost — the displacement of the residents, the transformation of a quiet place into a company town, the concentration of civic power in a corporation — was a genuine loss, whatever one thought of the enterprise that caused it. And these costs were incurred in the service of a private company's ambitions, driven by a commercial and personal vision, with the broader public interest — in the environment, in the community, in the fair balancing of costs and benefits — represented only weakly by the fragmented and often overmatched processes of regulation. SpaceX and its defenders argued that the costs were manageable and justified by the importance of the work, that the company invested in mitigation and monitoring, funded turtle patrols and habitat surveys, and that the benefits — the jobs, the technology, the advancement of spaceflight, the improbable revival of one of the poorest corners of Texas — outweighed the harms. Critics argued that the company moved too fast, cut corners, treated a fragile place and a small community as acceptable casualties, and used its wealth and power to override the interests of those in its way. Both views captured part of the truth.

You meet the result of all this on a map today, where a dot marked Starbase, Texas, sits at the end of State Highway 4, an incorporated city of a few hundred souls that did not officially exist a few years earlier, its name lifted from science fiction and made a legal fact. Drive out to it and you pass through the checkpoints and the parking lots, past the towers and the tank farms, to a place that is at once a functioning municipality and the private ground of a single company — a town with a mayor and a seal whose mayor answers, in his day job, to the man who owns the town. Whatever one concludes about the balance of harm and achievement, the sheer novelty of the thing is worth registering: in an age that had filed the company town away as a fossil of the industrial past, one of the most advanced enterprises on the planet had gone and built a new one, on a wildlife refuge, at the end of Texas, and named it for the stars.

There is a deeper resonance in the story of Boca Chica, one

that connects it to the whole history of frontier-building. The transformation of a wild, empty place into an industrial site in the service of a grand ambition, with costs borne by the land and by those who lived there — this is an old story, the story of railroads and mines and mill towns, of the remaking of landscapes and communities in the name of progress, with all the ambivalence that story has always carried. Boca Chica was the frontier town of the new space age, and like the frontier towns of the past, it embodied both the romance and the ruthlessness of the enterprise that built it — the thrilling ambition of reaching for Mars, and the real costs imposed on a fragile coast and a small community in the reaching. That the rockets rising from Boca Chica might one day carry humanity to another world did not erase the harm done to this one, and the honest chronicler must hold both in view: the extraordinary achievement of building the largest rockets ever made on a remote Texas coast, and the genuine costs — to the environment, to the community, to the balance between private ambition and public good — that the achievement exacted. Starbase, the town at the end of Texas, is a monument to both, and the traveler who drives that lonely road today, past the guarded gates and the gleaming towers where a quiet village used to be, is looking at the future and its price in the same frame.

Chapter 45: Flight One

On April 20, 2023, the largest and most powerful rocket ever built stood on its launch mount at Starbase, fully stacked for the first time — the Super Heavy booster below, with its thirty-three Raptor engines, and the Starship upper stage above, together standing some hundred and twenty meters tall, a gleaming steel tower against the flat Texas sky. It was a moment years in the making, watched by an enormous global audience, and it carried an almost unbearable weight of expectation. This was the vehicle meant to carry humanity to Mars, to return astronauts to the Moon under NASA's Artemis program, to launch the next generation of Starlink, to transform the economics of space. And now, for the first time, all thirty-three engines of the booster would fire together, unleashing something on the order of sixteen million pounds of thrust — roughly twice the power of the Saturn V that had carried Apollo to the Moon, more thrust than any single vehicle had ever produced at liftoff — and the whole colossal machine would attempt to fly. A first attempt three days earlier, on April 17, had been scrubbed in the final minutes by a frozen pressurant valve on the booster, and there was a note of mordant humor in the rescheduled date: the twentieth of April, 4/20, a number Musk had made into a running joke for years, now attached to the most consequential launch of his career.

The countdown reached zero, and the thirty-three Raptors

ignited. For a few seconds the vehicle sat on the pad amid a growing inferno, the engines building to full power, and then, slowly, ponderously, almost reluctantly, the largest rocket in history lifted off, climbing away from the mount on a column of fire and, ominously, on a vast billowing cloud of pulverized concrete and dust. That it flew at all was, in a sense, a triumph — the sheer engineering of igniting thirty-three engines in concert and lifting several thousand tons off the ground was a feat no one had accomplished at this scale, not the Americans with the Saturn V's five engines nor the Soviets with the doomed N1, whose thirty-engine first stage had exploded on all four of its attempts half a century earlier. But almost immediately, things began to go wrong, and the flight became a demonstration of both how far SpaceX had come and how much remained to be conquered. Several of the booster's engines failed or were shut down during the ascent, some of them at or near liftoff, robbing the vehicle of thrust and of control authority. The rocket climbed, but it climbed wounded. As it rose through the thickening resistance of the lower atmosphere and approached the moment when the booster was to separate from the upper stage, it began to lose control, pitching over and then tumbling, spinning end over end at the edge of space, unable to complete the separation. The vehicle, careening out of control high over the Gulf of Mexico, had to be destroyed.

Here the flight produced one of its most sobering findings, and one that later drew close scrutiny. Starship carried an automated flight-termination system — explosive charges meant to rupture the tanks and end the flight the instant the vehicle strayed beyond its safe corridor. The destruct command was issued as the rocket tumbled; and then, for an unnervingly long interval — some forty seconds — nothing happened, the wounded giant continuing its cartwheel through the upper air before the charges finally did their work and the vehicle erupted into a fireball roughly four minutes after launch, its debris raining down into the sea. That the termination system had been so slow to finish the job was a serious matter, not a trivial one, and SpaceX would be required to overhaul it before flying again. A rocket that does not reliably destroy itself on command is a rocket that cannot be trusted over any inhabited ground, and the forty-second lag was precisely the kind of failure that turns a spectacular test into a genuine safety problem.

Much depends, in judging a flight like this, on the expectations set before it, and here it is worth being fair to what SpaceX actually promised. Musk had spent the preceding weeks conspicuously lowering the bar. There was, he warned, perhaps a fifty percent chance the rocket would even reach orbit; more strikingly, he told audiences that he would count the flight a success if it merely cleared the launch pad without wrecking it — "if we get far enough away from the launch pad before something goes wrong," he said, "then I think I would consider that to be a success." This was partly genuine engineering realism, a company that had learned from the Falcon 1 years that first flights of new rockets usually fail, and partly the shrewd management of a global audience's expectations, so that anything short of total catastrophe could be framed as a win. Both things were true at once. It is a recurring feature of the SpaceX story that the company set its own scoring rules, and a skeptic is entitled to note how convenient those rules could be; but it is also true that by the genuine, technical standard of a maiden flight of the most complex rocket ever built, getting thirty-three engines lit and the vehicle off the pad and through the worst aerodynamic stresses of the lower atmosphere really was an achievement, whatever came after.

By the conventional standards of the aerospace industry, the flight was a failure: the rocket had not reached its objectives, had not achieved the near-orbital trajectory it was aiming for, and had been destroyed minutes after launch. But SpaceX, true to its philosophy, framed it very differently, and the framing was not entirely spin. The company's stated goal for this first flight had been modest by its own lights: primarily, to clear the launch pad — to get the enormous vehicle off the ground without destroying the launch infrastructure — and to gather data about how the vehicle behaved in flight. By that measure, the flight had partial success: it had lifted off, it had flown for several minutes, it had survived the punishing aerodynamic pressures of the lower atmosphere while tumbling, and it had generated an enormous quantity of data about the performance of the engines, the structure, and the flight dynamics of the largest rocket ever built — data that could be gained no other way than by flying. SpaceX celebrated the flight as a valuable test, exactly the kind of learning- through-flight its iterative method was built on, and it set immediately to analyzing the failures and preparing the next vehicle. Musk and the company had said all along that they expected the early flights to fail, that they were testing hardware to destruction to learn, and that success would come through iteration, not through getting everything right the first time. In that light, the fiery end of the first flight was not a catastrophe but a data point, painful but productive.

It is worth remembering, too, the strange communal texture of the moment, because it was unlike the way rocket failures had been experienced before. Millions watched the flight live on SpaceX's own webcast, and the company's young hosts, standing in front of a cheering crowd of employees, greeted each development — even the tumble, even the fireball — with whoops and applause, framing the destruction as a triumph in real time. To some viewers this was inspiring, the visible joy of a team that had genuinely cleared the bar it had set; to others it was jarring, a rocket cartwheeling to its doom over the Gulf while a crowd cheered as though at a fireworks show. Both reactions were understandable, and the dissonance between them captured something essential about how SpaceX had changed the emotional grammar of spaceflight: failure, staged openly and narrated cheerfully, had become part of the product, a spectacle the company invited the world to share rather than an embarrassment to be managed behind closed doors. No government program had ever televised its own rocket exploding and called it a good day; SpaceX did, and meant it.

There was, however, one aspect of the flight that could not be waved away as productive learning, and that provoked genuine criticism and lasting consequence: the damage to the launch site and the surrounding environment. The thirty-three engines, firing at full power just above the pad, had unleashed a force the mount was not adequately prepared to withstand. SpaceX had, in a decision that many observers found astonishing, launched without a flame trench and without a water-deluge system — the standard infrastructure that channels and floods away the searing exhaust of a large rocket, protecting the pad from the heat and the acoustic shock. Musk had gambled that a specially formulated, fondu-based high-strength concrete beneath the mount would hold; it did not. The exhaust tore into the pad, excavated a crater beneath the mount, and hurled chunks of pulverized concrete and steel outward — some fragments flung hundreds of meters, some reportedly weighing thousands of pounds — cratering the site, scattering debris across the environmentally sensitive coastal land including the adjacent wildlife refuge, sparking a brush fire in the state park, and raining a fine grit of pulverized concrete on the town of Port Isabel, some six miles away, where residents found their cars and windowsills coated in dust. The dust plume drifted for miles, and the episode drew sharp criticism from environmental groups and hard questions from regulators about the adequacy of SpaceX's environmental review and the wisdom of launching without proper pad protection. The Federal Aviation Administration opened a formal mishap investigation, and when it closed that investigation the following September it directed sixty-three corrective actions — redesigns of vehicle hardware to prevent leaks and fires, a rebuilt and hardened launch pad, additional testing of the flight-termination system, and more — all of which SpaceX had to complete before it would be licensed to fly again. The delay stretched across the better part of a year, during which the company not only reworked the rocket but rebuilt the pad entirely, this time with a massive steel flame deflector and a high-volume water- deluge system that could dump hundreds of thousands of gallons in the seconds around ignition.

The environmental and regulatory dimension of the Starship program deserves acknowledgment, because it represents a real cost and a real tension in the SpaceX story. Boca Chica is an ecologically sensitive place, a habitat for protected species and a refuge for wildlife, and the transformation of it into a rocket-testing and launch site — with the noise, the debris, the fires, the beach closures, and the occasional destruction of a prototype scattering fragments across the landscape — imposed genuine burdens on the local environment and the surrounding community. SpaceX and its defenders argued that the burdens were manageable and justified by the importance of the work, and that the company took mitigation seriously; critics argued that the company moved too fast, cut corners on environmental protection, and treated a fragile ecosystem as an acceptable casualty of its ambitions. Both views captured part of the truth, and the tension between the drive to build and fly quickly and the obligation to protect the environment and the community would persist throughout the program, a recurring source of friction, delay, and legitimate concern. It is part of the honest accounting of what the new space age costs, and to whom.

The rebuild that followed was itself a lesson in how SpaceX turned a bad day into a redesign, and two changes from that long grounding would shape everything that came after. The first was the reengineered pad — the steel flame deflector and the deluge system that would, on the very next flight, let the mount shrug off a launch that had cratered it before. The second was a change to the rocket itself that the tumbling failure of separation had made urgent: SpaceX adopted hot-staging, bolting a new vented interstage ring atop the booster so that the upper stage could light its engines while still attached and blast itself free, rather than relying on the gentle coasting separation that had failed so spectacularly as the wounded vehicle spun. It was, in a sense, an old Soviet technique revived, and Musk acknowledged as much; the point is that the catastrophe of Flight One did not merely destroy a rocket but reshaped the design of the next one, which is exactly what the iterative method was supposed to accomplish, and exactly what it did. The forty-second lag in the flight-termination system, the engine-outs, the shattered pad, the tumbling separation — each became a specific fix, folded into the next vehicle, so that the second flight would fail later and higher and for different reasons. That was progress, of a brutal and expensive kind, but progress nonetheless.

The first flight of Starship, then, was characteristically double-edged: a genuine achievement in getting the largest rocket in history off the ground and holding it together through a violent tumble, a valuable if fiery source of data, and, at the same time, a vivid display of how much remained to be solved and of the collateral damage the program could inflict. It was the beginning of a long, difficult, and very public campaign of test flights, each building on the last, each a mixture of progress and setback, that would unfold over the following years and that would, flight by flight, push Starship toward the capability it was designed for even as it repeatedly reminded the world that this was the hardest thing SpaceX had ever attempted. The next flights would go higher and achieve more — reaching space, reaching near-orbital velocity, and, in a moment of pure engineering theater that few who watched Flight One could have imagined, catching the returning Super Heavy booster out of the sky with the arms of the launch tower — but they would also bring their own failures and their own hard lessons, in a campaign whose ultimate outcome, even as this book goes to press, remains unfinished.

Chapter 46: Catching a Skyscraper

The flights that followed the fiery debut came at intervals of months, each preceded by the rebuilding of hardware, the analysis of the previous failure, and the grinding work of securing regulatory approval, and each pushing Starship a little further toward its goal. The second flight, on November 18, 2023, showed how much SpaceX had learned in the seven months since the first. This time the rebuilt launch pad, now armored with its steel flame deflector and its high-volume water-deluge system, survived the liftoff intact — the crater problem solved, the deluge slamming up through the exhaust and dampening the shock, the mount emerging scorched but whole. This time all thirty-three engines of the booster fired and kept firing through the ascent, a dramatic improvement over the engine-outs of the first flight. And this time the vehicle achieved something the first had not: a clean separation of the booster from the upper stage, accomplished with a new and slightly alarming technique called hot-staging, in which the upper stage lights its own engines while still attached to the booster, the exhaust venting through a specially built, vented interstage ring, the thrust itself shoving the two apart. Hot-staging is efficient — it wastes none of the coasting time a conventional separation loses — but it is tricky, demanding that a vehicle survive being blasted, at close range, by the engines of the stage above it, and its success on the second flight was a genuine advance. The flight still ended in the loss of both stages: the booster broke apart moments after separation, apparently as it attempted to turn back, and the upper stage was lost some eight minutes in, high above the ocean, when a leak and fire in its aft section triggered the flight-termination system. But it had flown farther, achieved more, and validated the key improvements. The iterative march was working, and its logic was becoming visible in the data: each flight failed later, and higher, than the one before.

The third flight, on March 14, 2024, went further still. The upper stage reached space and very nearly reached orbital velocity, coasting through the vacuum on a suborbital arc that carried it most of the way around the world before it was lost during the fiery reentry, its steel skin and heat shield not yet able to survive the full ferocity of the return. Along the way, the flight rehearsed capabilities Starship would need for real work: the opening and closing of its payload door — the "Pez dispenser" slot through which future Starlinks would be released — and a demonstration of transferring cryogenic propellant between internal tanks, a small but essential step toward the orbital refueling on which the entire Moon-and-Mars architecture depends, since a Starship bound for the Moon will have to be topped off in orbit by a fleet of tanker ships. Each flight was reaching a little closer to the full profile: launch, separation, the upper stage to orbital speed, and — the ultimate goal — both stages surviving reentry and returning to be reused. The fourth flight, on June 6, 2024, achieved a landmark that had eluded every attempt before it: the upper stage survived reentry. It survived it barely, and unforgettably. One of the forward steering flaps, caught in the worst of the plasma, began to burn and shred during the descent, its hinge glowing, chunks of it peeling away — and the onboard cameras, streaming live, captured the harrowing, almost unbearable images of the ship streaking through the incandescent sheath of reentry with a flap visibly disintegrating, melting before the eyes of millions, and yet somehow holding attitude. Against the odds the crippled ship rode it out, flipped upright, lit its engines, and set itself down in a controlled splashdown on the Indian Ocean. The booster, meanwhile, executed its own controlled descent to a soft splashdown in the Gulf — so precise, one SpaceX engineer marveled afterward, that it had come down "with half a centimeter accuracy." For the first time, both stages had survived their return, if only to the water rather than to a landing. Starship had proven it could come home.

The flap fire on the fourth flight is worth pausing over, because it shows how even a "success" fed the iterative machine. Starship's forward flaps sat, in the Block 1 design, near the leading edges of the vehicle, squarely in the path of the hottest plasma, and the fourth flight proved the point in the most graphic way possible, the hinge line glowing and the flap surface peeling into the airstream. The ship survived, but the lesson was unmistakable, and SpaceX responded by moving the forward flaps on later vehicles further leeward, tucking them out of the worst of the heat, and by reworking the heat-shield tiles that shielded the seams. This is the texture of what "learning through flight" actually means: not a slogan but a specific flap, in a specific place, redesigned because a camera showed it burning. Reentry was, in many ways, the single hardest problem the ship faced — harder, arguably, than the booster catch — because a returning Starship had to survive the full inferno of orbital velocity across its whole vast belly, with thousands of ceramic tiles each of which had to stay put, where a Falcon 9 booster came home from a gentler suborbital lob. That asymmetry is why the booster came to heel long before the ship did.

And then, on the fifth flight, on October 13, 2024, SpaceX did something that looked, even to jaded observers, like magic. The goal this time was not merely to bring the Super Heavy booster back to a soft splashdown but to catch it — to have the enormous booster, after separating and flying back toward Starbase, descend with extraordinary precision to the launch tower and be plucked from the air by two great mechanical arms mounted on the tower, the "chopsticks," which would close around the booster and take its weight as its engines shut off, snatching a returning skyscraper out of the sky. It was an idea so audacious that many had doubted SpaceX would really attempt it on an early flight, let alone succeed on the first try. Catching the booster with the tower, rather than landing it on legs as Falcon 9 did, was central to the plan for rapid reuse: a booster caught by the tower could, in principle, be lowered straight back onto the mount and readied to fly again in short order, without the weight and complexity of landing legs, enabling the fast turnaround the Mars vision demanded. But it required a precision almost beyond belief — bringing a vehicle the size of a twenty-story building, descending from the edge of space at thousands of kilometers an hour, to a near-stationary hover within centimeters of a fixed target, and closing mechanical arms around it, all autonomously, in a handful of seconds, with essentially no margin for error. SpaceX had built in a hard gate against hubris: thousands of separate health checks on the vehicle and the tower had to pass in the final minutes, and if any one of them failed the booster would automatically divert to a splashdown in the Gulf rather than risk destroying the tower. Only if everything was perfect would the catch be attempted.

On that October morning, the fifth Starship lifted off, the stages separated cleanly, and the Super Heavy booster turned and flew back toward the launch site while the upper stage continued on toward its own splashdown in the Indian Ocean. As the booster descended toward Starbase, slowing, steering itself with exquisite precision toward the waiting tower, the flight computers ran their final checks and gave the go. The booster came down through the last hundreds of meters, its engines reigniting for the landing burn, throttling to a near-hover as it slid in alongside the tower — and the great arms closed around it, catching the descending skyscraper in mid-air and holding it fast against the pins beneath its grid fins as its engines cut off, the booster hanging from the tower, caught, intact, exactly as designed, roughly seven minutes after it had left the ground. The reaction at Starbase was pandemonium — a roar of disbelief and joy that rivaled the first Falcon 9 landing nine years earlier, engineers screaming and embracing and pounding one another as they watched a feat that had existed, until that instant, only in animation and imagination. SpaceX had caught a rocket the size of a building out of the sky with a pair of mechanical arms. It was, by any measure, one of the most remarkable engineering achievements in the history of spaceflight, and it turned the seemingly fantastical plan for Starship's reuse from an animation into a demonstrated fact.

To grasp why the watching engineers came unglued, it helps to sit with the numbers behind those few seconds. A Super Heavy booster stands about as tall as a twenty-story building and, even emptied of most of its propellant, weighs on the order of two hundred tons; it comes screaming back toward the coast from tens of kilometers up, and it must arrive at the tower not roughly but exactly, decelerating from supersonic speeds to a dead hover in the right cubic meters of air, at the right instant, so that the arms can close on the pins beneath its grid fins. There is no second try and no human in the loop; the whole terminal maneuver unfolds faster than a person could react, steered by the vehicle's own computers weighing thousands of variables in real time. Falcon 9 had taught SpaceX to land a booster on a target, but it landed on its own legs, on a wide barge or a broad concrete pad, with meters of margin. To catch meant throwing away the legs and the margin both and betting the tower — an irreplaceable piece of infrastructure — that the guidance would be perfect. The company's willingness to make that bet, and the automated gates that would wave the booster off to the sea if it were not perfect, together expressed a confidence in its own control software that even admirers found audacious. That it worked on the first real attempt is the part that strains belief.

The booster catch was a genuine triumph, and it deserves to

be celebrated as such, a moment when SpaceX once again turned the apparently impossible into the demonstrated. It showed that the company's mastery of propulsive landing and precision control, honed over a decade and scores of flights on Falcon 9, could be extended even to a vehicle of Starship's colossal scale, and it validated a key element of the architecture that was supposed to make rapid reuse possible. It is worth being clear, though, about what the catch did and did not prove, because the achievement was so dazzling that it was easy to over-read. Catching the booster demonstrated a spectacular mastery of the descent and the terminal maneuver — but it did not, by itself, deliver reuse. No caught booster had yet been flown a second time; the harder economic proof, that a returned vehicle could be inspected, refueled, and launched again quickly and cheaply, still lay ahead, as did the far more daunting problem of bringing the upper stage — the ship that had to survive the full inferno of orbital reentry, flaps and all — back to a catch of its own. The footage of the catch became, almost instantly, one of the defining images of the era — the vast booster sliding down out of a clear sky and settling into the arms of the tower, replayed on every screen, shared past all counting, watched even by people who followed nothing else about spaceflight. It had the quality of a thing that should not be possible and yet was plainly happening, and that combination is rare enough that it lodged in the public imagination the way the first Moon landing footage or the first Falcon 9 touchdown had. For a company whose entire method depended on persuading the world that the fantastical was merely difficult, the catch was worth more than any brochure: it was proof, in seven minutes of live video, that the animations SpaceX had been showing for years were engineering plans, not dreams. The sixth flight, on November 19, 2024, continued the campaign, relighting a Raptor in the vacuum of space to prove the engine could be restarted for orbital maneuvers, and drawing an unusual visitor to Starbase in the president-elect, who watched the launch beside Musk; on that flight the booster catch was waved off in the final seconds when the automated checks flagged a problem with the tower, and the booster diverted, as designed, to a soft splashdown in the Gulf — a small but reassuring proof that the safety gate worked as intended. By the end of 2024, Starship had come a very long way from the tumbling wreck of its first flight: it was separating cleanly, reaching space, surviving reentry, and catching its booster with the tower. The pieces of the fully reusable system were coming together, one hard flight at a time. But the path ahead was not smooth, and the following year, 2025, would bring a stretch of failures that tested the company's confidence and reminded everyone that Starship remained a developmental vehicle — unproven, unpredictable, and still capable of blowing up in new and unexpected ways. The hard year was coming.

Chapter 47: The Hard Year

If 2024 had been the year Starship began to come together, 2025 was the year it seemed, for a stretch, to come apart. The source of the trouble was a new, larger, more capable version of the Starship upper stage — the second generation, designated Block 2 or Version 2 — which SpaceX introduced in an effort to increase the vehicle's performance, incorporating a redesigned structure, larger propellant tanks, a reworked forward flap arrangement, and other improvements meant to push Starship toward its full operational potential. But the redesigned vehicle brought new problems, and the early flights of the second-generation Starship in 2025 suffered a run of failures that was, even by the standards of a company that embraced failure as a teacher, sobering.

The seventh flight, on January 16, 2025, began with a genuine triumph and ended in a genuine debacle, and the juxtaposition was itself instructive. The booster performed flawlessly and flew home to be caught, for the second time, by the arms of the tower — proof that the booster catch was becoming not a stunt but a repeatable maneuver. But the new-design upper stage above it never got the chance to shine. Climbing toward space after a clean separation, the Block 2 ship developed a fire in its aft section — the region SpaceX engineers came to call the "attic," the enclosed volume around the engines and plumbing — fed by leaking propellant, and the fire spread until it severed communications and the vehicle broke apart, some eight and a half minutes into the flight. What made this failure different from the fiery losses of the developmental era was where the debris came down. The ship broke up high over the Atlantic near the Turks and Caicos Islands, and its remains — glowing streaks that tourists filmed from their balconies, and, in the days that followed, chunks of debris and scorched heatshield tiles that washed up on the beaches — fell across inhabited islands and busy air corridors. The Federal Aviation Administration activated what it calls a Debris Response Area and briefly halted or diverted air traffic near the falling wreckage; several airliners, caught in the holding pattern, radioed that they were running low on fuel. No one was hurt, and the property damage was minor — a single reported dent to a car — but the episode changed the texture of the criticism, because a failure that scatters flaming debris over vacationers and passenger jets is not the same kind of failure as a prototype crumpling on a remote Texas stand.

The eighth flight, on March 6, 2025, delivered an almost eerie repetition. Once again the booster came home and was caught cleanly by the tower; once again the new-design upper stage was lost during ascent, to trouble in its propulsion section, several engines shutting down before the ship tumbled and broke apart; and once again the debris rained from the sky over the Caribbean, this time streaking spectacularly over the Bahamas and Florida, prompting the FAA to ground flights at several Florida airports and to open, once more, a debris response zone. Two flights of the Block 2 vehicle, two losses of the same general kind, with the same alarming public consequence, was a pattern that demanded explanation, and it strongly suggested that the second-generation Starship had introduced a flaw — some resonance, some vibration environment, some weakness in the reworked structure and propulsion — that SpaceX had not yet fully understood. The company investigated, pointed to harmonic vibration and propellant leaks feeding fires in the aft section, made changes, and pressed on. But the ninth flight, on May 27, 2025 — notable in its own right as the first to reuse a flown booster, one of the very Super Heavys that had already been to space and back — brought no relief. The reflown booster was lost during its descent test, and the upper stage, though it reached space this time, lost attitude control during its coast, tumbling out of its planned orientation, and was passivated and given up, disintegrating on reentry within its planned corridor. And then, before the tenth flight could even be attempted, came perhaps the most jarring loss of all: on June 18, 2025, Ship 36, a Block 2 upper stage sitting on a test stand for a routine engine firing, detonated in a colossal explosion that lit up the night and destroyed the vehicle where it stood, the failure later traced to the rupture of a composite-overwrapped pressure vessel in the nose — a burst gas bottle, not an engine, but a whole ship lost all the same, and the test stand with it.

The investigations that followed each loss painted a picture of a vehicle undone not by any single blunder but by the treacherous physics of its own newness. On the seventh flight, SpaceX concluded, the Block 2 ship had shaken itself apart from the inside: the vehicle's structure had resonated with the thrumming of its engines at a frequency far stronger than the designers had predicted, a harmonic buildup that stressed the propulsion plumbing until propellant leaked and the aft "attic" caught fire. This is one of the oldest and most insidious hazards in rocketry — the tendency of a slender, powerful machine to find a resonant note and amplify it toward destruction — and it had bitten SpaceX precisely because the redesigned ship was a new instrument that had never been played at full volume. The eighth flight's loss, though it traced to a somewhat different failure in the propulsion section, rhymed with the seventh closely enough to suggest the company had not yet fully mapped the vibratory world its larger ship lived in. These were not the errors of carelessness so much as the unknowns of the frontier, but to a public watching debris streak over its beaches, the distinction offered cold comfort.

For SpaceX itself, the hard year exacted a cost that did not show up on any telemetry stream: a toll on confidence and on morale. This was a company whose brand was momentum, whose people were accustomed to the narrative of relentless forward progress, and to watch ship after ship break apart — and a whole vehicle detonate on the stand before it could even fly — was a genuinely disorienting experience for an organization that had grown used to winning. The reused booster of the ninth flight was meant to be a bright counter-story, a Super Heavy flying to space and back and then flying again, the reuse the whole architecture existed to prove; instead it was lost on its descent, the good news swallowed by the bad. There is a temptation, in retrospect, to smooth these months into just another dip on the way up, but that is not how they felt as they happened, and it is not how the company's critics — or, quietly, some of its own engineers — experienced them. The doubt was real, and the question of whether the Block 2 design had been a genuine misstep hung over Starbase through the summer.

This run of failures was the hardest stretch Starship had faced since the earliest days, and it drew intensified criticism and doubt. Skeptics who had always regarded the Mars vision as a fantasy pointed to the repeated explosions as evidence that SpaceX had overreached, that Starship was too big and too ambitious, that the company's move-fast-and-blow-things-up method had finally hit a wall it could not iterate its way past. The debris falling from the sky over populated islands and air routes gave the criticism a sharper, more legitimate edge, raising real safety and regulatory concerns and prompting the FAA, which licenses the launches and bears responsibility for public safety, to require investigations and impose constraints. The environmental and community concerns around Boca Chica intensified in step. And the sheer repetition of failure — flight after flight ending in the loss of the ship, a whole vehicle lost even on the ground — tested the patience of observers who had been promised, for years, that Starship was on the cusp of transforming spaceflight, and who now watched it repeatedly fail to complete a test flight. Even sympathetic voices began to ask whether the Block 2 redesign had been a misstep, whether SpaceX had pushed too far too fast, whether the confident timelines for lunar landings and Mars missions that depended on Starship were not, once again, wildly optimistic. NASA had staked its Artemis plan to return astronauts to the Moon on a version of this very vehicle, and every ship that broke apart over the Caribbean was, implicitly, a question mark drawn over that plan.

SpaceX's response was, characteristically, to keep flying and keep iterating, and to insist that the failures, however painful and public, were the normal cost of developing a radically new vehicle at the edge of the possible. There was truth in this. The history of the program's own earlier phases — the string of destroyed prototypes before SN15 landed, the tumbling wreck of the first integrated flight before the booster catch — had shown that SpaceX's method really could grind through failure to success, that the explosions really were, in the aggregate, productive, that the company really did learn from each loss and build a better vehicle. The question in 2025 was whether that pattern would hold for the second-generation Starship, or whether the new design carried problems that would prove more stubborn. And there was a deeper question, too, about the human and material and environmental costs of the method — about how many vehicles could be lost, how much debris could fall, how much regulatory and public patience could be spent, before the iterative approach became untenable at Starship's scale. Blowing up cheap steel prototypes at a remote test site was one thing; losing full vehicles on flights that scattered burning debris across international airspace and forced airliners to divert was another, and the calculus of acceptable failure grew more fraught as the vehicle grew larger and its failures more consequential.

The stakes of the stretch reached well beyond Starbase, and

that is part of why it drew such scrutiny. NASA had selected a version of Starship as the lander that would set the first Artemis astronauts on the Moon since Apollo, a decision that tied the American return to the lunar surface directly to the fortunes of the vehicle now breaking apart over the Caribbean; every failed flight was, implicitly, a slip in a timeline that Congress, the agency, and a watching geopolitical rival were all tracking. SpaceX's own answer to the crisis was, in the end, not to patch the second-generation ship indefinitely but to hasten past it — to pour its effort into a third-generation Starship, larger and more capable still, and into a new launch and catch complex built to handle it, treating the Block 2 vehicle less as a design to be perfected than as a rung to be climbed and left behind. It was iteration at the level of whole generations of hardware, the same restless logic that had thrown over carbon fiber for steel now applied to the ship itself, and it carried the same wager: that moving on fast was wiser than lingering to make a flawed thing perfect.

The hard year did, in the end, begin to turn. After the run of losses, SpaceX identified and addressed the problems that had been plaguing the second-generation vehicle — the leaks and the fires in the aft section, the vibration environments, the weaknesses that had defeated one ship after another — and the tenth flight, on August 26, 2025, returned Starship to success. The ship flew its full planned profile, opened its payload door and deployed a batch of eight Starlink mass simulators — the first time Starship had ever released a payload, a milestone that pointed past testing toward actual work — relit an engine in space, and rode its reentry down to a controlled splashdown, while the booster completed its own descent even after losing an engine on the way up. The eleventh flight, on October 13, 2025 — one year to the day after the first booster catch — succeeded as well, achieving its major objectives, and it marked a milestone of a different kind: it was the last flight of the second-generation Starship and the first-generation Super Heavy booster, the closing of a chapter, as SpaceX prepared to move on to a third-generation vehicle, larger and more capable still, and to a new launch pad built to handle it. The hard year had been survived. The iterative method had, once more, delivered — though at a cost in vehicles, in debris, in delay, and in public confidence that was real, and that the company's critics were right to note. Starship had proven, yet again, that it could recover from failure; but it had also proven that it remained a developmental vehicle, years from the routine reliability its grand mission required, and its timelines — for the Moon, for Mars, for everything — remained deeply uncertain. As 2025 gave way to 2026, SpaceX prepared to fly the next generation of its great rocket, and the world waited to see whether the new version would finally bring the breakthrough to full operational capability, or whether it would bring new failures of its own. The answer, when the twelfth flight came in the spring of 2026, would be characteristically mixed — a brilliant success for the upper stage, which flew its profile and splashed down intact in the Indian Ocean, and yet another loss of the booster, which broke apart over the Gulf — and it would leave the vehicle grounded once more pending an FAA investigation, and Starship, and the mission it carried, still poised between triumph and uncertainty.

Chapter 48: Iterate or Die

Running through the whole of SpaceX's history, from the exploding Falcon 1 prototypes to the shredded Starships of 2025, is a single philosophy so central that it deserves to be examined on its own: the philosophy of rapid iteration, of learning through failure, of building and testing and breaking and rebuilding at a pace and a tolerance for destruction that the traditional aerospace industry found reckless and that SpaceX found essential. This approach was the source of much of the company's speed and success, and also of its most vivid failures and its sharpest criticism, and understanding it — its logic, its costs, and its limits — is key to understanding the whole enterprise. Musk gave it a motto that he repeated, in various forms, for two decades: failure is an option here, he liked to say, and if things are not failing, you are not innovating enough. It was half creed and half provocation, and the company built its culture around it.

The philosophy can be stated simply. Rather than trying to design a perfect vehicle on paper before building it — analyzing every possibility, testing exhaustively on the ground to eliminate all risk before flight, the traditional aerospace approach born of the extreme cost of failure — SpaceX built hardware quickly, flew it, watched it fail, learned from the failure, and built an improved version, iterating rapidly toward a working design. Failure, in this approach, was not a disgrace to be avoided at all costs but a source of information to be harvested, a normal and expected part of development, even a sign that the company was pushing hard enough. The explosions that punctuated SpaceX's history — the prototypes lost at Boca Chica, the boosters crashed on drone ships, the Starships torn apart in the sky — were, in this framing, not disasters but iterations, each one a lesson that moved the design forward faster than any amount of cautious ground analysis could have. The company even embraced the failures publicly, and did so with a certain gallows wit; when a rocket blew up, SpaceX would speak, half-seriously, of a "rapid unscheduled disassembly," and it released a blooper reel of its own crashing boosters set to jaunty music, framing the wreckage as steps on the way to success — a candor that was partly genuine philosophy and partly shrewd public relations, and effective as both.

The logic of this approach rested on economics and on speed, and its origin story was not a boardroom theory but a near-death experience. SpaceX's first rocket, the small Falcon 1, failed on its first flight in 2006, failed again in 2007, and failed a third time in the summer of 2008, each failure consuming money the company did not have. Musk has said the company had funds for perhaps four attempts; the fourth, in September 2008, finally reached orbit, days before SpaceX would have run out of cash entirely. The lesson burned into the company by that experience was not that failure must be avoided at all costs — it plainly could not be — but that failure had to be made survivable, which meant it had to be made cheap. Iterating through failure is only viable if failure is affordable, if you can lose vehicles and learn from the loss without bankrupting yourself, and SpaceX worked relentlessly to drive its hardware costs down, from the mass-produced Merlin and Raptor engines to the stainless-steel Starship prototypes that cost a small fraction of what a carbon-fiber vehicle would have. Cheap hardware made failure affordable, and affordable failure made rapid iteration possible.

A concrete comparison makes the payoff vivid. Consider the vehicle built the old way, in the same years, for a similar purpose: NASA's Space Launch System, the giant rocket meant to carry Artemis astronauts to the Moon. SLS was engineered with the full traditional rigor — extensively analyzed, ground-tested, its Shuttle-derived engines and boosters qualified exhaustively before it ever flew — and it worked the first time, launching flawlessly on its debut at the end of 2022. But it took well over a decade and tens of billions of dollars to reach that first flight, each vehicle costs on the order of billions of dollars, and every one is thrown away after a single use. Starship was built the other way: flown before it was finished, blown up again and again, its early flights scattering wreckage across the Gulf and the Caribbean — and yet developed for a fraction of the cost, at a pace that let SpaceX fly a dozen full-scale test flights of the largest rocket ever built in the time a traditional program might have spent qualifying a single design on the ground. One method buys reliability up front and pays for it in decades and billions; the other buys speed and cheapness up front and pays for it in public explosions. Set the two side by side and the trade at the heart of the iterative philosophy stops being abstract: SpaceX was willing to fail loudly and often in exchange for moving perhaps ten times faster and cheaper than the alternative, and by flying early and learning from each loss it compressed development timelines the industry had thought immovable. The results spoke for themselves — the reusable rocket, the human-rated spacecraft, the mega- constellation, the largest rocket ever built — all achieved at a pace the establishment had deemed impossible, largely through the power of iteration.

The Falcon 9 landing campaign, years before Starship, had been the philosophy's first great vindication, and it is the cleanest illustration of iteration paying off. SpaceX tried, and failed, to land boosters again and again — a booster that ran out of hydraulic fluid for its steering fins and spun into the sea, a booster that came in too hard and toppled, a booster whose throttle valve responded too slowly and that struck the drone ship at an angle and exploded in a fireball Musk himself posted with the wry caption that it was a "rapid unscheduled disassembly." Each crash refined the next attempt, until, in December 2015, a Falcon 9 first stage settled upright onto a landing pad in Florida, and propulsive landing passed from stunt to routine so completely that, within a few years, SpaceX was flying individual boosters ten, fifteen, twenty times. The method had turned an industry certainty — that orbital rockets were disposable — into an anachronism. And crucially, it had worked even in the unforgiving arena of an operational vehicle: when a Falcon 9 broke apart climbing to orbit in the summer of 2015, undone by a failed strut in a helium bottle, and again when one exploded on the pad during fueling in 2016, its ruptured pressure vessel taking a satellite with it, SpaceX did not abandon the method but applied it — finding the specific fault, fixing it, and returning to flight, in each case, within months. Iteration, done with discipline, could absorb even failures that destroyed customers' payloads and still come out the far side flying more often and more cheaply than anyone else.

But the philosophy had real costs and real limits, and honesty requires reckoning with them. The failures were not free even when the hardware was cheap; they consumed time, provoked regulatory scrutiny, and, as the vehicles grew larger, imposed consequences far beyond the loss of the vehicle itself — the debris that fell from failed Starships across international airspace and forced airliners to divert, the environmental damage at Boca Chica, the strain on the communities and regulators who had to absorb the results of the company's willingness to fail. Blowing up cheap prototypes at a remote test site was one thing; losing enormous vehicles on flights that scattered burning debris over inhabited islands and busy air routes was another, and the calculus of acceptable failure grew more fraught as the stakes rose. There were domains, too, where the iterate-through-failure approach simply could not apply — human spaceflight, above all, where a failure means not a lost vehicle but lost lives. Here SpaceX had to do something genuinely difficult: layer onto its fast-moving culture the slow, exhaustive rigor of human-rating, proving its vehicles safe through analysis and ground testing before risking a crew. And it did. When it lost a Dragon capsule to an explosion during a ground test of its abort system in 2019, it did not shrug and fly the next one; it stopped, investigated, traced the fault to a leaking valve, redesigned the system, and flew astronauts only when it was satisfied, carrying its first crew to orbit in the spring of 2020 without a casualty. The company's success in reconciling these two modes — moving fast where failure was affordable, moving carefully where lives were at stake — was one of its underappreciated achievements, but it also revealed that iteration through failure was no universal solvent, that some problems demanded exactly the caution the philosophy was built to transcend.

It is worth correcting, too, a lazy version of the story that SpaceX's own showmanship encouraged: the idea that the company simply blew things up until they worked, that the method was destruction for its own sake. In fact the value was never in the explosion but in the instrumentation — the hundreds of sensors on every vehicle, the telemetry streamed back in the seconds before a failure, the disciplined engineering post-mortem that turned each loss into a specific, actionable lesson. A failure that teaches nothing is just a failure; SpaceX's insight was that a heavily instrumented failure, cheaply produced and rigorously analyzed, could teach more, and faster, than years of conservative ground testing. History even offered a cautionary counter-example the company knew well: the Soviet N1 Moon rocket, which failed on all four of its launches in part because its designers, lacking the means to test all thirty of its first-stage engines together on the ground, learned about their interactions only by flying — an uncontrolled, uninstrumented, ruinously expensive version of iteration that taught too little, too slowly, and helped lose the Moon race. The difference between that and SpaceX was not the willingness to fail but the machinery for learning from it.

The criticism of SpaceX's approach was not merely that it produced spectacular failures — those were, in the company's framing, the point — but that it reflected a culture that could shade into recklessness, that moved too fast, cut corners, and imposed costs on others in its pursuit of speed. Critics pointed to the environmental harms, the debris, the burnout of workers driven to punishing schedules, the occasional safety lapses, as evidence that the move-fast philosophy, whatever its engineering merits, could be careless of consequences and of people. Defenders responded that the failures were controlled and instructive, that the approach had produced achievements no cautious method could have, and that the costs, while real, were justified by the results and by the importance of the mission. Both sides had a point, and the truth was that the iterate-or-die philosophy was genuinely double-edged: a source of extraordinary speed and achievement, and also of real costs and genuine risks, inseparable one from the other, the same boldness producing both the triumphs and the harms.

The contrast was sharpest, and most instructive, when set against a competitor who chose the opposite creed. Jeff Bezos's Blue Origin, founded around the same time as SpaceX and with far deeper pockets behind it, adopted as its motto the Latin phrase Gradatim Ferociter — "step by step, ferociously" — and even put a heraldic crest on its patch, with tortoises reaching for the stars, a deliberate emblem of patient, methodical progress. It was a genuine philosophy, honorably held, and it produced careful, reliable hardware; but for the better part of two decades it also produced far less flight, and Blue Origin trailed SpaceX into orbit by years despite its resources. The two companies became, in effect, a natural experiment in engineering temperament — the tortoise and the hare of the new space age, except that here the hare, by failing fast and often, kept winning the races that mattered. One should not draw the lesson too crudely; caution has its place, and the tortoise's rockets did not scatter debris over vacationers. But the comparison sharpened the point that SpaceX's speed was not luck or mere audacity: it was the fruit of a specific, contestable choice about how to treat failure, and that choice, in domain after domain, paid.

The deeper significance of the philosophy is that it represented a genuine alternative to the way hard engineering had been done, and one whose influence reached far beyond SpaceX. The traditional aerospace approach — cautious, slow, failure-averse, exhaustively documented — had produced magnificent results but at enormous cost and glacial pace, and it had come to seem, by the twenty-first century, sclerotic, unable to move at the speed the moment demanded. SpaceX's iterate-through-failure approach offered a different model, faster and cheaper and more adaptive, and its success forced the whole industry, and observers far beyond it, to reconsider whether the old caution was always wisdom or sometimes merely timidity. You meet its influence now well outside the rocket business — in the rival launch companies racing to fly hardware early, in the defense start-ups promising to "build fast and test fast," in a whole generation of engineers who took from SpaceX the conviction that the way to build a hard thing is to build it, break it, and build it again. The approach was not right for every problem — human lives, above all, demanded the old rigor — but where it applied, it had proven extraordinarily powerful, and it had begun to reshape how a new generation approached the building of complex, difficult, expensive things. Iterate or die was more than a slogan; it was a philosophy of engineering and of enterprise, with real power and real costs, and it was, as much as any rocket, part of what SpaceX gave to the world — a demonstration that even the hardest problems might yield not to caution but to the relentless, failure-embracing pursuit of rapid improvement.

Chapter 49: Return to Flight

By the spring of 2026, SpaceX was ready to fly the third generation of Starship — the V3 vehicle, larger and more capable than anything that had come before, powered by the latest version of the Raptor engine, Raptor 3, which had been refined through years of development into a more powerful, simpler, and more reliable engine than its predecessors. The numbers told the story of that refinement: the Raptor 3, unveiled by the company in the summer of 2024, produced some 280 tonnes of thrust at sea level, its combustion chamber running at an extraordinary 350 bar of pressure, and it did so while shedding much of the plumbing that had cluttered its forebears. Where earlier Raptors had been wrapped in a tangle of external pipes, wiring, and a protective heat shield to survive the inferno around them, the Raptor 3 folded its cooling channels and secondary flow paths inside the engine itself, so that it needed no separate heat shield and no fire-suppression system at all — a piece of engineering that made the engine not only stronger but cleaner, cheaper, and faster to build. It is a useful corrective to a common assumption about rocketry: that progress means adding complexity. In fact the most important advance in the Raptor was subtraction — fewer parts, fewer welds, fewer things to go wrong — and it was that simplicity, as much as the raw thrust, that pointed toward the mass production a Mars vehicle would require.

The V3 Starship and Super Heavy represented a major evolution, incorporating everything SpaceX had learned across the eleven previous flights and the countless tests, and it was meant to move Starship decisively toward its operational goals: launching the largest Starlink satellites, demonstrating the orbital refueling essential for the Moon and Mars, and, eventually, carrying payloads and people. To fly the new vehicle, SpaceX had built a second launch pad at Starbase, Pad 2, designed to handle the larger third-generation rocket, and the twelfth flight would be the first to fly from it — a fresh start, in a sense, with new hardware, a new engine, and a new pad. That the company could contemplate such a wholesale reinvention only a year after its greatest triumph was itself a mark of how differently SpaceX worked from the institutions it had displaced; a traditional program would have frozen a successful design and flown it for a generation, but SpaceX treated even its successes as way stations, tearing up what worked in pursuit of what might work better.

To grasp what the company was building on, one has to recall the extraordinary thing it had already done. On October 13, 2024, on Starship's fifth flight, the Super Heavy booster — a stainless-steel column the height of a twenty-story building, its thirty-three Raptor engines together generating more than seventeen million pounds of thrust at liftoff, roughly twice the power of the Saturn V that carried Apollo to the Moon — had flown itself back to the very tower it had launched from and been plucked out of the descending air by a pair of mechanical arms, the so-called "chopsticks" of the launch structure the engineers had nicknamed Mechazilla. It was the first time in the history of spaceflight that an orbital-class booster had been caught rather than landed, and the control room at Starbase erupted in a way that startled even the people in it. The maneuver looked, to many viewers, like showmanship — a spectacular trick performed for the cameras — but the logic behind it was ruthlessly practical. Catching the booster on the tower meant the vehicle needed no heavy landing legs, saving weight that could be spent on payload, and it returned the booster directly to the place where it would be refurbished and reflown, cutting out the slow business of hauling a rocket back from a distant landing site. The catch was not a stunt; it was the linchpin of a plan to fly the same booster again within hours rather than months.

The road to that triumph had been strewn with wreckage, and it is worth remembering how violent the learning had been, because the smoothness of the catch can make the whole enterprise look easier than it was. The very first integrated flight, in April 2023, had barely cleared the pad before the ascending rocket, several of its engines already dead, tore itself apart over the Gulf — and the raw force of its thirty-three engines, firing from a launch mount that lacked any flame trench or water-cooled steel, had cratered the concrete beneath the pad and flung chunks of debris for hundreds of meters, forcing SpaceX to rebuild the launch site around a massive steel deluge system before it could fly again. Even in 2025, after the glory of the booster catch, the campaign kept stumbling: on the seventh and eighth flights, early that year, the newest version of the upper stage failed within minutes of launch on each occasion, breaking apart high above the Caribbean and streaking the twilight sky over the Turks and Caicos and the Bahamas with burning debris, briefly disrupting airline traffic and scattering fragments across islands and beaches. These were not quiet laboratory failures; they were public, spectacular, and occasionally alarming, and they framed the stakes of the V3 vehicle and its twelfth flight — a campaign that had proven it could do the near-impossible and had also, again and again, proven it could fail.

The twelfth flight, on May 22, 2026, was in many respects a triumph, and it demonstrated capabilities that pointed directly toward Starship's future as an operational vehicle.

The V3 Starship and Super Heavy lifted off from the new Pad 2, the Raptor 3 engines performing their work, and the vehicle flew through its ascent and separation. In space, the upper stage did something no previous Starship had done: it deployed payloads. The flight carried a set of Starlink satellite simulators — mass models standing in for real satellites — and the Starship successfully released all of them, demonstrating the payload-deployment capability that would be essential for its role launching the next generation of the constellation, along with a pair of modified real Starlink satellites that were used to photograph the Starship in space, capturing images of the great vehicle against the black of the void. Then the Starship upper stage did the other thing it needed to prove: it reentered the atmosphere, endured the ferocious heat and stress of the return, gathered critical data on the performance of its heat shield and its structure, guided itself with its flaps to a precise splashdown zone in the Indian Ocean, and executed its landing flip and a landing burn on two Raptor engines before splashing down — a controlled, successful return of the upper stage, the part of the vehicle that had proven so hard to bring home, now performing with growing confidence. For the Starship upper stage, the twelfth flight was a clear success, and a significant one.

That the upper stage came home at all was a hard-won victory, because reentry had been Starship's cruelest teacher. The vehicle returns from orbit protected by thousands of black ceramic tiles, each one hand-checked and individually fitted, and on flight after flight tiles had torn loose, letting the fierce plasma of reentry chew into the steel beneath; on one earlier flight, engineers had deliberately flown the ship with tiles removed in places, courting damage in order to learn exactly where the vehicle was most vulnerable. Getting a Starship through the atmosphere and down to a controlled splashdown, its flaps glowing but intact, was therefore not a routine milestone but the resolution of a problem that had destroyed several of its predecessors — and it was proof that the most stubborn obstacle to full reusability was, at last, yielding.

But the Super Heavy booster, on this flight, did not fare as well. After separating and beginning its return toward the Gulf, the booster lost control and broke apart, destroyed before it could complete its descent — a failure that stood in stark contrast to the booster catch that had thrilled the world a year and a half earlier, and a reminder that even capabilities SpaceX had once demonstrated could not yet be taken for granted with new hardware. The loss of the booster triggered the now-familiar regulatory response: the Federal Aviation Administration declared the failure a mishap, required SpaceX to conduct an investigation into what had gone wrong, and grounded Starship until the investigation was complete and the agency was satisfied that the cause had been understood and addressed. It was a pattern the company had lived through many times, and one that its admirers and its critics read in opposite ways — the admirers seeing a healthy culture of flying, failing, and fixing at a pace no rival could match, the critics seeing a vehicle that kept breaking, held to account by a regulator increasingly stretched by the sheer frequency of the mishaps it was asked to investigate. As this book goes to press in mid-2026, that is where matters stand: Starship's twelfth flight has flown, the upper stage has succeeded and the booster has been lost, and the vehicle is grounded pending the outcome of the investigation, its return to flight awaiting the resolution of the mishap. The story is, quite literally, unfinished.

This is a fitting place to pause and take stock of where Starship stands, precisely because it is unresolved. After more than a dozen flights and years of development, Starship has proven a great deal. It has demonstrated that the largest rocket ever built can fly. It has shown that the booster can be caught out of the sky by the launch tower, and that the upper stage can survive reentry and return to a controlled splashdown. It has deployed payloads in space and tested the systems it will need for operational missions. It has, in short, validated the core of its revolutionary architecture, proving that a fully reusable super-heavy launch system is achievable in principle. And yet it has not yet achieved the full, routine, reliable operation that its mission requires. It has not yet reached orbit and returned both stages to be reused. It has not yet demonstrated the orbital refueling on which the Moon and Mars missions depend. It has not yet carried a payload to orbit operationally, let alone a human being. It continues, even in 2026, to lose vehicles — the booster on the twelfth flight the latest example — and to face the regulatory groundings and investigations that failures bring. The vehicle that is meant to carry humanity to Mars remains, for now, a developmental rocket, brilliant and unfinished, closer than ever to its goal and still some distance from it.

It is worth pausing, too, over how much of that unfinished work is deliberate. SpaceX's development philosophy — inherited from the early days of Falcon and pushed to its extreme with Starship — treats each vehicle as disposable data, to be flown to the edge of failure and beyond so that the engineers can learn from what breaks rather than from what a computer model predicts might break. The company built Starships faster than it could fly them, lining up prototypes at Starbase so that the loss of one need not stall the campaign, and it accepted spectacular public failures as the price of speed. This approach horrified traditionalists, for whom a launch was a solemn, once-in-years event to be insured against every conceivable fault; but it had already produced the Falcon 9, the most reliable and most frequently flown rocket in history, and its defenders argued that the smoking wreckage of a test article was not evidence of a program in trouble but of a program moving fast enough to reach Mars in a human lifetime. Whether that gamble would pay off for Starship, as it had for Falcon, was the question the twelfth flight left hanging.

The uncertainty extends to the timelines that depend on Starship, and those timelines matter far beyond SpaceX, because the vehicle is not only the company's Mars rocket but also, under contract to NASA, the lander that is meant to carry American astronauts back to the surface of the Moon. Starship's development schedule thus bears directly on the fate of the Artemis program and on the new race to the Moon, and the repeated slips and failures of the Starship campaign have contributed to the slipping of those larger timelines, a subject we will take up when we turn to the Moon. There is a hard concreteness to this dependence that is easy to lose in the abstraction of schedules: a NASA astronaut cannot set foot on the lunar surface under the current architecture until a Starship has not only reached orbit and returned, but done so many times over in quick succession, gathering and transferring propellant along the way — so that every booster that breaks apart over the Gulf, every grounding, every month lost to an investigation, is also a month by which the American return to the Moon recedes.

For now, the essential point is that Starship, the culmination of everything SpaceX has built, the vehicle that carries the company's deepest ambitions and a good part of the new space age's hopes, is a work in progress whose outcome cannot yet be known. It may, in the years after this book is written, achieve everything its makers intend — routine flights to orbit, the launching of vast constellations, the return of humans to the Moon, and eventually the journey to Mars. Or it may continue to struggle, its grandest promises receding into the familiar future tense of "someday." The honest chronicler, writing in mid-2026, can only report that the vehicle has come astonishingly far, that it has proven the possibility of things once thought impossible, and that it has not yet crossed the finish line — and that the most important test of all, the orbital refueling that unlocks the Moon and Mars, still lies ahead. To that critical, unproven capability we now turn.

Chapter 50: Refuel in Orbit

There is a single technology, unproven at the scale required, on which the entire grand architecture of Starship depends — a capability so central that without it the Moon missions cannot happen and the Mars vision remains pure fantasy, and yet so difficult that no one has ever demonstrated it at the necessary scale. It is orbital propellant transfer: the refueling of a spacecraft in orbit, pumping propellant from one vehicle to another while both hurtle around the Earth in the weightlessness of space. Understanding why this matters, and why it is so hard, is essential to understanding both the promise and the precariousness of Starship's entire plan.

The problem arises from the tyranny of the rocket equation, the same brutal arithmetic that governs all spaceflight. Starship can lift an enormous payload — a hundred tons or more — to low Earth orbit. But to go beyond low orbit, to the Moon or to Mars, requires a great deal more energy, and therefore a great deal more propellant, than a Starship has left in its tanks once it reaches orbit. A Starship that launches from Earth and climbs to low orbit arrives there with its tanks nearly empty, having spent almost all its propellant just getting there. To then journey onward to the Moon or Mars, it needs to be refilled — its tanks topped back up with the hundreds of tons of propellant required for the onward voyage. And since you cannot launch a fully fueled Starship directly to the Moon in one shot — the physics does not allow it — the only way to send a Starship beyond Earth orbit is to launch it to orbit, and then launch additional Starships, "tanker" versions filled with propellant, to rendezvous with it and transfer their propellant into its tanks, refilling it in orbit until it has enough to continue.

The scale of that refilling campaign is where the plan collides with skepticism, and it is worth being precise, because the numbers have been a subject of genuine dispute. It is often said, loosely, that a Starship simply "tops off" in orbit before heading for the Moon, as if a single visit from a tanker would do — but in fact a single lunar mission may require a whole fleet of tanker launches, one after another, each ferrying up a partial load. SpaceX's own executives have suggested the figure might be around ten flights; NASA officials and independent auditors, examining the same architecture, have floated estimates ranging into the high teens, and some outside analysts have put it higher still. The reason for the range is itself instructive: no one yet knows exactly how much propellant each tanker can deliver, how much will be lost along the way, or how quickly the flights can be flown — and every one of those unknowns pushes the number up or down. A single mission to the Moon or Mars might therefore require not one launch but a dozen or more, all rendezvousing and transferring propellant in orbit, a complex orbital ballet that has never been performed.

Transferring propellant between spacecraft in orbit is far harder than it sounds, and the difficulty lies in the strangeness of fluids in weightlessness. On Earth, gravity keeps liquid propellant settled at the bottom of a tank, where pumps and outlets can draw it out cleanly. In the weightlessness of orbit, there is no "bottom"; liquid propellant floats freely, forming blobs and clinging to surfaces, sloshing and drifting unpredictably, and getting it to flow reliably from one tank to another — without gas bubbles, without losing propellant to the vacuum, without the ultracold cryogenic propellants boiling off in the process — is a genuinely hard engineering problem. A concrete image helps: imagine trying to pour a drink from one cup to another not merely in the dark, but with the liquid refusing to stay in either cup, breaking into wandering globes that touch the walls and each other and drift where they will — and now imagine that the liquid is methane and oxygen chilled to hundreds of degrees below zero, so cold that any stray warmth turns it instantly to vapor, and that the two cups are two spacecraft closing on each other at thousands of miles an hour. That is the essence of the task. It requires careful management of the fluids, techniques to settle and control them — often by nudging the whole vehicle gently forward so that the propellant sloshes to one end and behaves, briefly, as if gravity had returned — precise rendezvous and connection between the vehicles, and the handling of cryogenic liquids across the gap between two spacecraft.

SpaceX had, in fact, taken the first real step. On March 14, 2024, during Starship's third flight, the company carried out a modest but genuine demonstration, transferring a quantity of cryogenic propellant from one tank to another inside a single Starship — an "intertank" transfer that proved, at small scale, that the fluids could be moved and managed in orbit. It was a real accomplishment, and NASA had helped fund the work as part of a program of technology-development awards aimed squarely at cracking the refueling problem. But moving propellant between two tanks a few meters apart within one vehicle is a world away from the full operation — two separate Starships docking nose to nose in orbit and transferring hundreds of tons of cryogenic propellant from one to the other — which had not, as of the time of this writing, been demonstrated. It was the great unproven step, scheduled for its first integrated demonstration in 2026, and everything downstream of it — the lunar landings, the Mars missions — waited on its success.

The stakes of orbital refueling were made concrete by

SpaceX's contract with NASA to provide the lunar lander for the Artemis program's return of astronauts to the Moon. NASA had selected a version of Starship, called Starship HLS — for Human Landing System — to carry astronauts from lunar orbit down to the surface of the Moon and back up. But the Starship HLS mission architecture depended entirely on orbital refueling. To send a Starship to the Moon capable of landing astronauts and returning them to lunar orbit required so much propellant that the Starship would have to be refueled in Earth orbit by that series of tanker launches before it could depart for the Moon. This meant that a single Artemis lunar landing using Starship would require not one launch but a whole campaign of launches: the HLS Starship itself, plus numerous tanker flights to fill it, plus a "depot" Starship to store the accumulated propellant in orbit while the tankers came and went, all rendezvousing and transferring with a reliability and a cadence that SpaceX had not yet demonstrated. And the whole campaign was a race against the clock in a way that ordinary spaceflight is not, because cryogenic propellant does not wait. Left in orbit, warmed by sunlight and the residual heat of the vehicle, liquid methane and liquid oxygen steadily boil away, so the tanker flights cannot be spread comfortably over months; they must be flown in rapid succession, one nearly on the heels of the last, before the propellant already delivered evaporates. If orbital refueling proved harder than expected, or if the number of tanker flights required turned out to be very large, or if the flights could not be flown fast enough to outrun the boil-off, or if Starship's own development continued to slip, the whole timeline for returning astronauts to the Moon would slip with it. And indeed, the challenges of Starship and its refueling architecture were a major factor in the repeated delays to the Artemis schedule, contributing to the reframing of the program's timeline that we will examine shortly.

The dependence on orbital refueling represents both the boldness and the risk of the Starship approach. It is bold because, if it works, it unlocks capabilities that are otherwise impossible: a fully reusable vehicle that can, through orbital refueling, carry enormous payloads not just to low orbit but all the way to the Moon and Mars, at a cost that could transform deep-space exploration and settlement. It is risky because it stakes the entire architecture on a technology that remains unproven at scale, and because it makes every deep-space mission depend on a complex sequence of multiple launches and orbital operations, any one of which could fail. Critics of the Starship approach pointed to this dependence as a fundamental vulnerability, arguing that the requirement for many tanker launches and unproven orbital refueling made the architecture fragile and the timelines unrealistic. There was a certain irony in the criticism, for the alternative — a giant expendable rocket powerful enough to throw a lander straight to the Moon in a single shot, as Apollo had done with the Saturn V — was exactly the old, ruinously expensive approach that SpaceX had built its whole existence to escape. Refueling was the price of reusability; it was the mechanism by which a vehicle that landed back on Earth could nonetheless reach destinations a single tank of propellant could never take it. Defenders argued that orbital refueling, while hard, was a solvable engineering problem, that SpaceX had a track record of solving such problems, and that once solved, refueling would open the solar system in a way no other approach could. As of mid-2026, the question was genuinely open. The first demonstration of orbital propellant transfer between Starships was on the near horizon but not yet accomplished, and until it was, the grandest promises of the Starship program — the Moon, and beyond it Mars — rested on a foundation that had not yet been shown to bear the weight.

The idea of refueling in orbit was not, it should be said, a SpaceX invention, and its long and rather political history is a revealing footnote to the whole debate. Engineers had dreamed of orbital propellant depots for decades — filling stations in the sky where a spacecraft bound for the Moon or beyond could top off its tanks rather than carrying every drop from the ground — and around the start of the 2010s, internal NASA studies concluded that a network of such depots, supplied by ordinary commercial rockets, could accomplish deep-space missions without any need for a giant new government heavy-lift vehicle at all. That conclusion was inconvenient, because Congress had just directed NASA to build exactly such a giant rocket, the Space Launch System, and the depot studies were quietly set aside, their implications too threatening to a program that supported thousands of jobs across many states. The technology that might have made the monster rocket unnecessary was left on the shelf for a decade — until SpaceX, pursuing the same logic from the opposite direction, made orbital refueling the beating heart of its entire architecture. There is an irony worth savoring in the fact that the Starship now contracted to land NASA's astronauts on the Moon depends on precisely the capability that NASA's own bureaucracy had once found reason to bury.

It is worth noting how new this frontier really is, because it is easy to assume that a species which has flown to the Moon and built space stations must surely have mastered so basic a thing as moving fuel between two vehicles. In fact it never has, not at this scale. Spacecraft have topped up small quantities of storable propellant — the kind that sits stable at room temperature — aboard the Russian and Chinese space stations for decades, and robotic experiments have nudged the problem forward. But the transfer of large volumes of deep-cold cryogenic propellant, the high-performance stuff that boils at the merest hint of warmth, from one spacecraft to another in orbit, is a genuinely unprecedented act, attempted in earnest for the first time only now, by SpaceX, because only now has anyone had a reason grand enough to need it. When it is finally done, few members of the public will notice; there will be no crew aboard, no dramatic countdown, only two silver ships joined in the dark and a slow transfer of propellant registering as numbers on an engineer's screen. And yet that quiet, invisible event may prove to be the true hinge of the new space age, the moment the solar system stops being a place we visit and starts being a place we can reach at will.

This is the state of the central technology of the new space age's most ambitious vehicle: proven in pieces, unproven as a whole, essential to everything, and not yet done. It captures, in miniature, the larger condition of Starship and of the Mars dream it serves — a program of extraordinary achievement and extraordinary remaining risk, closer than ever to unlocking the solar system and still one hard, unproven step away. Whether orbital refueling succeeds, and how soon, will shape not only the fate of Starship and the Mars vision but the outcome of the new race to the Moon, in which Starship is meant to be America's lander, racing against a Chinese program with its own architecture and its own deadline. With Starship's story brought up to the uncertain present, we turn now from SpaceX itself to the wider field it competes in and, in large part, created — the constellation of rivals, challengers, and national programs that make up the new space race, beginning with the other billionaire who set out, at almost the same time as Musk, to build a private path to space, and whose company's fortunes in 2025 and 2026 would provide a vivid reminder that, for everyone, rockets remain hard.

Chapter 51: The Road to Mars

Everything comes back to Mars. It was the reason SpaceX was founded, the animating purpose behind the rockets and the satellites and the fortune, the destination toward which, in Musk's telling, all of it was aimed. And yet Mars remains, after nearly a quarter of a century, a promise rather than an achievement — no human has gone, no city has been built, and the vision of a self-sustaining human presence on the red planet remains as much a dream in 2026 as it was in 2002. It is worth confronting that vision directly, both in its grandeur and in its difficulties, because it is the horizon toward which the whole enterprise points, and because the honest reader deserves a clear-eyed account of what it would actually take, and whether it is likely ever to happen.

The vision, as Musk articulated it over the years, was of a genuinely self-sustaining city on Mars — not a small research outpost dependent on resupply from Earth, but a settlement of eventually a million people, capable of surviving and growing on its own, so that if something catastrophic befell the Earth, the light of human consciousness would continue on another world. He put numbers to it, and the numbers were staggering: to build a self-sufficient city of a million within a human lifetime would take, by his reckoning, something like a thousand Starships, flying in great fleets during the narrow launch windows that open only once every twenty-six months, when Earth and Mars swing close enough for the crossing — perhaps a hundred thousand people carried to the surface at each favorable alignment, over roughly two decades of launches, until the settlement reached the critical mass at which it could survive even if the ships from Earth stopped coming. This was the ultimate expression of the multi-planetary imperative: not merely to visit Mars, as Apollo had visited the Moon, but to settle it, permanently and independently, making humanity a two-planet species and thereby insuring against extinction. The scale of the ambition dwarfed anything humanity had attempted — the establishment of a self-sufficient civilization on a hostile alien world, requiring the transport of enormous quantities of people and material across interplanetary distances, and the solving of countless problems of survival, sustenance, and society on a planet where nothing grows and the very air is unbreathable.

It is worth dwelling on just how unbreathable, because Mars is routinely and badly misimagined. It is often pictured, in films and in the mind's eye, as a kind of cold desert — a redder, chillier Arizona where a settler might, in a pinch, survive with a warm coat and a bottle of oxygen. In fact the surface of Mars is nothing of the sort. Its atmosphere is barely one percent as dense as Earth's and composed almost entirely of carbon dioxide, so thin that a human standing unprotected on the surface would not slowly suffocate but die within moments as the low pressure boiled the water in their blood. The temperatures average far below freezing. The soil is not inert dirt but a fine dust laced with perchlorates — toxic, chlorine-based salts that are poisonous to humans and that coat the whole planet. And because Mars long ago lost the global magnetic field and thick atmosphere that shield the Earth, its surface is continually raked by cosmic rays and solar radiation, so that settlers might have to live underground or beneath thick shielding, venturing out only in armor, their exposure carefully rationed across a lifetime. Its gravity is a little more than a third of Earth's, with unknown long-term effects on human bones, hearts, and the development of children never yet conceived or born off the Earth. Growing food, producing breathable air, extracting water from the subsurface ice, manufacturing fuel and materials, building habitats, maintaining human health and reproduction across generations on such a world — each of these is a profound challenge, many of them unsolved, some perhaps unsolvable in any practical sense. And beyond the technical obstacles lie the human ones: who would go, and why; how a Martian society would be governed and sustained; whether human beings could thrive, psychologically and biologically, across generations on another planet; whether the enormous cost and risk could ever be justified. The vision of a million-person Martian city is not merely an engineering problem but a civilizational one, and its realization, if it is possible at all, lies decades or generations in the future.

The dream did not spring fully formed from Musk's head, and it helps to know where it came from. In the 1990s, an aerospace engineer named Robert Zubrin had laid out, in a book called The Case for Mars and through the Mars Society he founded, a detailed and influential argument that a human mission to Mars was not a fantasy but an achievable near-term goal — provided one abandoned the ruinously heavy approach of carrying everything from Earth and instead "lived off the land," manufacturing return propellant and other necessities from the Martian atmosphere and soil. It was Zubrin's vision of in-situ resource use, of a Mars mission made affordable by ingenuity rather than brute force, that helped shape the thinking of a generation of enthusiasts, Musk among them; the young entrepreneur, newly rich from the sale of PayPal, first turned toward space in part through contact with these ideas, even contemplating a modest philanthropic stunt — sending a greenhouse to Mars to reawaken public interest — before concluding that the real bottleneck was the cost of launch and founding a rocket company instead. Zubrin, for his part, became both a champion and a gadfly of the SpaceX enterprise, praising

Starship's promise while sharply criticizing what he saw as the grandiosity of the million-person city, and dismissing as overblown the very fears — of radiation, of perchlorates — that others raised as showstoppers. That a settlement so dependent on unproven biology and chemistry could divide even its most committed advocates is itself a measure of how much about Mars remains genuinely unknown.

The plan's real ingenuity — the piece that separates it from mere science fiction — lies in how it proposes to get the settlers home, and it repays a close look because it makes an abstract vision suddenly concrete. A rocket that had to carry all its own return propellant from Earth to Mars and back would be impossibly heavy; the mathematics simply forbid it. So the architecture, following the logic Zubrin had championed decades earlier, calls for the ships to manufacture their return fuel on Mars itself, out of the planet's own materials. This is why Starship burns methane and oxygen rather than the hydrogen or kerosene of most large rockets: methane can be made on Mars. Feed carbon dioxide — of which the thin Martian air is almost entirely composed — together with hydrogen split from subsurface water ice into a reactor, apply heat and a catalyst, and the century-old chemistry of the Sabatier reaction yields methane and water; run the water back through electrolysis and you have oxygen as well. In principle, a Starship could land on Mars nearly empty, spend the long wait until the next launch window quietly making its own propellant from the air and the ice, and then fly home on fuel that never left the planet. It is an elegant idea, and it is genuinely central to the plan — but it, too, remains unproven at anything like the necessary scale, a chemical plant the size of a small industrial facility that would have to run unattended, in the cold and the dust, tens of millions of miles from the nearest repair.

And the crossing itself is no small thing. Mars and Earth align favorably only once every twenty-six months, and even then the journey takes something on the order of six to nine months each way — half a year or more sealed inside a spacecraft, exposed for the whole of it to the cosmic radiation of deep space, beyond the protection of Earth's magnetic field, with no possibility of rescue and no way home until the planets align again. A settler who fell gravely ill, or a colony that suffered a catastrophic failure of its air or its power, could not simply be evacuated; help, if it came at all, would be many months away. This is the hard core of what "self-sustaining" must mean: not merely a base that can feed itself, but a society that can survive, unaided, every disaster that might befall it, across the long silences between ships. Whether human beings and human institutions can be made that robust on a world that is trying, chemically and physically, to kill them is a question no one can yet answer, and it is the question on which the whole grand vision ultimately turns.

Critics of the Mars vision — and there are many serious ones — argue that it is a fantasy, a distraction, even a kind of escapism, and their objections deserve a fair hearing. Some argue that the resources and effort devoted to dreaming of Mars would be better spent addressing the problems of the planet we already have — climate change, poverty, disease — and that the fantasy of a Martian escape hatch is a dangerous distraction from the hard work of caring for the Earth, perhaps even a moral hazard, offering the powerful an imagined exit from the consequences of terrestrial neglect. Others argue that Mars is simply too hostile, that the technical obstacles to a self-sustaining settlement are insurmountable or nearly so, that the vision underestimates the difficulty of sustaining human life and society on another world, and that it will never amount to more than, at best, a small and dependent outpost. Still others question the whole premise, doubting that humanity faces an extinction risk grave enough to justify the effort, or that a Martian settlement would actually provide meaningful insurance against it — noting, tartly, that even a Mars ravaged by every catastrophe humanity might visit upon the Earth would remain, in its cold and airless hostility, far more lethal than the worst post-disaster Earth. These criticisms are not foolish, and a clear-eyed account must take them seriously: the Mars vision may indeed be a fantasy, and the honest chronicler cannot assert that it will succeed.

And yet the vision, whatever its ultimate fate, has done real work in the world, and this is the paradox worth holding onto. The dream of Mars — however distant, however uncertain, however possibly impossible — drove the creation of genuinely revolutionary capabilities: the reusable rocket, the largest launch vehicle ever built, the transformation of the economics of spaceflight. Whether or not humanity ever reaches Mars, the pursuit of that goal has already changed what is possible in space, has reopened the frontier, has produced technologies and capabilities of enormous value regardless of whether the ultimate destination is ever attained. The Mars vision functioned as a forcing function, a goal so ambitious that pursuing it required solving problems and building capabilities that more modest ambitions would never have demanded. Consider that the reusable Falcon 9, which slashed the cost of reaching orbit and now launches more than half the world's payloads, was built by a company whose stated reason for existing was not cheaper satellite launch at all but the settlement of another planet; the practical revolution was, in a sense, a byproduct of the impractical dream. Musk had always framed the whole undertaking as a problem of cost: a ticket to Mars, he argued, would have to fall to something an ordinary determined person might one day afford — on the order of the price of a house rather than the billions that had made even the Moon a destination for governments alone — and it was that fixation on driving down the cost of reaching orbit and beyond, pursued for the sake of an unreachable planet, that produced the reusable rocket which now hauls the world's satellites. Even the Starlink constellation, the cash engine that funds everything else, was conceived in part as a way to pay for Mars. In this sense, the road to Mars has been productive even though the destination remains unreached, and perhaps unreachable. The rockets built to carry humanity to another world have, in the meantime, transformed spaceflight around this one, and that transformation is real even if the Martian city never rises.

Where, then, does the road to Mars stand in 2026? At its beginning, still, despite everything. Starship, the vehicle meant to carry humanity to Mars, remains a developmental rocket, not yet reliably reaching orbit, its critical orbital-refueling capability unproven, its timelines uncertain. No human has left low Earth orbit on a SpaceX vehicle; the Moon, a far nearer destination, remains ahead; Mars is a dream whose realization, if it comes at all, lies far in the future. The confident forecasts of earlier years — Musk had at various points spoken of cargo landings within a few years and human landings not long after — had slipped, as such forecasts always did, into the more distant future, and the grand vision of the million-person city, the multiplanetary species, remained exactly that: a vision, unrealized and perhaps unrealizable, receding into the familiar future tense of "someday." But the road toward it has already remade spaceflight, and the pursuit continues, driven by the same conviction that animated it from the start: that humanity's future lies, at least in part, beyond the Earth, and that the effort to get there, however uncertain its success, is worth making. Whether that conviction is wisdom or folly, whether Mars is destiny or fantasy, the road toward it has been one of the great engines of the new space age, and it runs, still, toward a horizon not yet reached.

Chapter 52: The Space Barons

Elon Musk was not the only technology billionaire who looked up at the sky and decided to spend a fortune trying to reach it. At almost the same time that Musk was founding SpaceX, another internet entrepreneur — richer than Musk, and in some ways more patient — was quietly starting his own rocket company, with his own grand vision of humanity's future in space. His name was Jeff Bezos, the founder of the online retailer Amazon, and his company was called Blue Origin. It is commonly assumed that Blue Origin was the latecomer, a project Bezos took up only after Amazon had made him one of the richest men on Earth, trailing along behind Musk's head start. In fact the opposite is closer to the truth: Bezos founded Blue Origin in the year 2000, a full two years before Musk founded SpaceX, and he did so with characteristic quiet, telling almost no one, buying up land in West Texas under shell-company names and gathering a small band of engineers to think, from first principles, about how humanity might live in space. For two decades the rivalry between Musk and Bezos, between SpaceX and Blue Origin, would be one of the defining subplots of the new space age — two of the world's wealthiest men, two very different temperaments, two competing visions of how humanity should expand beyond Earth, and two companies whose fortunes would diverge dramatically before, in the

2020s, Blue Origin finally began to close the gap, only to be reminded, in the most brutal way, that rockets remain hard for everyone.

Bezos had dreamed of space since boyhood, and his vision, like Musk's, was expansive, but it pointed in a different direction. Where Musk fixed on Mars, on making humanity a multi-planetary species with a second home on another world, Bezos imagined something else: not the abandonment of Earth for another planet but the movement of heavy industry and, eventually, of large populations off the Earth into vast artificial habitats in space. The idea was not original to him, and he never pretended it was; it came from Gerard O'Neill, the Princeton physicist whose lectures and 1976 book The High Frontier had captivated Bezos as a teenager, and whose students had once calculated, only half in jest, that the solar system could support a human population in the trillions if only people were willing to live in orbit rather than on planets. O'Neill's vision was of enormous rotating cylinders, miles long, spun to generate artificial gravity, their interiors landscaped with rivers and forests and towns, in which millions of people might one day live — and it was this, not a Martian city, that Bezos adopted as his lodestar, imagining a future in which the Earth would be zoned, in effect, for residence and light industry, a garden preserved by moving the polluting and resource-hungry activities of civilization into space. Bezos's goal, as he put it again and again, was a future of "millions of people living and working in space," and his path to it ran through the Moon and through the gradual construction of the infrastructure — cheap, reliable, reusable launch above all — that such a future would require. Blue Origin's motto was "Gradatim Ferociter," Latin for "step by step, ferociously," and its coat of arms bore a pair of tortoises reaching for the stars — an emblem of Bezos's professed philosophy: patient, methodical, unhurried, building carefully toward long-term goals rather than racing to spectacular milestones.

For many years, Blue Origin's methodical patience looked,

from the outside, like slowness bordering on paralysis, especially in comparison with SpaceX's furious pace. While SpaceX was reaching orbit, landing rockets, carrying astronauts, and building a satellite constellation, Blue Origin — despite being founded earlier and backed by Bezos's even greater fortune, funded substantially by his selling of roughly a billion dollars of Amazon stock a year — moved far more slowly and stayed far closer to the ground. Its principal achievement for most of that period was New Shepard, a small suborbital rocket named after Alan Shepard, the first American in space. New Shepard was not an orbital vehicle; it could not reach the speeds needed to circle the Earth. Instead it flew straight up, past the boundary of space some hundred kilometers high, giving its passengers a few minutes of weightlessness and a view of the Earth against the black of space before descending back down, the capsule landing under parachutes and the booster landing propulsively, both reused. New Shepard began carrying passengers in July 2021 — the first crewed flight, on the twentieth of that month, carried Bezos himself, his brother Mark, the eighty-two-year-old aviator Wally Funk, and an eighteen-year-old Dutch student named Oliver Daemen, making Funk, on that single flight, the oldest person ever to reach space and Daemen the youngest. Funk's presence was its own quiet act of justice: she had been among the "Mercury 13," a group of women who passed the same tests as the original male astronauts in the early 1960s only to be denied the chance to fly, and now, six decades later, she finally crossed the boundary she had been kept from. New Shepard became a vehicle for a certain kind of space tourism, offering the very wealthy a brief taste of spaceflight, and its reusable booster landings demonstrated genuine capability. But it was a far cry from SpaceX's orbital feats, and it left Blue Origin, for two decades, in the position of the tortoise to SpaceX's hare — except that the hare showed no sign of stopping to rest.

The rivalry between the two men could be sharp, and it broke into the open at the very moment both companies first learned to bring a rocket home. In November 2015, Blue Origin landed a New Shepard booster upright after a suborbital hop, and Bezos, breaking a long public silence, marked the occasion with the first tweet he had ever sent, calling a reused rocket "the rarest of beasts." Musk was quick to point out the distinction his rival's celebration elided — that flying straight up to the edge of space and back was a far easier feat than reaching the tremendous speeds of orbit and returning from them. A few weeks later, in December 2015, SpaceX landed the first stage of an orbital Falcon 9 after a mission that had actually delivered satellites to orbit, and Bezos returned the needle with a tweet of his own: "Congrats @SpaceX on landing Falcon's suborbital booster stage. Welcome to the club!" — a pointed refusal to grant the distinction Musk had drawn. The barbs sat atop a real commercial quarrel; the two companies had already tangled in the patent office, where Blue Origin held a patent on the idea of landing a rocket on a ship at sea, which SpaceX challenged and largely got invalidated, clearing the way for the drone-ship landings that would become routine. Beneath the sniping lay a genuine philosophical divide about how fast to move and how much to risk, and for years the scoreboard ran heavily one way.

The real test of Blue Origin's ambitions was New Glenn, its long-awaited orbital rocket, a large, partially reusable heavy-lift vehicle named after John Glenn, the first American to orbit the Earth. New Glenn was designed to compete directly with SpaceX's Falcon rockets — a big, capable launcher whose first stage, like Falcon 9's, would land and be reused, and which could carry heavy payloads to orbit for commercial, government, and national-security customers, as well as launching the satellites of Amazon's own planned internet constellation, a Starlink rival called Project Kuiper. New Glenn had been in development for the better part of a decade, long delayed, a symbol of Blue Origin's slow pace — but in January 2025 it finally flew, reaching orbit on its very first attempt on the sixteenth of the month, a genuinely difficult feat that few rockets have ever managed on a maiden flight, and one that announced Blue Origin's arrival as a real orbital competitor. The company did not manage everything at once: on that first flight it tried and failed to land the reusable first stage on a ship at sea, losing the booster during its descent, a reminder that the recovery trick SpaceX had made look routine was anything but. Its second flight, in November 2025, went further, successfully deploying a pair of NASA science probes bound for Mars and, crucially, landing the reusable first stage at sea, on a landing vessel in the Atlantic — demonstrating the recovery and reuse capability that was essential to competing with SpaceX on cost. After two decades of patient, ferocious steps, Blue Origin seemed at last to be hitting its stride, flying an orbital rocket, landing its booster, and beginning to challenge SpaceX's dominance.

And then, in a matter of weeks in the spring of 2026, Blue Origin was reminded that in this business, hard-won progress can be undone in an instant. The third flight of New Glenn, in April 2026, suffered a partial failure: though the rocket flew, its payload was left in the wrong orbit, an "off-nominal" orbit that compromised the mission. Worse was to come. In May 2026, a New Glenn rocket exploded during a static-fire test on the ground at its launch complex at Cape Canaveral — a catastrophic failure that destroyed the vehicle and severely damaged the launch pad, Blue Origin's only operational orbital launch site, with early assessments suggesting it would take a year or more to rebuild. In a single explosion, Blue Origin lost not just a rocket but its ability to launch at all until the pad could be reconstructed, a devastating setback that threatened to stall the company's momentum just as it had begun to build. Around the same time, Blue Origin announced that it would pause New Shepard's suborbital flights for a couple of years, redirecting resources toward its lunar-lander program in support of NASA's Artemis effort. That program was itself substantial and real: Blue Origin was building not one but two Moon landers — an uncrewed cargo vehicle called Blue Moon Mark 1, and a much larger crewed lander, Blue Moon

Mark 2, which NASA had selected in 2023 as a second Human Landing System alongside SpaceX's Starship, meant to carry astronauts to the surface on the Artemis V mission later in the decade. A full-scale mockup of the crewed lander's cabin had already been delivered to NASA for astronaut training, a tangible sign that Blue Origin's lunar ambitions were more than paper. But the decision to pause New Shepard, reasonable as it was, underscored how thinly even a Bezos-funded company had to spread itself.

That lunar rivalry had a bruising prehistory of its own, and it is worth recalling because it shows how the competition between the two barons spilled beyond engineering into the courts. When NASA first awarded the contract to build the Artemis lunar lander, in April 2021, it chose SpaceX alone, citing a budget too thin to fund two providers and a Starship bid that came in at less than half the price of its rivals. Blue Origin was incensed. The company protested the decision to government auditors, who rejected the complaint, and then sued NASA in federal court, arguing the process had been unfair — a fight that froze SpaceX's lunar-lander work for roughly half a year while the case ground on. Blue Origin lost, but the episode delayed the very program both companies professed to serve, and it hardened the impression, in some quarters, of a company more practiced at litigation than at flight. It was only later, when NASA found the money to add a second lander, that Blue Origin won its place at the table on the merits — a reminder that the road to the Moon now ran not only through the physics of rocketry but through the politics of contracts and the patience of judges.

The company's leadership vowed to fly again before the end of the year, and Blue Origin's deep pockets and genuine capabilities meant it would surely recover in time. But the episode was a vivid, humbling illustration of the central truth of the whole enterprise, the one that no amount of money or ambition can repeal: rockets are hard, failure stalks everyone, and the reliability that SpaceX had achieved through years of painful iteration was not something a competitor could simply buy or will into existence. There is a temptation to read the two barons as a morality tale — the hare who could not sit still against the tortoise too cautious to move — but the truth is less tidy and more instructive. SpaceX's speed had come at the cost of many public explosions, and Blue Origin's caution had spared it that spectacle while costing it years; and in the end both companies discovered the same thing, that the only way past the hardness of rockets is through it, flying and failing and flying again, whether one does so quickly and loudly or slowly and in silence. Blue Origin remained a serious contender, and its rivalry with SpaceX would continue to shape the new space race — including, as we will see, in the competition to build the landers that would return humans to the Moon. But in the middle of 2026, the contrast between the two space barons remained stark: one company launching more than half the world's payloads and preparing for a historic public offering, the other picking itself up from an explosion that had leveled its only launch pad.

Chapter 53: The Challengers

Beyond the two space barons and their giant rockets, the new space age spawned a whole ecosystem of challengers — companies large and small, established and upstart, national and commercial, each pursuing its own piece of the transformed launch market that SpaceX's cost revolution had opened up. Some sought to compete directly with SpaceX; others carved out niches; still others were the legacy incumbents, struggling to adapt to a world in which the rules they had known for decades no longer applied. Together they made up the competitive landscape of the new space race, a field far more crowded and dynamic than the sleepy, government-dominated launch business of the pre-SpaceX era. That very crowding was itself a monument to what SpaceX had wrought: by proving that private companies could build rockets cheaply and that reusability could work, SpaceX had inspired and enabled a wave of new entrants, even as it dominated them.

Among the most successful of the newcomers was Rocket Lab, a company founded by the New Zealand engineer Peter Beck, which took a different approach from SpaceX by focusing initially on the small end of the market. Rocket Lab's Electron rocket was a small launcher, designed to carry modest payloads — small satellites, the kind that had proliferated as electronics shrank and the cost of building satellites fell — to orbit frequently and affordably. In an era when satellites were getting smaller and more numerous, there was a real market for a dedicated small launcher that could fly a small satellite to a specific orbit on its own schedule, rather than forcing it to hitch a ride as a secondary payload on a big rocket bound elsewhere. Electron found that market and flew regularly, making Rocket Lab one of the few new launch companies to achieve consistent commercial success — and Beck's engineers pursued their own clever, lower-cost path to reusability, not catching the whole booster on a tower as SpaceX did but attempting to snag the small Electron first stage in mid-air with a helicopter as it drifted down under a parachute, before ultimately settling on fishing the boosters out of the ocean and refurbishing them. And Rocket Lab did not stay small: it developed a larger, reusable rocket called Neutron, aimed squarely at the medium-lift market that Falcon 9 dominated, positioning itself as perhaps the most credible of the companies trying to follow SpaceX's path to reusable, cost-effective launch.

There is a small, telling story in that pivot, and it captures how thoroughly SpaceX had rewritten the industry's assumptions. Beck had for years insisted, publicly and emphatically, that Electron would never be reusable and that Rocket Lab would never build a large rocket — going so far as to declare that he would eat his own hat if the company ever reversed course on reusability. When it did reverse course, unveiling Neutron in 2021, Beck kept his word in the most literal way he could manage: he put a Rocket Lab cap into a blender, reduced it to a paste, and ate a spoonful of it on camera, explaining afterward that at his company "everybody lives and dies by do what you say you're going to do," and that the reversal was a debt he needed to pay before he could move forward. It was a moment of theater, but also of genuine humility, and it stood as a small emblem of a larger truth: that reusability, once dismissed across the industry as a fantasy that could never pay for itself, had become so obviously the future that even its most vocal skeptic was willing to eat his hat to join it. The old conventional wisdom — that rockets were like ammunition, built to be spent once — had collapsed, and Beck's blended cap was its epitaph.

Rocket Lab's success was, in its way, as instructive as SpaceX's, because it showed a second path to survival in the transformed industry — not matching SpaceX head-on but building a whole company patiently around a proven niche. Electron reached orbit for the first time in 2018 and then flew again and again, becoming one of the most frequently launched rockets in the United States, second only to Falcon 9; and rather than live on launch fees alone, Beck's company expanded into building the satellites themselves — spacecraft buses, components, solar arrays, reaction wheels — so that it could sell not just the ride to orbit but the thing being carried, a strategy of vertical integration borrowed straight from the SpaceX playbook. Rocket Lab even reached beyond Earth orbit, lofting a small NASA spacecraft toward the Moon and planning a privately funded probe to Venus, an audacious ambition for a company that had begun by lobbing shoebox satellites into low orbit. That a firm founded in New Zealand — a country with no prior rocket industry to speak of — could become the clearest challenger to the most valuable space company on Earth was itself a measure of how completely the barriers to entry had fallen.

Not everyone who tried was so fortunate, and the field was littered with the wreckage of ambition. The stock-market enthusiasm of the early 2020s had showered money on a whole generation of launch startups — companies promising 3D-printed rockets, ultra-cheap small launchers, novel engines, all racing to claim a piece of the opening market — and most of them discovered that building a rocket that reaches orbit is one of the hardest things a human organization can attempt. Some flew a few times and failed; some never flew at all; many burned through their funding and quietly folded. For every Rocket Lab that found a durable footing, several vanished, and the survivors learned the same hard lesson the incumbents were learning from the other direction: that in the economics SpaceX had created, there was very little room for a launch company that could not offer either a genuinely differentiated service or a genuinely competitive price.

The legacy incumbents, meanwhile, faced an existential challenge, and their responses varied. In the United States, the traditional launch establishment had been embodied for years by United Launch Alliance, a joint venture of Boeing and Lockheed Martin that had long held a near-monopoly on launching the nation's most important national-security and government payloads, flying reliable but expensive expendable rockets. It is easy, and unfair, to caricature these companies as bumbling dinosaurs; in truth ULA's Atlas V and Delta rockets were among the most reliable machines ever flown, entrusted with the most valuable and irreplaceable spacecraft the nation possessed precisely because they almost never failed. Their problem was not competence but economics. SpaceX's arrival shattered ULA's monopoly, forcing it to compete on price for the first time and, eventually, to develop a new rocket, Vulcan Centaur, meant to be cheaper and more capable than its predecessors. Vulcan was a solid, capable vehicle, and it entered service in the mid-2020s, but it was fundamentally an expendable rocket in an age when SpaceX was reusing its boosters dozens of times — a single Falcon 9 first stage had by then flown and landed well over twenty times — and it struggled to match SpaceX on cost and cadence. In Europe, the Ariane rocket family, long the mainstay of commercial launch through the company Arianespace, faced the same pressure; its new Ariane 6, developed to succeed the workhorse Ariane 5, arrived years late and, again, as an expendable rocket in a reusable age, expensive and slow by comparison with Falcon 9, leaving Europe struggling to maintain independent access to space in the face of SpaceX's dominance. These incumbents were not incompetent — they built good rockets, and they served important needs, particularly the strategic imperative of maintaining sovereign launch capability independent of any single provider, a need thrown into sharp relief when Europe's long-standing reliance on Russian Soyuz rockets evaporated after the invasion of Ukraine and left the continent, for a spell, with almost no way to reach orbit on its own. But they were, in the harsh economics of the new era, at a structural disadvantage, flying expendable vehicles against a competitor that had learned to fly its rockets again and again, and they found themselves perpetually chasing a company that had rewritten the rules.

There was also a wave of competition in the domain that had become SpaceX's financial engine: satellite constellations. Starlink's success attracted rivals who sought to build their own mega-constellations of low-orbit internet satellites, hoping to capture a share of the vast market SpaceX had pioneered. The most significant was Amazon's Project Kuiper, backed by Jeff Bezos's fortune and Amazon's resources, which aimed to deploy thousands of satellites to provide global internet, competing directly with Starlink — the Musk-Bezos rivalry extended from rockets to the sky itself. Kuiper faced the daunting task of building and launching its constellation years behind Starlink, which already had a vast head start and a self-reinforcing advantage in launching its own satellites on its own rockets; there was a certain irony in the fact that Amazon, unable to build satellites fast enough on its own launchers alone, was obliged at one point to buy launches from SpaceX itself, paying its arch-rival to help loft the constellation meant to compete with it. Other constellations were planned or building as well, including ventures backed by national governments wary of depending on an American company for critical connectivity. The European Union pursued its own satellite-internet ambitions, and China, in particular, moved to build state-backed mega-constellations — projects with names like Guowang and the Thousand Sails, or Qianfan, constellation, each envisioned at many thousands of satellites — both to serve its own population and to offer an alternative to Starlink for other nations, extending the geopolitical competition of the new space age into low Earth orbit. The sky, once nearly empty, was becoming a contested domain, crowded not only with SpaceX's satellites but with the constellations of its rivals and of nations determined not to cede the orbital high ground to a single company or country — and crowded enough, astronomers and space-traffic experts warned, to raise real fears about collisions and the clutter of orbital debris.

The race in orbit ran against a clock as much as against a rival, for the right to fill the sky came with deadlines. Regulators had learned to attach deployment requirements to the licenses that granted a company its slice of low Earth orbit — Amazon, for instance, was obliged to place roughly half of its more than three thousand Kuiper satellites into orbit by a date in 2026, or risk losing its authorization — and those deadlines turned constellation-building into a forced march, with companies committing to launch cadences they had not yet proven they could sustain. Amazon, unable to loft its satellites fast enough on the rockets it had contracted, at one point found itself in the awkward position of buying launches from SpaceX, paying the very rival whose Starlink it aimed to unseat. China's efforts, less bound by such market pressures, advanced with the steady push of the state; its Guowang and Qianfan constellations, each planned in the many thousands of satellites, were conceived not only as commercial services but as instruments of national policy, offering countries across the developing world an alternative to depending on an American company for the connective tissue of the modern economy. And as the launches multiplied, a quieter alarm grew louder — from astronomers who watched the streaks of satellites scar their long-exposure images of the night sky, and from experts in orbital debris who warned that a sky crowded with tens of thousands of spacecraft raised the real and rising risk of collisions, each of which could scatter a cloud of shrapnel that threatened everything else in its path.

The crowded, competitive field that had grown up around

SpaceX told a story with two morals, pointing in opposite directions. On one hand, it demonstrated the reality of SpaceX's dominance: despite all the challengers, SpaceX remained far ahead, launching more than half the world's payloads, operating the largest constellation by a wide margin, and setting the pace that everyone else struggled to match. No competitor had yet come close to matching SpaceX's combination of cost, cadence, and capability, and the company's lead, if anything, had widened over the years rather than narrowed — a fact that troubled even some of SpaceX's admirers, who worried that a single private company had come to hold so central a place in the world's access to space, launching the satellites of its rivals and its government alike, that its dominance had begun to look less like a triumph of competition than like a new kind of monopoly. On the other hand, the very existence of so many challengers — new launchers, reusable-rocket programs, rival constellations, national efforts — testified to how thoroughly SpaceX had transformed the industry, opening a domain that had been the preserve of a few governments to a flourishing ecosystem of private and public competitors, and inspiring a global race to build the capabilities SpaceX had pioneered. Whether that race would eventually produce genuine rivals to SpaceX's dominance, or whether SpaceX would maintain and extend its lead, remained one of the central open questions of the new space age. But the most consequential competition of all was not commercial but national, and it centered on the oldest and most evocative prize in spaceflight — the Moon, to which two great powers, the United States and China, were now racing, in a contest that revived the language and the stakes of the original space race of the twentieth century.

Chapter 54: The Small-Launch Boom

Not all of the new space age was built around giant rockets and mega-constellations. One of its most significant and least appreciated transformations happened at the other end of the scale, in the world of small satellites and small launchers, and it flowed directly from the same forces SpaceX had unleashed: falling costs, shrinking electronics, and the opening of space to new actors. The result was a boom in small satellites — thousands of them, built cheaply, launched frequently, and used for purposes that would once have required spacecraft costing a fortune — and a corresponding boom in the companies racing to launch them.

The small-satellite revolution had its roots in the relentless miniaturization of electronics, and in one surprisingly humble decision made not by a rocket company but by two university professors. In 1999, Bob Twiggs of Stanford and Jordi Puig-Suari of California Polytechnic State University, frustrated that students rarely got to build a real satellite before they graduated, proposed a standard so small and cheap that a class could finish one within a degree: a cube ten centimeters on a side — about the size of a Rubik's cube — weighing barely more than a kilogram, which they called the CubeSat. The idea was almost quaint, a teaching tool; instead it became a global standard, because it happened to arrive just as computers, sensors, cameras, and communications gear were growing small, cheap, and capable enough to fit inside such a box and still do useful work. Suddenly it was possible to build satellites the size of a shoebox or a loaf of bread, or smaller, that could image the Earth, relay data, conduct experiments, and test technologies — spacecraft that universities, startups, and even hobbyists could build and fly for a tiny fraction of what a traditional satellite cost. The barrier to putting something useful in orbit collapsed, and a wave of new actors — small companies, research institutions, developing nations, entrepreneurs — entered the space business. The number of satellites launched each year soared, driven not only by SpaceX's Starlink but by this proliferation of small spacecraft built by a diverse and growing community.

The single clearest illustration of what small satellites could do was the company Planet, which set out to photograph the entire land surface of the Earth every single day — a feat no government spy agency, with its handful of enormous and exquisite satellites, had ever attempted. Planet achieved it not with a few large satellites but with a swarm of small ones it called "Doves," each a three-unit CubeSat about the size of a loaf of bread and weighing some four kilograms, packed with off-the-shelf optics and electronics, and deployed by the dozen. Where a traditional reconnaissance satellite might cost hundreds of millions of dollars and be treasured as a national asset, a Dove was cheap enough that Planet could afford to keep launching fresh flocks of them, upgrading the whole constellation as the electronics improved and shrugging off the loss of any individual satellite. The result was a live, daily portrait of the changing planet, used for everything from tracking deforestation and crop health to counting cars in retail parking lots for financial analysts — a capability that, a generation earlier, would have belonged only to superpowers, now sold as a subscription service.

The sheer scale of the change is hard to overstate. For the

first several decades of the space age, the world launched only a few dozen to perhaps a hundred-odd satellites in a given year, each a costly and carefully guarded machine. By the middle of the 2020s that trickle had become a flood: in some single years of the decade, more spacecraft reached orbit than had been launched in the first several decades of spaceflight combined, the overwhelming majority of them small. Starlink accounted for a great share of the total, but hundreds of other operators — companies, universities, government agencies, and nations that had never before had a space program — filled in the rest, and the population of active satellites circling the Earth swelled from a few thousand to many times that in the span of a few years. Space, in raw numerical terms, had gone from an exclusive club to a crowded commons in less than a decade.

And what those small satellites could do reached, in time, into ordinary life on the ground. Capabilities that had once belonged only to the intelligence agencies of superpowers — the ability to photograph any point on the Earth, to watch a border or a port or a burning forest from above — became, in effect, a commercial product any newspaper, aid group, or commodity trader could buy. When wars broke out or disasters struck, images from private constellations appeared in the world's newspapers within hours, revealing troop movements, flooded towns, and the smoke of distant fires; commodity analysts studied satellite photographs of oil-storage tanks and crowded parking lots to gauge the health of economies before the official figures were published; environmental groups used orbital sensors to catch methane leaks and illegal deforestation in the act. None of this had been possible, at any price, a generation earlier. The small-satellite revolution did not merely put more objects in orbit; it democratized a kind of god's-eye view of the planet that had once been among the most tightly held secrets of the Cold War.

The small satellites needed to reach orbit, and this created a market for launch, but a market with particular needs. A small satellite could hitch a ride as a secondary payload on a large rocket, sharing the launch with a bigger primary satellite, and SpaceX did a booming business in these "rideshare" missions, packing dozens or hundreds of small satellites onto a single Falcon 9 and delivering them to orbit cheaply — a kind of bus service to space, carrying many small payloads at once. The scale of it could be startling: in January 2021, on the first of its dedicated Transporter rideshare flights, SpaceX launched a single Falcon 9 carrying 143 satellites at once, shattering the previous world record of 104 set a few years earlier by an Indian rocket, and landing the booster to fly again besides. SpaceX turned this into something like a scheduled service. Its Transporter flights departed on a regular cadence to the popular sun-synchronous orbit favored by Earth-imaging satellites, and a web of specialist middlemen — aggregators with names like Exolaunch, D-Orbit, and Spaceflight — bundled the payloads of dozens of small operators onto each one, handling the paperwork and the deployment hardware so that a customer needed only to buy a slot. The prices they quoted would have seemed fantastical a generation earlier: a small satellite could reach orbit for a few hundred thousand dollars, and a single CubeSat for far less, sums within reach of a university department or a modestly funded startup rather than a national government. It was, in the most literal sense, a bus service to space — regular departures, a fixed fare, a shared ride — and it did for orbit something like what the scheduled airline had once done for the sky.

But rideshare had a limitation: the small satellites had to go where the primary payload was going, on the primary's schedule, which did not always suit them — a satellite that needed a particular orbit or a particular launch date could find itself stranded, waiting for a bus that never went quite where it wanted. This created an opening for dedicated small launchers — rockets sized to carry a small satellite to precisely the orbit it wanted on its own schedule — and a wave of companies rose to fill it, developing small rockets aimed at the small-satellite market. The most successful was Rocket Lab, whose Electron rocket carried small payloads to orbit frequently and reliably, but there were many others, a whole ecosystem of small-launch startups, some succeeding, many failing, all chasing the dream of cheap, responsive, dedicated launch.

The small-launch boom was, in a sense, the democratization of space in action — the opening of orbit to a wide and growing range of actors who could never have afforded access at the old prices and with the old vehicles. A university could build and fly a satellite; a startup could deploy a constellation; a developing nation could establish a space program on a modest budget. The uses multiplied: Earth-observation companies built constellations of small imaging satellites that photographed the entire planet daily, enabling applications from agriculture to disaster response to financial analysis; data-relay companies connected remote sensors and devices through small-satellite networks; a proliferation of experimental and technology-demonstration satellites tested new ideas cheaply in orbit. The space economy broadened and diversified, no longer dominated solely by a few governments and large operators but populated by a vibrant ecosystem of small companies and small satellites, enabled by the falling costs and the new access that the transformation of launch had brought. This diversification was one of the genuine goods of the new space age, spreading the benefits and the participation of space activity far more widely than ever before.

Yet the small-launch boom also had a shadow, and it was cast, ironically, by SpaceX itself. The dedicated small launchers faced a formidable competitor in SpaceX's rideshare service, which could carry small satellites to orbit so cheaply — by packing many onto a single large reusable rocket — that it undercut the economics of the dedicated small launchers, which had to bear the full cost of a launch for a small payload. The reckoning was brutal, and the wreckage was real. Virgin Orbit, the small-launch venture spun out of Richard Branson's empire, which launched its rockets from beneath the wing of a modified Boeing 747, suffered a failed launch from Cornwall in January 2023 — a mission optimistically named "Start Me Up" that instead saw its rocket falter in the second stage and fall short of orbit — and within months the company had run out of money and filed for bankruptcy, its aircraft and assets sold off for parts. Astra, another highly touted startup, managed to reach orbit only twice in seven attempts, losing NASA payloads including a set of hurricane-tracking satellites when a rocket failed, and eventually abandoned the small rocket that had been its whole reason for being. Many of the small-launch startups struggled or failed, unable to compete with the low prices of rideshare on a cheap reusable Falcon 9, and the small-launch market proved harder and more brutal than its enthusiasts had hoped. SpaceX's dominance, in other words, extended even into the small-launch niche, its cheap reusable rockets and rideshare service capturing much of the small-satellite launch business and squeezing the dedicated small launchers. The very forces that had opened space to small satellites — falling costs, driven above all by SpaceX — also made it hard for the small-launch companies to survive, caught between the appeal of their dedicated service and the overwhelming cheapness of riding along on a SpaceX rocket.

Those companies that survived tended to do so by refusing to be pure small-launch companies at all. Rocket Lab endured because it had diversified into building satellites and spacecraft components, so that launch was only one of its revenue streams; others leaned on a customer that valued something rideshare could not offer — the military, which prized "responsive launch," the ability to place a satellite into a chosen orbit on short notice, without waiting for a rideshare bus to fill up and depart, so that a constellation damaged in a crisis could be quickly replenished. National-security demand, in other words, kept a corner of the dedicated small-launch market alive precisely because it was willing to pay a premium for schedule and control that the cheapest option could not provide. But that was a narrow and specialized refuge, not the vast open market the boom's early enthusiasts had imagined, and it left the broad commercial dream of cheap, frequent, dedicated small launch largely unrealized — swallowed, like so much else, by the economics of the reusable giant.

The small-launch boom thus illustrated, from a different angle, the central dynamics of the new space age: the democratization of access that falling costs enabled, the proliferation of new actors and new uses, the diversification of the space economy — and, running through it all, the dominance of SpaceX, whose cost advantage reached into every corner of the market, enabling the boom even as it squeezed the competitors trying to serve it. You meet the fruits of that boom now without knowing it, every time a weather app draws on fresh orbital imagery, a shipping company tracks a container across an ocean, a farmer checks the health of a field from a phone, or a news report shows satellite pictures of a flood or a war taken that very morning — all of it made possible by cheap, numerous small satellites that did not exist a generation ago. The world of small satellites was one of the new age's brightest developments, opening space to a wide and growing community and putting the capabilities of orbit within reach of universities, startups, and nations that could never have afforded them before. And it was, at the same time, one more arena in which SpaceX's overwhelming advantage shaped the outcome, its cheap reusable rockets both the enabler of the small-satellite revolution and the great competitor that the small-launch companies could not escape. The boom was real, the democratization was real, and so was the shadow of the giant that had made both possible.

Chapter 55: The Tourists

For most of the history of spaceflight, the people who went to space were a tiny, elite cadre of professional astronauts, selected from thousands of applicants, trained for years, and sent aloft on government missions of national importance. Fewer than six hundred human beings had left the Earth in the six decades since the first did so. Space was, in the most literal sense, the least accessible place a person could go, reserved for a handful of the exceptionally qualified and the exceptionally lucky. One of the quieter but genuinely significant transformations of the new space age was the beginning of the end of that exclusivity — the opening of spaceflight, however tentatively and however unequally, to people who were not professional astronauts: to paying customers, to private adventurers, to a widening circle of humanity that could, for the first time, buy its way to space.

It is often assumed that private space tourism began in 2021, with the widely televised joyrides of billionaires in their own rockets, but the first paying passenger had in fact reached orbit two decades earlier, and the story of how he did so reveals how bitterly the old guard resisted the change. In April 2001, an American businessman named Dennis Tito — a onetime NASA engineer who had left to make a fortune in finance — paid a reported twenty million dollars to the cash-strapped Russian space program and rode a Soyuz capsule to the International Space Station, spending nearly eight days in orbit and becoming the first person in history to fund his own trip to space. NASA was furious. The agency regarded the notion of a wealthy tourist wandering its space station as an affront and a safety hazard, and its officials fought the flight, at one point refusing to train Tito at the Johnson Space Center; only Russian insistence, and Russia's need for the money, carried the day. Tito would later call the experience the greatest moment of his life. His flight, and the handful of other wealthy passengers who followed him to the station over the next few years, made the essential point that the door, however narrowly, could be opened — that a private individual, given enough money and enough will, could go where only governments had sent people before.

The more visible and more modest form of the new tourism was suborbital — brief flights to the edge of space and back, offering a few minutes of weightlessness and a view of the Earth's curve against the black of space, without the vastly greater expense and difficulty of reaching orbit. Two companies led this business, and their rivalry was itself a subplot of the new space age. Blue Origin's New Shepard, as we have seen, carried passengers on suborbital hops, launching them straight up past the boundary of space and returning them under parachutes, the whole flight lasting only about ten minutes. Its competitor was Virgin Galactic, the company founded by the British entrepreneur Richard Branson, which took a different technical approach: a rocket-powered spaceplane, carried aloft under a large carrier aircraft and then released to fire its rocket engine and climb to the edge of space, before gliding back to a runway landing. Branson, like Bezos, flew on his own vehicle, and the two billionaires' race to be first to ride their own spacecraft to space, in the summer of 2021, was a widely publicized contest that captured the era's blend of genuine achievement and billionaire spectacle — Branson edging ahead by nine days in July, only for Bezos to note, pointedly, that Virgin

Galactic's spaceplane did not climb as high as the internationally recognized boundary of space that New Shepard crossed. Virgin Galactic went on to begin regular commercial service in 2023, at last carrying paying customers and researchers after nearly two decades of development — and after a fatal accident in 2014, when one of its rocket planes broke apart in a test flight and killed a pilot, a reminder written in blood that even the modest business of touching the edge of space was genuinely dangerous. The company sold its seats for hundreds of thousands of dollars, but it flew too rarely and at too great a cost to turn the profit it had promised for twenty years, and the gap between the dream it had marketed and the halting reality it delivered hung over the whole suborbital enterprise.

What lifted these suborbital flights above mere spectacle were the passengers who rode them, and the histories they carried. On Bezos's own first flight sat Wally Funk, then eighty-two, one of the "Mercury 13" women who had passed the astronaut tests in the early 1960s and been denied the chance to fly because of their sex; six decades late, she finally crossed the line she had been kept from. That autumn, New Shepard carried the actor William Shatner — Captain Kirk of Star Trek, then ninety years old — who returned shaken and tearful, describing not the exhilaration everyone expected but a profound grief at the sight of the thin, fragile film of the atmosphere and the cold black void pressing against it; he later wrote that the overwhelming feeling was one of sadness, "a funeral" for the planet he had just seen from above. And in 2024 the same rocket carried Ed Dwight, also ninety, who in the early 1960s had been chosen by President Kennedy as the nation's first Black astronaut candidate, only to be pushed out and never flown; his belated journey, brief as it was, closed a wound that had been open for sixty years. These were not idle novelties. They were, in their way, small acts of historical repair, and they gave the suborbital business a human weight its critics often missed.

For critics there were, and reasonably so. Suborbital tourism was, in truth, a limited thing — a costly novelty for the wealthy, a few minutes of weightlessness for a very high price — and critics fairly questioned whether it represented meaningful progress toward humanity's future in space or merely an expensive amusement for those who could afford it. The passengers did not orbit the Earth; they did not stay in space; they simply touched its edge and came back. But the suborbital business had a real significance beyond the experience it sold. It demonstrated that private companies could carry members of the public to space safely and repeatedly; it began to build the infrastructure, the operations, and the public acceptance of commercial human spaceflight; and it planted the idea, in the public mind, that going to space was no longer the exclusive province of governments and their astronauts — that it was becoming, however slowly and however expensively, a thing that ordinary people, or at least wealthy ones, might do. It was a first step, and first steps matter.

The more consequential development was orbital space tourism and private orbital spaceflight, which SpaceX, with its Crew Dragon, made possible in a way no one else could. As we saw in the story of the civilian missions — Inspiration4, the Axiom flights, Polaris Dawn — SpaceX carried private crews to orbit and to the space station, demonstrating that private human spaceflight could go far beyond a few minutes at the edge of space to genuine orbital missions lasting days, with all the capability and complexity that implied. The first of these, Inspiration4, in September 2021, was itself a landmark: the first time an all-civilian crew, with no professional astronaut aboard, orbited the Earth. It was funded and commanded by the entrepreneur Jared Isaacman, who bought all four seats and gave three of them away — one to a healthcare worker and childhood-cancer survivor named Hayley Arceneaux, who flew with a titanium rod in her leg from her own treatment and became, at twenty-nine, the youngest American to orbit the Earth and the first person to fly to space with a prosthesis — turning the mission into a fundraiser for St. Jude Children's Research Hospital that raised hundreds of millions of dollars. These missions cost vastly more than suborbital hops — tens of millions of dollars per seat — and so were, if anything, even more the preserve of the ultra-wealthy and their beneficiaries. But they represented a far deeper opening of spaceflight, proving that the full experience of orbital spaceflight, once reserved entirely for government astronauts, could be provided commercially to private individuals. And they pointed toward an imagined future in which orbital destinations — commercial space stations, built by companies like Axiom and others to replace or supplement the aging International Space Station — would host private visitors, researchers, and workers, a genuine human economy in low Earth orbit rather than a handful of government outposts.

The most striking of these private orbital missions pushed further still, blurring the line between tourism and genuine exploration. In September 2024, Isaacman flew a second time, on a mission called Polaris Dawn that he funded and commanded, carrying a crew that included two SpaceX engineers, Sarah Gillis and Anna Menon. The flight climbed to an apogee of some fourteen hundred kilometers — the farthest any human being had traveled from the Earth since the Apollo astronauts returned from the Moon, carrying the crew through the fringes of the radiation belts — and then, at a lower altitude, did something no private mission had ever attempted: a spacewalk. Because the Crew Dragon has no airlock, the entire cabin had to be vented to the vacuum of space, exposing all four occupants to it at once in newly designed SpaceX pressure suits, while Isaacman and then Gillis climbed partway out through the open hatch. Gillis, a violinist, would later play her instrument in orbit for a children's charity, a small grace note that captured the mission's blend of the daring and the human. Nor was the space station left out of the new economy: the company Axiom had begun flying private crews to the International Space Station itself, its first mission in April 2022 commanded by a veteran former NASA astronaut and carrying three paying customers who reportedly spent tens of millions of dollars each for a stay of more than two weeks in orbit — a foretaste of the commercial stations Axiom and others planned to build once the aging International Space Station was retired.

None of this was safe, and it is worth resisting the easy language of "joyrides," because spaceflight remains among the most dangerous things a human being can do, and the passengers who bought their way aloft were strapping themselves atop controlled explosions with the same slim margins that had killed professional astronauts before them. That SpaceX had by 2026 carried dozens of people to orbit and back aboard Crew Dragon without losing a single one was a genuine and hard-won achievement, not a guarantee. And the enterprise drew sharp criticism on other grounds — that it was obscene for the very rich to spend tens of millions of dollars on a few days in orbit while so many on the ground went without; that the flights burned prodigious quantities of propellant for the pleasure of a handful; that calling any of it "democratization" was a marketing gloss on what remained, for now, a luxury of the ultra-wealthy. The criticisms were fair, and they coexisted uneasily with the genuine wonder the flights produced, the sense — voiced by passenger after passenger — that seeing the Earth whole and small against the dark had changed them. Both things were true at once.

Whether this vision of a broadly accessible human presence in space would ever be realized remained, like so much in the new space age, an open question, and one shadowed by the same tension between promise and inequity that ran through the whole enterprise. The optimistic case held that space tourism and private spaceflight were following the familiar path of every transformative technology — beginning as expensive playthings of the rich, then, as costs fell and capabilities grew, becoming accessible to ever wider circles, until what had been the preserve of a few became available to many, as had happened with air travel, with computing, with countless other technologies that began as luxuries and became commonplace. Air travel itself, after all, had once been the province of the daring and the rich, its early passengers a novelty photographed for the newspapers; within a lifetime it became a bus with wings. If launch costs kept falling, driven by reusability and competition and scale, and if vehicles like Starship delivered the order-of-magnitude reductions they promised, then the number of people who could go to space might grow from hundreds to thousands to, eventually, far more, and a genuine human expansion into space might slowly take shape. The pessimistic case held that space would remain fundamentally expensive and dangerous, that the physics and the costs would never fall far enough to make it broadly accessible, and that the "democratization" of space was a marketing slogan disguising what was really just a new luxury for the super-rich, of little relevance to the vast majority of humanity. Both cases had merit, and the truth would depend on developments not yet knowable — above all on whether the cost revolution SpaceX had begun could be carried far enough to change not just who launched satellites but who could leave the Earth. What was certain was that the door, firmly shut for sixty years, had been pushed ajar, and that private citizens were, for the first time, flying to space in growing numbers. Where that would lead, no one could yet say. But the opening of spaceflight to private human beings was real, and it was one more way in which the new space age had begun to change humanity's relationship with the frontier above.

Chapter 56: The Return to the Moon

In December 1972, the astronauts of Apollo 17 left the surface of the Moon, and no human being has set foot there since. For half a century, the Moon — the destination that had defined the original space race, the place where humanity had made its greatest leap — sat untrodden, a silent monument to a burst of ambition that had flared and faded. It is often said, loosely, that humanity "lost the ability" to go to the Moon, as though the knowledge had somehow evaporated, and there is a grain of literal truth in it — the tooling was scrapped, the assembly lines dismantled, the engineers scattered and grew old — but the deeper truth is starker and more sobering: humanity did not forget how to go to the Moon; it decided to stop. The Apollo program, having won the race to the Moon and demonstrated American technological supremacy in the Cold War, was cancelled while it still had rockets standing ready — the final three planned missions, Apollo 18, 19, and 20, struck from the schedule, their Saturn V rockets, each one a functioning marvel that had cost a fortune to build, hauled off to lie on their sides as lawn ornaments outside museums in Houston, Florida, and Alabama, where they can still be visited today. The political will and the funding that had sustained the effort evaporated the moment the goal was achieved. The generation that had watched Neil Armstrong step onto the lunar surface grew old, and the Moon receded into history, a thing humanity had done once, long ago, and then stopped doing. The dream of returning — of going back to stay this time, of building a lasting human presence on another world — persisted in speeches and studies and cancelled programs, but for decades it came to nothing.

The new space age revived that dream, and gave it, at last, a real vehicle and a real urgency. The revival took shape as NASA's Artemis program — named, fittingly, for the Greek goddess of the Moon, the twin sister of Apollo — an effort to return American astronauts to the lunar surface and, this time, to establish a sustained presence, with a base, regular missions, and the Moon as a stepping stone toward Mars. Artemis was built around a government rocket and spacecraft of the traditional kind: the Space Launch System, or SLS, an enormous expendable rocket derived from Space Shuttle technology, and the Orion crew capsule, developed over many years at great expense to carry astronauts to lunar orbit and back. The first Artemis mission, Artemis I, flew in November 2022, sending an uncrewed Orion capsule on a twenty-five-day journey around the Moon and back atop the SLS in a successful test of the basic hardware. It was a genuine achievement, the first flight of the system that was meant to carry humans back to the Moon, and it demonstrated that the rocket and the capsule worked. But Artemis, for all its ambition, was also a creature of the old way of doing things. The SLS was expensive, expendable, and slow to build — a rocket that, by the estimates of NASA's own inspector general, cost something on the order of four billion dollars per launch and could fly only once every year or two, the product of the traditional cost-plus contracting that SpaceX had spent two decades disrupting, its engines in some cases literal leftovers from the Space Shuttle, refurbished and then thrown away into the ocean after a single use. And it depended, for the crucial final step of actually landing astronauts on the Moon, on the new way, on SpaceX.

For NASA, unable to build a lunar lander of its own within its budget and timeline, had turned to the commercial sector, as it had for cargo and crew, and had selected SpaceX's Starship — in its Human Landing System variant, Starship HLS — to carry astronauts from lunar orbit down to the surface of the Moon and back. There was a strange and telling asymmetry in the arrangement that is worth pausing over: the lander SpaceX proposed was itself a vehicle far larger than the entire Apollo Lunar Module, indeed larger than the Saturn V that had launched the whole Apollo stack — a stainless-steel skyscraper of a spacecraft descending to the dusty lunar surface to deliver two astronauts, where Apollo's spidery little lander had done the same job at a tiny fraction of the size. Later, NASA added a second lander provider, Blue Origin, with its Blue Moon lander, following the same logic of redundancy that had shaped the Commercial Crew Program. This meant that the return of humans to the Moon depended not only on the government's SLS and Orion but on the private companies' landers — above all on Starship, whose development, as we have seen, remained fraught with difficulty and dependent on the unproven technology of orbital refueling. The Artemis architecture thus stitched together the old and the new: a traditional government rocket to carry the crew to lunar orbit, and a commercial vehicle to land them, with all the promise and all the risk that the commercial approach entailed. And the fate of the whole program came to hinge, in large part, on whether SpaceX could get Starship working — reliably reaching orbit, mastering orbital refueling, and demonstrating that the HLS lander could safely carry astronauts to the lunar surface — a set of milestones that Starship's troubled test campaign had repeatedly pushed further into the future.

The architecture grew more elaborate still, for Artemis was never conceived as a simple sprint to the surface in the manner of Apollo. NASA and its international partners planned a small space station, the Lunar Gateway, to be assembled in an elongated orbit around the Moon — built with contributions from Europe, Japan, Canada, and the United Arab

Emirates, and meant to serve as a way station where crews and landers could gather before descending and to which they could return. To some, the Gateway embodied the collaborative, sustainable spirit that distinguished the new lunar effort from the flags-and-footprints dashes of the 1960s; to others, including some veteran engineers, it was a costly detour, a station orbiting the Moon that a mission bound for the surface had to visit for reasons more political and industrial than technical. Either way, it added yet another set of pieces that had to be built, launched, and made to work together before an astronaut could stand on the lunar soil, and another set of schedules that could slip.

Before any human returned, robots went ahead to scout the ground, and their fortunes offered a sobering preview of how hard the Moon remained. NASA, extending its commercial model to the lunar surface, had begun paying small private companies to deliver instruments to the Moon, and the first attempts humbled everyone: in January 2024 one company's lander suffered a crippling propellant leak on the way out and never reached the Moon at all, while the following month another, built by a Houston firm, became the first American spacecraft to touch the lunar surface since Apollo — a genuine milestone — only to come down too fast, catch a landing leg, and topple onto its side, finishing its mission tilted and half-crippled. These missions aimed, pointedly, at the Moon's south pole, and the reason is the same one that draws both America and China to that forbidding, crater-shadowed region: in the permanent darkness at the bottoms of its polar craters, where sunlight has not fallen for billions of years, lies water ice. That ice is the most valuable resource on the Moon, for it can be melted into drinking water, split into breathable oxygen, and refined into the hydrogen and oxygen of rocket propellant — turning the lunar south pole, in the eyes of every spacefaring power, into the single most strategically important patch of ground beyond the Earth, and the prize around which the new race to the Moon is really organized.

The human face of the program made the stakes vivid. In 2023, NASA named the four astronauts who would fly Artemis II, the first crewed mission, a loop around the Moon and back without landing: the commander Reid Wiseman, the pilot Victor Glover, the mission specialist Christina Koch, and the Canadian Jeremy Hansen. The crew was, by design, a portrait of how much had changed since Apollo — Koch set to become the first woman ever to travel to the vicinity of the Moon, Glover the first Black astronaut to do so, and Hansen the first person of any nation other than the United States to journey to the Moon at all, a marker of how the new lunar effort was meant to be international and inclusive where the old one had been a contest between two superpowers. But naming a crew is not the same as flying it, and the astronauts found themselves waiting as the schedule around them kept slipping.

As a result, the Artemis timeline slipped, again and again, and by 2026 the schedule had been substantially reframed. The mission that was to carry the first astronauts back to the lunar surface, Artemis III, had originally been envisioned for the mid-2020s, but the delays in Starship's development, along with challenges with the spacesuits — themselves contracted out to a commercial provider and running behind — the Orion capsule, whose heat shield had eroded unexpectedly during the Artemis I reentry and required lengthy investigation, and other elements, pushed it back. By 2026, NASA had reconceived the near-term steps: rather than an immediate landing, the plan called for a demonstration in Earth orbit, testing the rendezvous and docking of the crew vehicle with the lander systems, with the actual crewed lunar landing slipping to a later mission and a later date — into the latter part of the decade. The reframing was a candid, if disappointing, acknowledgment that the pieces were not yet ready, that the landers in particular were still under development and not yet human-rated, and that the confident timelines of a few years earlier had run into the hard reality of how difficult the task remained. Returning humans to the Moon, it turned out, was nearly as hard the second time as the first, and the reliance on Starship, brilliant but unfinished, meant that the schedule was hostage to the fortunes of a vehicle that was still, in 2026, losing boosters and getting grounded by regulators.

The delays and difficulties of Artemis raised uncomfortable questions, and they gained a sharp edge from the fact that the United States was no longer the only nation racing to the Moon. The prospect of an American return to the lunar surface, so long delayed and now dependent on the uncertain progress of Starship, had to be measured against the steady, determined advance of a rival that had set its own deadline and was moving toward it with the focused resources of a state that did not answer to the vagaries of congressional funding or the schedule slips of private contractors. The competition had a diplomatic dimension as well, and it, too, echoed the old bloc politics of the twentieth century. The United States had gathered a growing coalition of nations around a set of principles for lunar exploration it called the Artemis Accords — a framework for peaceful, cooperative, and transparent activity on the Moon, to which dozens of countries had by the mid-2020s put their names. China and Russia, for their part, had declined to join and were assembling a rival grouping around their own proposed International Lunar Research Station, courting their own partners among the nations of the developing world. The Moon, untrodden for half a century, was thus becoming the object not merely of two national programs but of two competing coalitions, each with its own rules, its own partners, and its own vision of who would write the norms for humanity's return — a contest as much about the future governance of the frontier as about the engineering of reaching it.

China had announced its intention to land astronauts on the Moon before the end of the decade, by 2030, and it was building the specific hardware to do it: a new heavy rocket, the Long March 10, and a pair of new spacecraft with poetically chosen names — the crew ship Mengzhou, "Dream Vessel," and the lunar lander Lanyue, "Embracing the Moon." Its architecture was, in a sense, more conservative and more Apollo-like than America's: two Long March 10 launches, one carrying the crew ship and one the lander, meeting in lunar orbit, with the astronauts transferring to the lander for the descent — an approach that, whatever its own risks, depended on no orbital refueling and no fleet of tankers, and could be tested in discrete, verifiable steps. China moved methodically toward it, testing engines, spacecraft, and landing systems on the ground and vowing, in the words of its program officials, to "spare no effort." The possibility that China might reach the Moon around the same time as, or even before, the United States returned — that the nation which had won the original space race might be matched or beaten in the new one — gave Artemis a competitive urgency that echoed the Cold War, and it turned the return to the Moon from a matter of exploration into a matter of national prestige and strategic competition. The contrast was not lost on American officials, who watched a rival with a simpler plan and an unwavering deadline close the distance while their own effort waited on a revolutionary vehicle that was still learning to fly. To understand the full shape of the new race to the Moon, then, we must turn to the other great power now reaching for the lunar surface — to China's space program, its ambitions, and its deadline, and to the revived great-power competition that has brought the language of the space race roaring back after half a century.

Chapter 57: The Lunar Economy

Why go back to the Moon at all? Apollo went to win a race and prove a point, and having done so, it stopped. The new push to the Moon rests on a different rationale — not a one-off demonstration but the establishment of a lasting presence, and behind that ambition lies a vision of the Moon not merely as a destination to visit but as a place to use, a source of resources and a stepping stone that might, over time, support a genuine economy in space. Whether that vision is realistic or fanciful, it is the logic behind Artemis and behind the broader interest in the Moon, and it is worth examining with some care, because it is easy to mistake the enthusiasm for the substance, and easy, too, to dismiss the substance because of the enthusiasm.

The single most important reason the Moon has become interesting again is water — specifically, water ice, believed to exist in significant quantities in the permanently shadowed craters near the lunar poles, regions so deep in shadow that they never see sunlight and remain frigidly cold, cold enough to trap water ice for billions of years. It is worth pausing on just how cold. The floors of these craters, tilted forever away from the low polar sun, are among the coldest known places in the entire solar system — colder, at roughly forty degrees above absolute zero, than the surface of Pluto

— and it is precisely that unrelenting cold that makes them cosmic cold traps, catching and holding water molecules delivered over eons by comets and micrometeorites that would long since have boiled away anywhere the sun could reach. The evidence for this water is not merely inferred but directly, and rather dramatically, measured. In October 2009, NASA deliberately crashed a spent Centaur rocket stage into a permanently shadowed crater called Cabeus, near the lunar south pole, and flew a small trailing spacecraft, LCROSS, straight through the debris plume the impact threw up, sampling it with its instruments in the final minutes before it, too, slammed into the surface. The plume it flew through contained something like a hundred and fifty kilograms of water, vapor and ice together, blasted out of the darkness — water that amounted to several percent by mass of the excavated soil. It was a violent, ingenious way to answer a question, and it turned the presence of lunar water from a hope into a fact.

Water, on the Moon, is treasure. It can be split, by electrolysis, into hydrogen and oxygen, providing breathable air for astronauts and — crucially — propellant for rockets, so that the Moon might serve as a filling station in space, a place to produce fuel rather than hauling it all up from the Earth at enormous cost. This is the concept of in-situ resource utilization: using the resources found in space rather than carrying everything from Earth, and it is central to the vision of a sustainable presence on the Moon and beyond. The economic logic is worth making concrete. Every kilogram of propellant burned in space today has to be lifted, at great expense, out of Earth's deep gravity well; a spacecraft departing for Mars from low Earth orbit must carry with it all the fuel for the journey, and the fuel to accelerate that fuel, in the punishing exponential arithmetic that governs all rocketry. But the Moon's gravity is only a sixth of Earth's, and a spacecraft topped off with propellant made from lunar ice — mined, electrolyzed, and liquefied at a polar depot — would escape the Moon far more cheaply than it could ever escape the Earth. In that vision the Moon becomes not a dead end but a node, a place to refuel on the way to somewhere else, and the water in Cabeus and its sister craters becomes the first commodity of an economy beyond the Earth.

The vision extends beyond water. The Moon has other potential resources — minerals, metals, and rare materials, including, some speculate, elements valuable for terrestrial industry or for use in space. It offers a stable platform for scientific instruments, including telescopes that might, on the Moon's far side, be shielded from the radio noise of the Earth. And it serves, in the grander visions, as a stepping stone and a proving ground for the journey to Mars — a place to develop and test the technologies, the techniques, and the experience of living and working on another world, closer to home and easier to reach than Mars, before attempting the far greater challenge of the red planet. To support this activity, NASA and its partners envisioned infrastructure in and around the Moon, including a small space station in lunar orbit called Gateway, meant to serve as a staging point for missions to the surface, and, over time, a base on the surface itself, a lasting human presence rather than the brief visits of Apollo. The vision, in its fullest form, was of a lunar economy — a sustained human and robotic presence, drawing on local resources, supporting science and exploration and perhaps commerce, and serving as a stepping stone to Mars.

It is often said, in the popular telling, that Artemis is simply Apollo again — flags and footprints, a second race to plant a boot and come home. In fact the whole point of the enterprise is the opposite: not the sprint but the settling, not the visit but the outpost, and this difference explains almost everything about why it has proved so much harder and slower than the original. Apollo could afford to ignore the poles, land in the sunlit equatorial plains where conditions were benign, and stay only hours or days, because it never meant to come back to stay. The new effort must reach the very hardest terrain on the Moon — the rugged, shadowed, deep-frozen polar highlands where the ice is thought to lie — and must build there the machinery of a permanent presence, and that is a different order of problem entirely, one that no amount of political will can simply buy its way through in a few years.

Not every resource touted for the Moon deserves equal credence, and it is worth being candid about which claims rest on evidence and which on enthusiasm. The most persistently oversold is helium-3, a rare isotope deposited in the lunar soil by the solar wind and frequently invoked as the fuel of some future fusion economy — a prize supposedly worth going to the Moon to mine. The trouble is that the fusion reactors that would burn it do not exist, may never exist in practical form, and even if they did would require the processing of staggering tonnages of regolith to yield a usable quantity; helium-3 is less a business plan than a talking point. Water, by contrast, is real, measured, and immediately useful, which is precisely why the serious planning has converged on it and quietly set the more exotic dreams aside. There is a genuine irony here worth savoring: for decades after Apollo, the scientific consensus held the Moon to be utterly, bone-dry — the returned samples seemed to confirm a world without water, baked and dead — and it was only the orbital and impact measurements of the twenty-first century, LCROSS foremost among them, that overturned that verdict and revealed a Moon whose poles hoarded ice all along. The place that had seemed the driest body humanity had ever visited turned out, at its shadowed poles, to be one of the more interesting reservoirs of water in the inner solar system, and that reversal, as much as any rocket, is why the Moon came back into view.

How much of this vision is realistic, and how much is aspiration outrunning reality, is a fair question, and the honest answer is that much remains uncertain. The water ice at the lunar poles is believed to exist, but its quantity, accessibility, and the practicality of extracting and processing it are not yet established, and much of it, rather than lying as a convenient sheet of frost, is thought to be bound up within the regolith at concentrations of a few percent, mixed through the frozen, abrasive soil in ways that would make extraction a formidable industrial undertaking in the coldest, darkest, most hostile environment humans have ever tried to work in. The engineering of lunar resource utilization — mining ice from permanently shadowed craters, processing it into propellant, doing so reliably and economically on that hostile surface — is unproven at every step. The broader lunar economy — the sustained presence, the infrastructure, the commerce — depends on solving these challenges and on the continued investment and commitment of governments and companies over the long term, which is far from guaranteed given the history of lunar ambitions that flared and faded. The Gateway station and the lunar base remained, as this book was written, plans and partial developments rather than established realities; indeed, by early 2026 NASA had signaled that it would pause Gateway's development entirely, redirecting its energy toward a surface base, a reshuffling that captured how fluid and contested the architecture still was. The lunar economy was a vision, plausible in its logic but unproven in its execution, and whether it would be realized depended on technical, financial, and political developments not yet determined.

The scale of the commitment, and of the slippage, can be read in the accounts. Since the inception of the Human Landing System program in 2019, NASA had obligated nearly seven billion dollars to lander development and was projected to spend well over eighteen billion through the end of the decade, and still the timelines kept moving. A single critical technical demonstration — the transfer of supercold propellant between two Starship vehicles in orbit, the linchpin of the entire refueling scheme on which the lunar landing depended — had been scheduled for March 2025 and then slipped a full year, to 2026, and NASA's own auditors flagged the maturity of these cryogenic technologies as a top risk to the whole endeavor. Cryogenic fluid management — keeping propellants cold enough to stay liquid, and moving them from one tank to another in the weightlessness of space without boiling them away or letting them slosh uncontrollably — sounds like a detail. It is in fact one of the hardest unsolved problems standing between the present and the Moon, and it is a measure of how the grandest visions rest, in the end, on the mastery of the unglamorous.

SpaceX's role in the lunar vision was, as everywhere, central. Its Starship, in the Human Landing System variant, was meant to carry astronauts to the lunar surface, and its enormous payload capacity — if Starship worked and orbital refueling succeeded — could deliver to the Moon cargo and infrastructure on a scale that could transform what was possible there, potentially enabling the bases and the resource operations that the lunar economy required. A fully working Starship, capable of delivering a hundred tons or more to the lunar surface at relatively low cost, would be a game-changer for lunar ambitions, making possible a scale of activity that the small landers of the past could never support; the entire Apollo program landed, across six missions, a total mass on the Moon that a single Starship was designed to exceed in one flight. But this depended, once again, on Starship's success and on the mastery of orbital refueling, both uncertain, and the repeated delays in Starship's development contributed to the slipping of the lunar timelines. The lunar economy, like so much of the new space age, hung on the fortunes of Starship — brilliant, unfinished, and not yet proven — and on whether the grand visions could be translated into the hard reality of operations on another world.

Where does one meet this vision today, short of the Moon itself? One meets it in the quiet redirection of the world's space agencies toward a single strip of cratered highland at the lunar south pole, the target now of American, Chinese, Indian, and commercial ambition alike, chosen for one reason above all others: the ice. One meets it in the language of "depots" and "propellant transfer" and "in-situ resource utilization" that has migrated from academic papers into budget documents and contract awards. And one meets it, most tangibly, in the robotic scouts already being sent ahead —

the small commercial landers, the mapping orbiters, the instruments designed to sniff out exactly where the water lies and in what form — the reconnaissance of a frontier that the architects of the lunar economy hope to inhabit. One meets it, too, in the quiet reorientation of an entire industry's vocabulary and capital toward a strip of shadowed craters that, a generation ago, no one had any reason to care about — a shift in attention that is itself a kind of down payment on the belief that the Moon will someday pay its way. The return to the Moon, then, was driven not by the desire for another flags-and-footprints demonstration but by the vision of a lasting presence and a genuine economy in space, grounded in the promise of lunar resources, above all water ice, and in the Moon's role as a stepping stone to Mars. Whether that vision would be realized — whether the water would be found and used, the bases built, the economy established — remained uncertain, dependent on solving hard technical problems and sustaining long-term commitment, and hostage, in significant part, to the success of Starship. But the vision was the reason the Moon had become interesting again, the logic behind Artemis and the new race, and it represented a genuine ambition to make the Moon not merely a place to visit but a place to use and, perhaps, eventually to inhabit — a first step, in the grandest telling, toward a human presence not just on the Moon but throughout the solar system, of which the return to the Moon was meant to be the beginning.

Chapter 58: The Red Dragon

While the United States rediscovered the Moon through a mixture of government programs and private companies, another nation was pursuing the lunar surface with the focused determination of a rising power that saw spaceflight as central to its destiny. China, which had launched its first astronaut into orbit only in 2003 — decades after the United States and the Soviet Union — had, in the intervening years, built one of the most capable and fastest-advancing space programs in the world, and by the 2020s it had emerged as the principal challenger to American leadership in space, the rival that gave the new space race its great-power dimension and revived, after fifty years, the language and the stakes of the original contest between superpowers.

The first flight itself is worth recalling, because it fixes how quickly China moved. On the fifteenth of October 2003, a former fighter pilot named Yang Liwei rode a Long March rocket out of the Gobi Desert aboard the Shenzhou 5 capsule, orbited the Earth fourteen times over twenty-one hours, and returned to become the first person China had ever sent into space — an achievement that made China only the third nation in history, after Russia and the United States, to launch a human being into orbit under its own power. Yang later recounted an eerie detail from the flight: a mysterious knocking sound, like someone tapping the hull with a wooden mallet, that recurred without explanation and that no engineer could afterward reproduce or fully account for. It was a small, human, unsettling episode at the dawn of a program that would, within two decades, land rovers on the far side of the Moon. That trajectory — from a single nervous astronaut listening to phantom knocking, to a methodical march toward a crewed lunar landing — is the measure of what China accomplished, and of why the rest of the world came to take it so seriously.

China's approach to space was, in almost every respect, the opposite of SpaceX's. Where SpaceX was a private company driven by an individual's vision and disciplined by the market, China's program was a state enterprise, an instrument of national policy, funded and directed by the government and the ruling Communist Party, pursued with the patience, the resources, and the strategic purpose of a state that had made space a priority of national development. China's space program advanced methodically across every domain: it launched a series of increasingly sophisticated robotic missions to the Moon, including landing rovers on the lunar surface and, in a genuine first, returning samples from the far side of the Moon — the hemisphere permanently turned away from Earth, never before sampled — a feat that demonstrated capabilities matching or exceeding those of any other nation. That mission, Chang'e 6, touched down in June 2024 in the Apollo Basin, within the vast and ancient South Pole–Aitken impact basin on the lunar far side, scooped up close to two kilograms of rock and soil, and flew it back to a landing in Inner Mongolia — the first material ever brought to Earth from the hemisphere no human eye on the ground can see. The far side poses a problem that captures the sophistication required: because the body of the Moon itself blocks all direct radio contact with Earth, China first had to station a relay satellite, aptly named Queqiao, "Magpie Bridge," in a position beyond the Moon from which it could pass signals around the obstacle, without which no lander or ascent vehicle on the far side could have been commanded at all. It was the kind of patient, systems-level groundwork that characterized the whole program.

The far-side sample return did not appear from nowhere; it was the culmination of a lunar campaign built rung by deliberate rung, each mission a controlled advance on the last, and the sequence is worth laying out because it reveals the method behind the achievement. In 2013, Chang'e 3 set China's first lander and its Yutu — "Jade Rabbit" — rover on the Moon, making China only the third nation ever to soft-land there. In January 2019, Chang'e 4 accomplished something no one had done before, touching down on the far side itself and deploying a rover that would become the longest-lived in history. In 2020, Chang'e 5 brought home roughly a kilogram and three-quarters of lunar soil from the near side — the first fresh samples returned to Earth by anyone since a Soviet probe in 1976, ending a forty-four-year drought in lunar sample return. And then, in 2024, Chang'e 6 extended that capability to the unreachable far side. Landing, roving, near-side return, far-side return: each mission demonstrated a discrete new competence, tested it, and banked it, so that by the time China spoke of landing astronauts, it had already proven, robotically, most of the pieces the crewed landing would require. This was not the improvisation of a latecomer but the patient accumulation of a program that knew exactly where it was going.

It built and operated its own space station, Tiangong, a permanently crewed orbital outpost that gave China an independent human presence in space. That China had a station of its own at all was, in part, a consequence of exclusion: in 2011 the United States Congress passed the so-called Wolf Amendment, which effectively barred NASA from bilateral cooperation with Chinese entities, shutting China out of the International Space Station and leaving it little choice but to build its own — which it did, and which stood to become the only operational crewed space station once the aging International Space Station was retired around the end of the decade. It is one of the quieter ironies of the new space age that a policy meant to isolate China's program instead spurred it to full independence, and that the prospect now loomed of China possessing, at least for a time, the sole permanent human outpost in orbit — a striking reversal of the old order, and one the framers of the exclusion can hardly have intended. Tiangong was, by the middle of the decade, no prototype but a fully operational outpost, continuously inhabited by rotating three-person crews who conducted spacewalks and experiments in a station China built, launched, and ran entirely on its own — the tangible proof that a nation locked out of the world's shared laboratory had simply built a rival one, and now possessed the deeper capability of sustaining human beings in orbit indefinitely by itself.

At the center of China's ambitions in the 2020s was a declared goal that echoed the great contest of the twentieth century: to land Chinese astronauts on the surface of the Moon before the end of the decade, before 2030. To accomplish this, China was developing the necessary hardware with characteristic determination — a powerful new rocket, the Long March 10, designed to carry crews toward the Moon; a new crewed spacecraft, named Mengzhou, "Dream Vessel," to carry the astronauts; and a lunar lander, named Lanyue, "Embracing the Moon," to take them down to the surface. The program was advancing through a deliberate sequence of tests and demonstrations — a zero-altitude escape test of the Mengzhou capsule, landing-and-takeoff verification of the Lanyue lander, and, in early 2026, a successful low-altitude flight test of the Long March 10, each a rung on the ladder toward the landing — pursued with the same methodical patience that characterized all of China's space efforts. Chinese officials expressed confidence that they would meet their deadline, vowing to "spare no effort" to land astronauts on the Moon before 2030. Whether China would actually achieve a crewed lunar landing before the United States returned was genuinely uncertain — both programs faced technical challenges and possible delays — but the very possibility that China might reach the Moon first, or at nearly the same time, transformed the strategic landscape and injected into the new space age a competitive urgency that had been absent since Apollo.

It is often said, in the West, that China's space program is a secretive imitator, following where others have led and copying what others have built. The record complicates that easy judgment. The far-side sample return was not a copy of anything; no other nation, including the United States, had ever done it, and it required capabilities — the relay satellite, the far-side landing, the automated ascent and rendezvous — that no one else had assembled. China's methodical, publicly announced, on-schedule progression of tests, each building deliberately on the last, looked less like a copycat sprint and more like the disciplined, well-funded state engineering that had once carried Apollo to the Moon. The condescension with which China's rise was sometimes greeted in Washington was, in this light, a form of complacency, and the officials who took the challenge seriously were the ones reading the program correctly.

The revival of a great-power race to the Moon carried echoes of the original space race, but the differences were as significant as the similarities, and they illuminated how much the world had changed. The original space race, between the United States and the Soviet Union, had been a contest between two superpowers using their government space programs as instruments of Cold War competition, a demonstration of ideological and technological superiority conducted through national agencies with essentially unlimited political backing. The new race had a similar great-power dimension — the United States and China, the two leading powers of the twenty-first century, competing for prestige, influence, and strategic advantage in space — but it was complicated and enriched by the presence of the commercial sector, above all SpaceX. The American effort was not a pure government program like Apollo but a hybrid, depending on private companies for critical capabilities, its rockets and landers a mixture of government and commercial hardware, its progress tied to the fortunes of firms like SpaceX and Blue Origin. China's effort, by contrast, remained a state enterprise in the classic mold, more directly comparable to the Apollo-era programs, though China too was beginning to cultivate its own commercial space sector in imitation of the American model. The contrast raised a genuine strategic question, much debated in Washington: was the American approach, with its reliance on private companies and its vulnerability to their delays and failures, a source of strength — harnessing the innovation and cost-efficiency of the market — or a source of weakness, introducing uncertainty and dependence into a matter of national importance? Would the dynamism of SpaceX and the commercial sector give the United States a decisive edge, or would China's focused, well-funded, patiently executed state program prove the more reliable path to the Moon?

China's ambitions did not stop at a single landing. Together with Russia and a growing roster of partner states — among them Pakistan, Venezuela, Belarus, South Africa, Egypt, and others — China had begun to organize an International Lunar Research Station, a planned permanent base at the lunar south pole conceived as a counterweight to the American-led Artemis coalition, its membership pointedly drawn from nations outside the Artemis Accords. The Moon, in other words, was acquiring not one bloc but two, each gathering allies, each advancing its own norms and its own architecture, in a competition that reached from the engineering of rockets into the shaping of the rules under which the Moon would be used.

China had also begun, in conscious imitation of the American model, to cultivate a commercial space sector of its own — a fleet of privately capitalized startups encouraged and steered by the state, tasked with reproducing the cost-cutting dynamism that SpaceX had unleashed. The results were not merely derivative. In July 2023, a Beijing-based company called LandSpace launched its Zhuque-2 to orbit and in doing so became the first entity anywhere in the world to fly a rocket burning methane and liquid oxygen — the very propellant combination SpaceX had chosen for Starship — reaching that milestone ahead of SpaceX, ahead of Blue Origin, ahead of every American firm racing toward the same goal. It was a pointed reminder that the assumption of permanent Western primacy in commercial spaceflight was exactly that, an assumption, and that China's imitation of the American model might, in places, outrun the original. Chinese firms were pursuing reusable boosters, their own small-satellite constellations, and their own launch cadence, and while none yet approached SpaceX's scale, the direction was unmistakable: China intended to compete not only state-to-state but company-to-company, on the very terrain where the United States had assumed its lead was safest.

These questions extended beyond the Moon to the broader competition for advantage in space, which had become, by the 2020s, a recognized arena of great-power rivalry with military and economic as well as symbolic dimensions. Space was increasingly understood as critical infrastructure — for communications, navigation, surveillance, and, potentially, warfare — and the nations that dominated it would hold significant advantages. China's rapid rise as a space power, its growing constellation of satellites, its independent human-spaceflight capability, and its lunar ambitions all reflected an understanding that leadership in space was central to leadership in the twenty-first century, and the United States responded by treating the competition with China as a strategic priority. The new space race was thus not only, or even primarily, about the Moon; it was about which powers, and which systems — state or commercial, authoritarian or democratic — would shape humanity's expanding presence beyond Earth, and command the advantages that presence conferred. And it was into this contest that SpaceX's extraordinary dominance injected a factor without precedent: a single private company, controlled by a single individual, that launched more than half the world's payloads and operated the majority of its satellites, whose capabilities had become essential to American power even as its concentration raised profound questions. To the meaning of that dominance — its sources, its scale, and its dangers — we now turn.

Chapter 59: The Rest of the World in Space

The new space race is often told as a story of two or three players — SpaceX, Blue Origin, China — but this narrows a far richer and more global reality. The transformation of spaceflight in the early twenty-first century was not confined to the United States and China; it drew in nations across the world, some with long space histories and some newly ambitious, each pursuing its own path into a domain that had come to seem essential to national development, prestige, and security. To understand the full shape of the new space age is to look beyond the headline rivalries to the broader globalization of space, in which more nations than ever before were reaching for orbit and beyond. A useful measure of the change is simply this: for the first sixty years of the space age, only three powers — the Soviet Union, the United States, and China — had ever set a spacecraft down softly on the Moon. In the span of a single year, 2023 to 2024, that exclusive club expanded by two, as first India and then Japan joined it, and the enlargement of that small fraternity in so short a time says much about how widely the capability to operate in deep space had spread.

India emerged as one of the most significant of these players, its space program a source of national pride and a model of ambitious achievement on a modest budget. The Indian

Space Research Organisation had, over decades, built a capable and cost-effective program, launching its own satellites, sending missions to the Moon and Mars, and developing a reputation for accomplishing sophisticated feats at a fraction of the cost of the established space powers. In August 2023, its Chandrayaan-3 mission set its Vikram lander down near the lunar south pole, making India the fourth nation ever to achieve a soft landing on the Moon and — more remarkably — the first in history to land near the south pole, the very region of suspected water ice that had drawn the world's attention. The achievement was the sweeter for following failure: Chandrayaan-2 had crashed into the Moon on its own landing attempt in 2019, and the engineers who watched Vikram touch down four years later were, in a very human sense, redeeming a loss. What astonished outside observers most was the price. The entire mission cost on the order of seventy-five million dollars — less than the production budget of several Hollywood films about space — a figure that made India's accomplishment not merely a matter of national pride but a genuine demonstration that world-class deep-space capability need not carry a superpower's price tag. India kept costs low through frugal engineering, the adaptation of proven technology, and a deep bench of highly skilled engineers, and it turned that frugality into a competitive identity, positioning itself as a provider of low-cost launch and space services to the world and cultivating a growing commercial space sector in imitation of the American model. India's rise reflected both its growing technological capability and its determination to claim a place among the leading spacefaring nations, and it added a significant player to the new space age.

India's genius for frugality had announced itself even before the Moon. In 2014 its Mangalyaan probe reached orbit around Mars, making India the first nation ever to reach the red planet on its very first attempt — a feat the United States, the Soviet Union, and Europe had all failed to manage on their first tries — and it did so for about seventy-four million dollars, less than the hundred-million-dollar budget of the Hollywood film "Gravity," and roughly a ninth of what NASA had spent to send its own contemporaneous orbiter to Mars. The comparison became a global talking point, and India's prime minister seized on it, praising his scientists for demonstrating "a new paradigm of frugal engineering, and the power of imagination." Behind the slogan lay a serious point about the economics of the new space age: capability that had once been the exclusive preserve of superpowers spending lavishly could now be had, by a determined nation with skilled and inexpensive engineers, for the price of a studio picture. India went on to build toward crewed spaceflight of its own, and its frugal successes rewrote, for a watching world of aspiring space nations, the assumed relationship between ambition and budget.

Japan and Europe, long-established space powers, continued their own substantial programs, though both grappled with the competitive pressure that SpaceX's transformation of launch had created. Japan maintained a capable program of launch, exploration, and satellite operations, contributing to the International Space Station and conducting ambitious robotic missions of striking finesse. Its Hayabusa2 probe traveled to the distant asteroid Ryugu, fired a projectile into its surface to expose fresh material, and returned to Earth in December 2020 with about five grams of primordial asteroid dust — a tiny mass, but one of incalculable scientific value, a sample of the raw material from which the solar system was made. In January 2024 Japan's SLIM lander, nicknamed the "Moon Sniper," made Japan the fifth nation to soft-land on the Moon, and did so with a precision no one had achieved before, touching down within roughly a hundred meters, and by some measures a mere ten meters, of its intended target — where earlier lunar landings had aimed at ellipses kilometers across. The mission was also a lesson in imperfect triumph: SLIM came to rest tilted at an awkward angle, one of its engines having failed in the final descent, its solar panels pointed the wrong way, and for a time it seemed the craft might die in the dark; yet when the sun's angle shifted days later, the lander revived and returned its data, a small parable of the resilience that space engineering demands. Japan pursued all this even as it faced the challenge of competing with SpaceX's cheap reusable launch.

Europe, through the European Space Agency and the Arianespace launch provider, remained a major space power with a broad program of science, Earth observation, navigation, and exploration, but it struggled with the same competitive pressure. Its new Ariane 6 rocket, meant to secure Europe's independent access to space, finally flew its maiden mission in July 2024 — years later than planned, and, as an expendable vehicle designed in an age that SpaceX had already ended, unable to match the American company on cost. There was a poignancy in the timing: Europe had spent a decade and billions of euros building a rocket that was thoroughly conventional at the very moment reusability was rewriting the economics of the field, and the anxiety this produced was real and openly discussed. The concern about strategic autonomy — the fear of becoming dependent on SpaceX or on any foreign provider — drove European efforts to sustain its own launch capability and to develop its own satellite constellations, even at higher cost, for the sake of independence. Europe, in short, was paying a premium not for capability but for sovereignty, having concluded that reliance on a single American company for something so fundamental was a risk no serious power could accept. Japan wrestled with the same anxiety in its own key: its new flagship rocket, the H3, meant to give the country an affordable, sovereign path to orbit, failed on its debut in 2023 before succeeding in 2024, a stumble that underscored how hard it had become for even an advanced industrial nation to field a competitive launcher against a rival that was flying, landing, and reflying rockets on a nearly daily rhythm.

There was, moreover, a dimension of the global scramble that went beyond rockets and satellites to the choosing of sides. As the United States and China each assembled a coalition around a rival vision of the Moon — the

American-led Artemis framework on one hand, the Chinese-and-Russian lunar-base project on the other — the world's other space nations found themselves courted, and their decisions to sign with one camp or the other became small acts of geopolitical alignment. India, the UAE, Japan, and dozens of others attached themselves to the Artemis principles; a different roster gathered around the alternative. The globalization of space thus did not dissolve the great-power rivalry into a placid commons; it enlarged the rivalry, giving each side allies, clients, and constituencies, and turning the question of how the Moon would be used into one that implicated not two powers but scores of nations, each with a stake, a preference, and a signature on one accord or another.

Beyond the established powers, a striking feature of the new space age was the entry of nations that had never before had significant space programs, drawn in by the falling costs and the growing importance of space. The United Arab Emirates offered the most vivid example. Having built its space effort largely from scratch and at remarkable speed, the UAE sent its Hope probe into orbit around Mars in February 2021, becoming the first Arab nation to reach another planet, only the fifth entity ever to reach Mars, and just the second, after India, to succeed on its very first attempt — a domain in which roughly half of all missions had historically failed. The Emirates pursued space quite deliberately as a tool of national development and diversification, a way to inspire its young people toward science and to signal its aspiration to lead in advanced technology, and it went on to plan lunar missions and to fly its own astronauts. Other nations across the Middle East, Asia, Africa, and Latin America established or expanded their own space efforts, launching satellites for communications, Earth observation, and national prestige, participating in a domain that falling costs had made accessible to a far wider range of countries than ever before. The democratization of space extended, in this sense, to nations as well as to companies and individuals: the transformation of launch economics opened orbit not only to universities and startups but to states that could never have afforded a space program in the era of expensive expendable rockets. Space became, more than ever before, a global endeavor, pursued by a growing community of nations, each with its own ambitions and its own reasons for reaching upward.

South Korea offered a telling example of how fast a determined newcomer could climb. In 2022 it launched its first homegrown orbital rocket, the Nuri, becoming one of only a handful of nations able to loft a satellite of its own making on a booster of its own design, and in the same year it dispatched a lunar orbiter, Danuri, to circle the Moon — a nation that had possessed no independent launch capability at all a few years earlier vaulting, almost in a single stride, into the ranks of deep-space actors. The pattern repeated, in various keys, across the world: new national space agencies stood up, first satellites launched, memoranda signed to join one lunar coalition or the other, ambitions declared that would have seemed fantastical a generation before. What made all of this possible was, at bottom, the same force reshaping everything else in the domain — the collapse in the cost of access — for a nation that need no longer build and operate a ruinously expensive launch program to participate could instead buy its way to orbit, and the barrier to entry that had once confined spaceflight to two superpowers had fallen low enough for dozens to step over it.

There was, too, a quieter enabler behind much of this proliferation, and it pointed back to the very company whose dominance the wider story concerned. Many of these newer entrants reached orbit not on rockets of their own but by buying rides on SpaceX's Falcon 9, whose rideshare missions carried the small satellites of dozens of nations, companies, and universities to orbit for a fraction of the old cost — a single launch sometimes dispensing more than a hundred spacecraft into the sky like seeds from a pod. The globalization of space and the concentration of launch in one company's hands were thus not opposing facts but intertwined ones: cheap access broadened the field of who could participate, even as it narrowed the field of who provided the access. You meet this paradox today every time a nation that has never built a rocket nonetheless operates a satellite — its national ambition realized atop another power's hardware.

This globalization of space had consequences both hopeful and fraught. On the hopeful side, it spread the benefits and the participation of space activity far more widely, gave more nations a stake in the peaceful use and exploration of space, and enriched the whole endeavor with a diversity of actors, ambitions, and approaches. A world in which many nations reached for space was, in some respects, a world in which space was more genuinely a common human endeavor, less the preserve of a few superpowers. On the fraught side, the proliferation of national space programs, constellations, and ambitions intensified the competition for the finite resources of orbit and of the Moon, added to the crowding of the skies, and raised the stakes of the weak governance that struggled to manage a domain now pursued by so many. The globalization of space made the questions of how orbit and the Moon and the solar system would be shared, governed, and kept peaceful more pressing than ever, as the number of actors with the capability and the ambition to operate in space grew, and as the domain became ever more crowded and contested. The new space age was, in the end, not merely a contest among a few giants but a global transformation, drawing in nations across the world and reshaping humanity's collective relationship with the frontier above — a transformation of which SpaceX was the catalyst and the dominant force, but which extended far beyond any single company or country, and which belonged, increasingly, to the whole of humanity.

Chapter 60: The Price of Orbit

To grasp the full scale of what SpaceX had accomplished, and the concentration of power it represented, consider a single set of numbers from the middle of the 2020s. In 2025, SpaceX conducted well over a hundred and sixty orbital launches — a figure that, by itself, exceeded the launch totals of most nations and, indeed, of the entire rest of the world combined. In that year, roughly half of all the orbital launches conducted anywhere on Earth, by any country or company, were SpaceX's; the company launched more mass into orbit than everyone else on the planet put together. To make the cadence concrete: SpaceX flew its five-hundredth Falcon 9 in the summer of 2025, and across the year the rocket was lifting off, on average, more than once every three days, a rhythm of operations that the aerospace industry had, for its entire prior history, regarded as physically impossible for a single vehicle. It operated the largest satellite constellation in history, with well over ten thousand Starlink satellites in orbit, amounting to something like two-thirds of all the active spacecraft in the sky. It was the only entity, public or private, in the Western world capable of routinely carrying astronauts to orbit. And it had achieved a launch cadence and a cost structure that no competitor could approach, flying its reusable boosters again and again — one booster, tail number B1067, reaching its twenty-ninth flight in July 2025 — at a price per kilogram to orbit that was a fraction of what the industry had charged before SpaceX existed. By any measure, a single private company had come to dominate access to space to a degree without precedent in the history of the domain.

That single booster deserves a moment's attention, because it is the most eloquent rebuttal to a misconception that the industry clung to for years. It was long and confidently said, by SpaceX's competitors and by many sober analysts, that rocket reuse would never pay — that the weight of landing hardware, the cost of inspection and refurbishment, and the hidden damage of the punishing return through the atmosphere would eat up whatever was saved, so that a "reusable" rocket would be reusable only in the way that a burned match is reusable, at more trouble than it was worth. B1067 answered them by flying twenty-nine times. A rocket stage that the old orthodoxy said should be thrown away after a single use had instead flown to the edge of space and returned to a pinpoint landing twenty-nine separate times, each flight costing a small fraction of a new booster, and in doing so it turned the theoretical case for reusability into an operational fact. The refurbishment that skeptics said would be prohibitive had, through relentless engineering and sheer repetition, become closer to the turnaround of an airliner than the rebuild of a spacecraft.

The consequence in hard numbers is worth stating, because it is the foundation of everything else. In the era of the Space Shuttle, delivering a kilogram of payload to low Earth orbit cost, by the fuller accountings, tens of thousands of dollars — a figure so high that a bag of groceries' worth of mass to orbit ran to the price of a luxury car. Falcon 9, flying reused, brought the advertised cost down toward a couple of thousand dollars a kilogram, and SpaceX's own internal cost, for its own Starlink launches, lower still. That is a reduction of better than an order of magnitude, achieved not by a laboratory breakthrough but by the mundane, relentless act of not throwing the rocket away — and it is why competitors flying expendable vehicles found themselves not merely behind but structurally unable to compete on price, no matter how capable their hardware. Nor was launch SpaceX's only monopoly. Its Crew Dragon capsule had, since 2020, been the sole means by which the United States could carry its own astronauts to orbit, ending years of humiliating dependence on Russian Soyuz seats — and it remained, through the middle of the decade, the only Western spacecraft certified to do so, a second monopoly nested inside the first. Boeing, awarded a parallel contract at the same time and more money to fulfill it, had struggled for years to deliver a working alternative, leaving SpaceX to carry not only its own crews but, on occasion, to serve as the nation's only lifeline to the space station it had helped build.

This dominance was the direct fruit of the reusability revolution that had been the company's founding wager. By recovering and reflying its rockets, SpaceX had driven its own cost of launch far below that of any competitor still building expendable vehicles, and this cost advantage had translated into a commercial dominance that fed on itself. Because SpaceX could launch cheaply, it won the lion's share of the commercial launch market; because it launched so often — driven both by its commercial customers and by its own insatiable Starlink program — it kept its costs low and its operations sharp, refining its reusable rockets and its manufacturing to ever-greater efficiency; and because it operated Starlink, it had a captive, high-volume demand for its own launches that guaranteed a high cadence regardless of the commercial market. Each element reinforced the others in a virtuous cycle — cheap launch, high cadence, the Starlink engine — that competitors, lacking one or more of the pieces, could not replicate. A rival might build a reusable rocket, but without a Starlink-scale internal customer it could not match SpaceX's cadence; a rival might build a constellation, but without its own cheap reusable rockets it could not match SpaceX's economics. SpaceX had assembled a combination of capabilities that was extraordinarily difficult to challenge, and its lead, rather than eroding as competitors emerged, had if anything widened. The gap can be stated plainly: where SpaceX flew more than a hundred and sixty times in a year, its nearest Western commercial rivals were still measuring their annual launches in the single digits or the low tens, and the distance was not closing. Jeff Bezos's Blue Origin, founded two years before SpaceX and lavishly funded by the richest fortune on Earth, had spent two decades reaching the threshold of orbital flight while SpaceX built an empire; the traditional United Launch Alliance, long the Pentagon's favored contractor, fielded a new rocket that was capable but expendable and expensive; Rocket Lab carved out a real business at the small end of the market and worked toward a larger, reusable vehicle; and Europe, India, and China each pressed ahead with programs of their own. None of them was foolish or incapable — several were formidable — and yet each confronted the same brutal arithmetic: to match SpaceX they needed not one breakthrough but the whole interlocking system at once, the reusable rocket and the high cadence and the captive constellation, and assembling all three against a competitor already a decade down the road was a task that money alone could not accelerate. It was a lead measured not merely in rockets or dollars but in accumulated experience — the tens of thousands of small lessons learned by flying, landing, and reflying, again and again, that no rival could simply purchase and that only time and repetition could confer.

The benefits of this dominance were real and substantial, and they should not be minimized. SpaceX's cheap, frequent, reliable launch had transformed what was possible in space, enabling the deployment of vast constellations, lowering the cost of every kind of space activity, and giving the United States and its allies an abundance of launch capacity that was a genuine strategic asset. The cost revolution had opened space to new actors — companies, universities, developing nations — that could never have afforded access at the old prices, democratizing, to some degree, the ability to put things in orbit. And SpaceX's capabilities had restored

American leadership in a domain where it had been faltering, ending dependence on Russia for crew launch, providing capabilities that rivaled or exceeded any nation's, and giving the United States a commercial space sector that was the envy of the world. In many respects, SpaceX's dominance was a triumph — for the company, for American technological leadership, and for the expansion of human activity in space. The reusable rocket that the industry had dismissed as impractical had become the foundation of a new space economy, and SpaceX had built it.

But the very scale of that dominance raised concerns that were equally real, and that grew more pressing as the concentration deepened. The most basic was the danger of depending so heavily on a single company for something so critical. When one private firm launches half the world's payloads, operates most of its satellites, and provides the only Western means of carrying astronauts to orbit, the failure of that firm — through a technical catastrophe, a financial collapse, a decision by its leadership, or any other cause — would leave a vast and dangerous gap that no one else could quickly fill. This was not a hypothetical. In July 2024, after more than three hundred consecutive successful missions, a Falcon 9 upper stage failed on a routine Starlink flight, and the consequence was immediate and revealing: the Federal Aviation Administration grounded the entire fleet, and because that fleet was carrying so large a share of the world's launches, a single anomaly on a single rocket halted much of humanity's access to orbit until SpaceX identified the cause and won clearance to fly again a couple of weeks later. The interruption was brief, and SpaceX's recovery was impressively fast, but the episode demonstrated in miniature exactly the vulnerability that the concentration created: a grounding of the Falcon 9 fleet, of the kind that had happened before and would happen again, could bring much of the world's launch activity to a standstill. A disruption to Starlink could sever communications that millions, and militaries, had come to depend on. The concentration that made SpaceX so powerful also made the world dangerously dependent on it, creating a single point of failure in a domain of growing importance, precisely the kind of vulnerability that prudent nations try to avoid. And this dependence was compounded by the fact that SpaceX was controlled, to an unusual degree, by a single individual whose judgment, priorities, and conduct could shape the behavior of an entity on which so much now rested — a concern dramatized by the Starlink-in-Ukraine episodes and by the broader question of the founder, to which we will come.

That the July 2024 anomaly was not a freak was underscored within weeks, when a Falcon 9 booster toppled and was lost on landing after an otherwise successful flight, and again by the ordinary background rhythm of a machine flown so often that even a very low failure rate must, statistically, produce the occasional mishap. This is the paradox at the heart of SpaceX's reliability: the very cadence that made it indispensable also guaranteed that it would, from time to time, stumble, and because so large a share of the world's launch now rode on one company's hardware, each stumble carried consequences out of all proportion to its rarity. A launch provider that flew a dozen times a year could absorb a mishap quietly; a provider that flew more than a hundred and sixty times, and carried with it the schedules of governments, militaries, and commercial operators across the globe, could not. The world had, without quite deciding to, arranged its access to space so that a single company's bad week became everyone's problem — an efficiency in ordinary times, and a fragility in extraordinary ones.

There was, too, the question of competition and monopoly. SpaceX had achieved its dominance largely through genuine merit — by building better, cheaper rockets than anyone else — and there was nothing inherently wrong with a company winning by being better. But dominance, however earned, brings market power, and market power brings the potential for abuse: the ability to raise prices once competitors are driven out, to use control of launch to advantage one's own constellation over rivals', to shape the terms of access to space in one's own interest. It is worth remembering that SpaceX itself had once been the insurgent making precisely this argument. In its early years the company had gone to court to challenge the cozy arrangement by which the United States military bought its launches, protesting that a protected incumbent was foreclosing competition — and now SpaceX was the incumbent, the giant whose scale others found impossible to match, and the argument it had once made against others could, in time, be made against it. Whether SpaceX would use its dominance benignly or exploitatively, whether the competition from Blue Origin, Rocket Lab, China, and others would be sufficient to discipline it, whether governments would need to intervene to ensure fair access and guard against the risks of concentration — these were open questions, and they were the natural accompaniment of the company's success. The new space age had, in the space of two decades, gone from a domain dominated by government programs to one dominated by a single private company, and the world had not yet worked out how to think about, let alone govern, that transformation. The concentration of orbital power in SpaceX's hands was one of the defining facts of the new era, and one of its defining problems, promising both the dynamism of private enterprise and the dangers of dependence on a single, individually controlled company operating in a domain of national and even civilizational importance. Nowhere were those stakes higher than in the domain where space and national security met — the militarization of the orbital high ground, where SpaceX had become indispensable to the armed forces of the world's leading power.

Chapter 61: The Economics of Abundance

For the whole of the space age, the defining condition of spaceflight was scarcity. Access to orbit was scarce and expensive, rationed by cost, limited to a few flights a year, available only for the most valuable payloads and the most important missions. Everything about how humanity used space was shaped by that scarcity: satellites were built to last for decades because launching a replacement was so costly; missions were few and precious; the possibilities were bounded by the high price of getting anything off the ground. The deepest consequence of SpaceX's revolution — deeper, in a sense, than any single achievement — was to begin to replace that scarcity with abundance, to make access to space cheap and frequent enough that the old constraints started to fall away, and to open possibilities that scarcity had foreclosed. Understanding this shift from scarcity to abundance is key to grasping the full significance of what SpaceX wrought.

The scale of the price collapse is worth stating in plain figures, because it is the engine beneath everything else. In the era of the Space Shuttle, lifting a single kilogram to low Earth orbit cost on the order of tens of thousands of dollars

— by many accountings north of fifty thousand — a price that made every gram a subject of agonized negotiation among engineers and accountants. The reusable Falcon 9 drove that figure down toward a couple of thousand dollars a kilogram, and Starship, if it worked as designed, aimed lower still, toward the low hundreds and eventually, in the most optimistic reckonings, below. An order-of-magnitude reduction had already been achieved, and a further one was the explicit goal. It is difficult to name another domain of engineering in which the fundamental cost of the activity fell by a factor of ten within a single career, and harder still to overstate what such a fall does to the range of the thinkable, because once a cost drops far enough, it does not merely make old activities cheaper — it changes the very question of what is worth attempting at all.

Consider how scarcity had shaped the satellite. In the era of expensive launch, a satellite was a precious, long-lived, ultra-reliable object, built at enormous cost to last fifteen or twenty years, because launching a replacement was so expensive that a satellite had to justify its launch by serving for a very long time. This drove satellites to be large, expensive, conservative, and slow to incorporate new technology, since they were frozen at the technology of their launch for decades. A traditional communications satellite bound for the high geostationary orbit might cost hundreds of millions of dollars and take the better part of a decade to design and build, a jewel so valuable that its loss on launch was a corporate catastrophe. Abundance inverted this logic completely. When launch became cheap and frequent, as Starlink demonstrated, satellites could be cheap, mass-produced, short-lived, and constantly replaced with newer, better versions — a constellation in perpetual renewal, always incorporating the latest technology, treating individual satellites as expendable and the system as the reliable whole. SpaceX built its Starlink spacecraft not one at a time in a clean-room cathedral but on a production line turning out several a day, each designed to work for only about five years before deorbiting and burning up, to be succeeded by a better model. The old orthodoxy held that a satellite must be exquisite because it must be permanent; abundance replaced the exquisite and permanent with the cheap and renewable, and found that for many purposes the renewable system was not merely acceptable but superior — younger, more current, self-healing, never obsolete. This was a profound change in how satellites were conceived and built, made possible by the shift from scarce, expensive launch to abundant, cheap launch, and it pointed to how abundance could transform every aspect of space activity.

The possibilities that cheap, abundant launch opened extended far beyond satellites. Mega-constellations like Starlink, connecting the world from orbit, were economically impossible under scarcity and became possible under abundance. Ambitious science missions, once rationed by the high cost of launch, could become more frequent and more capable. The delivery of large amounts of cargo and infrastructure to the Moon or Mars, prohibitive under scarcity, became conceivable under the abundance that a fully reusable Starship promised. Space manufacturing — producing in orbit materials or products that benefit from weightlessness or vacuum — long a dream frustrated by the cost of access, became more plausible as launch costs fell. Space tourism and private human spaceflight, impossible when a launch cost a fortune, began to emerge as access cheapened. And the grandest visions — a lunar economy, a Martian settlement, a genuine human expansion into space — depended entirely on the shift from scarcity to abundance, on driving the cost of access down far enough that moving people and material into space in large quantities became affordable. Abundance was the precondition for everything the new space age imagined, the foundation on which all the grander possibilities rested.

Some of these possibilities are more concrete than the grand visions suggest, and worth naming precisely, because they show how a change in launch cost reaches into industries that have nothing obviously to do with rockets. Consider manufacturing in orbit. There are materials that can be made better, or only, in the near-perfect vacuum and the weightlessness of space — certain exotic optical fibers, for instance, that in the absence of gravity form without the tiny imperfections that gravity introduces on Earth, promising signal quality unattainable on the ground; certain crystals and pharmaceutical compounds that grow larger and purer without a gravitational field to distort them. For the whole of the space age such ideas remained laboratory curiosities, strangled at birth by the arithmetic: no product, however superior, could justify a manufacturing process whose raw input — the trip to orbit and back — cost tens of thousands of dollars a kilogram. Lower that cost by an order of magnitude and the calculation changes; a factory in orbit ceases to be science fiction and becomes a spreadsheet. The same logic applies, more speculatively, to the notion of point-to-point travel — using a rocket to carry passengers or urgent cargo between two points on Earth in under an hour, over the pole and back down — an idea that is absurd at the old cost of flight and merely audacious at the new one. Whether these particular businesses will materialize is uncertain; the point is structural, that each becomes thinkable only once the cost of access falls, and that a whole penumbra of such possibilities hangs, waiting, on the far side of cheap launch.

There is a deeper economic principle at work here, one that the pioneers of every transformative technology have discovered in turn, often to their surprise. When the cost of something falls dramatically, total consumption of it does not merely hold steady at a lower price — it explodes, as uses that were never worth the old price come flooding in. Cheaper steel did not mean the world bought the same amount of steel for less; it meant the world built skyscrapers and bridges and railways it had never contemplated. Cheaper computation did not shrink the computing bill; it put a computer in every pocket and then in every appliance. The economists call this induced demand, and it is why the promoters of cheap launch are not merely selling discounts on existing activity but betting on the eruption of activity that does not yet exist — the mega-constellation that only made sense once launch was cheap, the Earth-observation business that only closed its accounts at the new price, the private astronaut who only became imaginable when a seat cost a fortune rather than a nation's treasury. If Starship delivered the second order-of-magnitude reduction it promised, the reasonable expectation was not a quieter space industry doing the same things for less, but a louder and stranger one, doing things no one had yet thought to propose.

The economic logic of this transformation deserves emphasis, because it is easy to miss amid the spectacle of rockets and constellations. The cost of launch had been, for the whole of the space age, the gating constraint on what humanity could do in space — the bottleneck through which everything had to pass, and which limited the possibilities to what could bear the enormous expense. By attacking that constraint — by driving the cost of launch down through reusability, by an order of magnitude and potentially more — SpaceX did not merely make the existing activities cheaper; it changed what was possible, opening a space of possibilities that the old costs had foreclosed. This is the classic pattern by which a transformative technology reshapes the world: not merely by doing the old things more cheaply, but by making possible new things that were previously unthinkable. The railroad did not merely make travel cheaper; it created new industries, cities, and ways of life. Cheap electricity did not merely light homes; it enabled a century of invention. The analogy repays a closer look, because in each case the decisive change was not a matter of degree but of kind. When the cost of moving goods overland fell far enough, it became sensible to grow wheat a thousand miles from where it would be eaten, to build a city where there had been only prairie, to organize an entire continental economy around the assumption of cheap transport — none of which was a plan anyone had drawn up in advance; all of it emerged, unbidden, once the constraint lifted. Cheap access to space, if it is truly achieved, may do the same — enabling activities, industries, and possibilities in space that no one can yet fully imagine, precisely because the scarcity that foreclosed them is only now beginning to lift, and because the most consequential uses of a cheap new capability are rarely the ones its pioneers set out to enable.

There is, of course, a crucial caveat, and it must be stated clearly, because it is the point on which the skeptics are right and the enthusiasts too easily glide past. Cheap launch is necessary for abundance but not sufficient, and the full abundance that the grandest visions require has not yet been achieved. SpaceX drove launch costs down dramatically with Falcon 9's reusable boosters, but the truly transformative abundance — the order-of-magnitude-and-more reduction that would make space genuinely cheap, that would enable the lunar economy and the Martian settlement and the space manufacturing — depended on Starship, on a fully and rapidly reusable super-heavy vehicle, and Starship remained, as this book was written, unfinished, its promise unrealized, its boosters still occasionally lost and its critical in-orbit refueling still undemonstrated. It is worth being precise about what "not yet" means here: the economics of abundance were, in 2026, partly realized and partly still a promise — real enough to have transformed satellites and enabled Starlink and lowered the cost of access substantially, but not yet carried to the ultimate conclusion that the grandest visions required. Whether that ultimate abundance would be achieved — whether Starship would work and drive costs down far enough to open the full space of possibilities — remained the central open question, on which the future of the new space age depended.

And yet even the partial abundance already achieved had begun to produce sights that would have been unimaginable a decade earlier, and they are the clearest evidence that the shift is real and not merely projected. In 2021 a SpaceX Dragon carried four private citizens — not a professional astronaut among them — into orbit for several days, the first all-civilian orbital flight in history; a few years later, on a mission called Polaris Dawn, a private crew flew higher than any human had gone since Apollo and conducted the first commercial spacewalk, in suits SpaceX had designed for the purpose. Private companies began ferrying paying crews to the International Space Station on chartered missions. None of this was tourism in the trivial sense; it was the early, halting commercialization of human spaceflight itself, an activity that for sixty years had been the exclusive province of national governments spending public fortunes, now beginning — tentatively, expensively, but unmistakably — to open to private hands. That it happened at all was a function of falling cost, and it offered a small, concrete preview of what a fuller abundance might bring.

But the direction was clear, and the shift already underway was profound. For the first time in the space age, the fundamental constraint of scarcity was beginning to lift, and the possibilities that abundance would open — some already realized, many still to come — represented a genuine transformation in humanity's relationship with space, from a domain rationed by cost and reserved for the few to one that cheap access might open to activities, industries, and ambitions previously unimaginable. You meet the leading edge of this shift already, in the most ordinary of places: in a satellite dish on a farmhouse roof in a country that never had broadband, fed by a constellation that abundance alone made buildable; in the falling price quoted to a university that wants to fly an experiment; in the casual assumption, unthinkable a generation ago, that a private citizen might one day buy a seat to orbit. The economics of abundance were the deepest meaning of SpaceX's revolution, the underlying shift that all the visible achievements expressed, and if the abundance were carried far enough — if Starship succeeded and the costs fell far enough — the consequences might reshape not just an industry but the trajectory of human civilization, opening the solar system to human activity in a way that scarcity had always prevented. That was the promise, still partly unfulfilled, of the transformation SpaceX had begun: not merely cheaper rockets, but an age of abundance in space, and all that such abundance might make possible.

Chapter 62: The High Ground

From the earliest days of the space age, space and national security were intertwined. The rockets that carried the first satellites into orbit were descendants of ballistic missiles built to deliver nuclear warheads; the first satellites included spacecraft designed to spy on adversaries from above; and the strategic value of the orbital high ground — for reconnaissance, for communications, for navigation, for early warning of missile attack — was understood by military planners from the outset. But for decades this military dimension of space operated largely in the background of public awareness, conducted through classified government programs, its importance quietly growing as the world's militaries came to depend ever more heavily on the satellites overhead. The new space age brought this military dimension to the foreground, both because the strategic stakes had grown and because the commercial revolution SpaceX had wrought had transformed the military use of space as thoroughly as it had transformed the civilian.

Modern militaries had come to depend on space to an extraordinary degree. Satellites provided the navigation signals that guided everything from troops to precision weapons; the communications that linked forces across the globe; the reconnaissance and surveillance that watched adversaries and assessed threats; the early warning that detected missile launches; the weather data that shaped operations. The dependence is easy to state abstractly and worth making concrete. A single satellite-guided bomb, of the kind that has become routine in modern warfare, is essentially a cheap iron weapon strapped to a receiver that listens for the faint timing signals of the navigation constellation overhead and steers itself to within a few feet of its target; deprive it of those signals and it reverts to a dumb, inaccurate lump of metal. The entire edifice of precision warfare, on which the armed forces of the leading powers had staked their doctrine, rested on the continued functioning of a few dozen navigation satellites, along with the communications and reconnaissance spacecraft that fought around them. A modern military deprived of its satellites would be crippled — unable to navigate, communicate, or see as it had learned to do — and this dependence made space itself a domain of potential conflict, a place where a war might be won or lost, and where adversaries might seek to blind or deafen one another by attacking the satellites on which each relied. The recognition of space as a war-fighting domain led the United States, in December 2019, to establish a Space Force, the first new branch of its armed forces in more than seventy years, dedicated wholly to the domain — a step that acknowledged, formally, what had long been true: that space was now central to national defense, a high ground that had to be contested and secured.

The ways to contest that high ground extend well beyond the crude spectacle of blowing a satellite apart, and it is the quieter methods that most preoccupied military planners. A satellite need not be destroyed to be defeated: its signals can be jammed, drowned out by noise from the ground; its navigation timing can be spoofed, fed false data to mislead the weapons that trust it; its sensors can be dazzled by lasers; and, most unsettling of all, it can be shadowed in orbit by a so-called co-orbital vehicle, an "inspector" satellite that sidles up to another nation's spacecraft and lingers within reach, a capability that both China and

Russia were observed to be developing and testing. Against this spreading array of threats, the old model of a few enormous, exquisite, irreplaceable military satellites looked increasingly like a liability — a small number of high-value targets, each catastrophic to lose. The response, championed by a new Space Development Agency, was to disperse capability across large numbers of small, cheap, frequently replaced spacecraft, a proliferated architecture whose resilience lay precisely in its numbers — and that response, of course, depended on cheap and frequent launch to build and sustain it, which pointed straight back to SpaceX.

Into this militarized domain, SpaceX injected the same transforming force it had brought to civilian space: cheap, frequent, reliable launch, and the capabilities that flowed from it. The military became one of SpaceX's most important customers, relying on the company to launch its most sensitive and valuable satellites — the reconnaissance spacecraft, the communications and navigation satellites, the assets on which national security depended — at a fraction of the cost and with a frequency that the old launch providers could not match. SpaceX's reliability and cadence made it indispensable to the military's access to space, and its reusable rockets gave the United States a responsive launch capability — the ability to put assets into orbit quickly and affordably — that had strategic value in itself. But SpaceX's role went beyond launch. The company's Starlink constellation, and its military variant Starshield — announced in late 2022 and soon under contract to the Space Force — offered the armed forces something new and valuable: a vast, distributed, resilient network of satellites providing communications that were extremely difficult for an adversary to destroy or disrupt, precisely because they were spread across thousands of spacecraft rather than concentrated in a few. The logic of resilience was the mirror image of the old vulnerability: a handful of large, exquisite military satellites presented a handful of targets, each of whose loss would be grievous, whereas a constellation of thousands could absorb the loss of dozens without meaningful degradation, its very redundancy a form of armor. The demonstration of Starlink's value in the war in Ukraine had shown military planners around the world what such a constellation could do — keeping a nation's forces connected under bombardment when its ground infrastructure had been shattered — and the United States and its allies moved to incorporate these capabilities into their own strategic planning, making SpaceX not just a launch contractor but a provider of critical military infrastructure.

Yet Ukraine also revealed the shadow side of that dependence, and it is worth correcting a widespread misunderstanding of what actually happened. It is often said that Elon Musk "switched off" Starlink over Crimea to thwart a Ukrainian attack. That is not quite what occurred. In September 2022, faced with a Ukrainian request to extend Starlink's coverage to the waters off Russian-occupied Crimea to support a sea-drone strike on the Russian fleet at Sevastopol, Musk declined to activate service in that area — service that had never been switched on there in the first place — reasoning, by his own later account, that enabling an attack on the naval base of a nuclear-armed power could make him and his company complicit in a dramatic escalation of the war. The distinction between refusing to extend a capability and "turning off" one already in use is real, but for the national-security officials watching, it hardly softened the underlying alarm. Here was a private businessman, on his own personal judgment, making what amounted to a battlefield decision in a war between nations — a decision about which military operations would be possible and which would not — and there was no established mechanism by which the government could compel, override, or even reliably predict him. The episode was, for the officials responsible for American strategy, a chilling demonstration of exactly the vulnerability that the concentration had created.

This deepening entanglement of SpaceX with national security intensified all the concerns that the company's dominance had already raised, and added new ones specific to the military domain. The dependence of the armed forces on a single private company for launch and for critical communications created a strategic vulnerability of exactly the kind that military planners are trained to avoid: a single point of failure, a capability concentrated in one firm controlled by one individual, whose decisions could affect military operations, as the Ukraine episodes had shown in the starkest terms. The idea that a private citizen might, through his control of a commercial constellation, influence the conduct of military operations — enabling or restricting capabilities according to his own judgment, beyond the control of the government that depended on them — was deeply troubling to national-security officials, and it drove efforts to establish clearer arrangements, dedicated government-controlled services like Starshield, and greater redundancy. But the underlying reality was difficult to escape: SpaceX had built capabilities that the military wanted and needed, that no one else could provide as well or as cheaply, and that dependence, once established, was hard to unwind. The company's dominance in civilian space had made it dominant in military space as well, with all the strategic advantages and all the strategic risks that entailed.

The depth of that entanglement is easy to underestimate, because so much of it is classified, but its outlines were plain. SpaceX launched the nation's spy satellites, its missile-warning spacecraft, its navigation and communications assets; it built, under contract, a large proliferated network of reconnaissance satellites for the National Reconnaissance Office, the secretive agency that operates America's eyes in orbit — so that the country's ability to watch its adversaries from above now rested, in significant part, on the same company that launched its astronauts and connected its farmhouses. The military prized, too, the promise of responsive launch: the ability, in a crisis, to replace a satellite shot down or disabled by an enemy within days rather than years, a capability that only a company flying at SpaceX's cadence could plausibly offer, and one that turned raw launch tempo into a strategic asset in its own right. Every one of these threads strengthened the armed forces and, in the same motion, deepened their dependence, knitting the world's most powerful military ever more tightly to a single private firm — indispensable, irreplaceable in the near term, and controlled by one man. The unease this produced was not confined to the United States. Watching how completely one American company had come to underpin American military power, other nations drew the obvious lesson and moved to build resilient constellations of their own — Europe, for one, committing to a sovereign secure-communications network precisely so that it would not, in a crisis, find its forces dependent on infrastructure it did not control. Dependence on SpaceX had become, in this way, a strategic consideration for the whole Western alliance, and the scramble to hedge against it was itself a measure of how central the company had grown.

The militarization of space raised, too, the darker prospect of conflict in orbit — of weapons designed to attack satellites, of the deliberate creation of debris to deny an adversary the use of space, of a war that extended into the domain on which so much of modern life and modern warfare depended. This was no longer a hypothesis. In November 2021, Russia fired a missile that destroyed one of its own defunct satellites, the roughly two-ton Cosmos 1408, shattering it into more than fifteen hundred trackable fragments and many thousands of smaller ones — debris that spread across low Earth orbit, forced the astronauts aboard the International Space Station to shelter in their return capsules in case of a strike, and, in the months that followed, compelled SpaceX's own Starlink satellites to maneuver out of the way of the wreckage. The Russian test, destructive as it was, was not even the worst of its kind. In 2007 China had destroyed one of its own aging weather satellites, Fengyun-1C, at a far higher altitude, producing the single largest debris cloud in the history of spaceflight — thousands of trackable fragments, many of which will remain a hazard in orbit for decades or centuries, long after the political point of the test has been forgotten. India joined the club in 2019 with a deliberately low-altitude test designed to let its debris fall quickly back to Earth, a tacit acknowledgment of the lesson China had taught. The pattern across all of them was the same: a demonstration meant to project strength that in fact seeded the commons with shrapnel dangerous to friend and foe and bystander alike. It was a vivid illustration of a peculiarly self-defeating danger: China, Russia, the United States, and India had all, at one time or another, demonstrated the ability to smash a satellite, and yet every such demonstration scattered long-lived shrapnel that threatened the demonstrator's own spacecraft as surely as anyone else's. The growing strategic importance of space, the intensifying competition between the great powers, and the concentration of critical capabilities in constellations like Starlink all raised the stakes of a potential conflict in orbit — a conflict that could, in the worst case, not only cripple militaries but, through the cascading generation of debris in the runaway chain reaction that scientists called the Kessler syndrome, render parts of near-Earth space unusable for everyone, a catastrophe for the whole of humanity's presence beyond the Earth. The new space age had made space more valuable, more crowded, and more contested than ever before, and it had done so faster than the world's institutions and laws could adapt, leaving the governance of the domain — the rules of the road, the limits on weapons, the management of debris and traffic, the arrangements for security and cooperation — lagging far behind the reality. SpaceX had not created these tensions, but its transformation of space had accelerated and sharpened them, and the company sat at the center of them, indispensable and concentrated, a private firm woven into the security of nations. It was against this backdrop of dominance, dependence, and unresolved danger that the deepest and most idiosyncratic risk of all came into focus — the question of the founder himself, whose singular control over so much power, and whose increasingly erratic and polarizing public conduct, made the concentration of capability in SpaceX not merely an institutional concern but a deeply personal one.

Chapter 63: Space Law and the Commons

As the new space age filled the skies with satellites, reopened the race to the Moon, and made orbit a domain of commerce and competition and conflict, it ran ever harder against a body of law and governance that had been written for a different era and that struggled to keep pace with the transformed reality. The rules that govern human activity in space — who may do what, who owns what, who is responsible for what — were largely established in the 1960s, at the dawn of the space age, and they reflect the assumptions of that time, when space was the domain of two superpower governments and commercial activity was unimaginable. The gap between those old rules and the new reality became one of the quieter but more consequential problems of the transformed space age.

The foundational document is the Outer Space Treaty, opened for signature in January 1967 and in force by that October — a Cold War-era agreement that established the basic principles still nominally governing space, and that by the mid-2020s had been ratified by more than a hundred and ten nations, including every major spacefaring power, China and Russia among them. It declared that space should be free for exploration and use by all nations, that no nation could claim sovereignty over the Moon or other celestial bodies, that space should be used for peaceful purposes and not host weapons of mass destruction, and that nations bore responsibility for their activities in space, including those of their private companies. That last principle repays a moment's thought, because it produces a consequence that surprises many: under the treaty, a private company's actions in space are, in law, the responsibility of its home government, so that every Falcon 9 that rises from Florida is, in the eyes of international law, an act of the United States, for which the United States answers to the world. There is, in the legal architecture, no such thing as a purely private actor in orbit; behind SpaceX stands, inescapably, the government that licenses and thereby vouches for it. These principles were noble and, for their time, sufficient, when the only actors were a handful of governments and the activity was limited. But they were written for a world that no longer exists. They said little about commercial activity, about the ownership of resources extracted from the Moon or asteroids, about the management of orbital traffic and debris, about the crowding of low Earth orbit by mega-constellations, about the myriad questions raised by a space age of private companies, mega-constellations, and resource ambitions. The old law provided a framework, but a framework with enormous gaps, and into those gaps rushed the new realities that the treaty's authors could never have foreseen.

The Outer Space Treaty did not stand entirely alone; it was the first and broadest of a small cluster of agreements negotiated in the following decade, and the fate of the others is instructive. A Liability Convention, concluded in 1972, made a launching state absolutely liable for damage its space objects caused on Earth, and it was invoked in earnest exactly once: in 1978, when a Soviet nuclear-powered reconnaissance satellite, Cosmos 954, fell out of orbit and scattered radioactive debris across the frozen expanse of northern Canada, prompting a costly cleanup and, eventually, a partial Soviet payment toward the bill — the only time in history a nation has paid compensation under the treaties for the wreckage of a spacecraft. That single case, tidy as it was, hardly prepared the law for a sky filled with tens of thousands of privately owned satellites whose collisions would generate not one identifiable culprit but a diffuse cloud of shared fault. More telling still was the Moon Agreement of 1979, which tried to go further than the Outer Space Treaty by declaring the Moon and its resources the "common heritage of mankind," to be shared under an international regime — a phrase that alarmed the very nations capable of reaching the Moon, none of which ratified it. The Moon Agreement stands today as a near-dead letter, a treaty almost no one who matters has signed, and its failure is itself a lesson: the spacefaring powers would accept rules against seizing territory, but not rules that gave others a claim on what they might one day extract. That unresolved tension — between the Moon as everyone's common inheritance and the Moon as a place to be mined by whoever gets there first — sat at the center of every dispute that followed.

Consider the question of resources, around which one of the more stubborn misconceptions has grown. It is often said that because no one may own the Moon, it must be illegal to mine it — that the treaty's ban on appropriation forecloses any commercial exploitation of lunar or asteroid materials. In fact the treaty says no such thing. It prohibits national appropriation — the planting of a flag and the drawing of a border, the claiming of celestial territory as sovereign soil — but it is silent on whether a company that extracts water ice or minerals may own what it has dug up. The distinction is precisely the one that governs the high seas, and the analogy is the one that legal scholars reach for again and again: no nation may own the Atlantic Ocean, yet a fishing vessel plainly owns the fish it hauls aboard. Whether the same logic extends to lunar ice — whether one may keep what one extracts from a body no one may claim — the treaty did not clearly say, and the ambiguity created real uncertainty over the legal basis for the resource ambitions that were central to the plans for lunar and Martian activity. Into that vacuum stepped national legislatures. The

United States, in its Commercial Space Launch Competitiveness Act of 2015, declared that American citizens and companies were entitled to own and sell the resources they extracted from space; Luxembourg, improbably, made itself a hub of the nascent asteroid-mining industry by passing a similar law of its own; and the United States promoted a set of principles, the Artemis Accords, intended to establish norms for peaceful and cooperative lunar activity, including the extraction of resources.

The Artemis Accords tell their own story of how contested this terrain had become. Launched in October 2020 with just eight founding signatories — Australia, Canada, Italy, Japan, Luxembourg, the United Arab Emirates, the United Kingdom, and the United States — they grew within a few years into a coalition of dozens of nations, an American-led effort to write the rules of lunar behavior through voluntary accord rather than binding treaty. But they were national laws and voluntary accords, not universal agreements, and they were contested by other powers, notably China and Russia, which pursued their own approaches and questioned the legitimacy of the American-led framework. China and Russia declined to sign, criticizing the Accords as an attempt to impose Washington's preferred rules outside the proper channels of the United Nations, and organized instead their own coalition around a rival lunar base project, gathering their own roster of partner states — a competing bloc, with a competing vision, and pointedly no overlap in membership. The governance of space resources, in short, remained unsettled, a matter of competing national laws and rival norms rather than clear and universal rules, and the potential for conflict — over lucrative sites on the Moon, over the right to extract and own resources — was real, sharpened by the fact that the most valuable real estate, the ice-bearing craters of the south pole, was finite, and that two rival coalitions were converging on the same narrow ground.

The problem of orbital traffic and debris was even more

pressing, and even less adequately governed. As tens of thousands of satellites filled low Earth orbit — Starlink and its rivals, the mega-constellations of companies and nations — the risk of collisions and the accumulation of debris grew, threatening the usability of the orbital environment. Yet there was no effective global authority to manage orbital traffic, to coordinate the placement of satellites, to require the removal of debris, to prevent the crowding of orbit beyond safe limits. What governance existed was piecemeal and national. The United States, through its Federal Communications Commission, tightened its rule in 2022 to require that satellites in low orbit be brought down within five years of the end of their missions, down from a long-standing and widely ignored guideline of twenty-five — a meaningful step, but one that bound only spacecraft licensed in America, and left the rest of the world's operators to their own patchwork of standards. The management of the orbital commons was thus left to a patchwork of national regulators, voluntary guidelines, and the goodwill and self-interest of the operators, with no binding global framework to ensure that the shared resource of orbit was not despoiled by the uncoordinated actions of many actors each pursuing its own interest. The classic tragedy of the commons loomed: each operator, benefiting from placing its satellites in orbit, had little incentive to bear the costs of protecting the shared environment, and the collective result threatened to be a low Earth orbit so crowded and so cluttered with debris as to become dangerous or unusable — a catastrophe for all, produced by the rational self-interest of each. The governance needed to prevent this — binding rules on debris, traffic management, limits on congestion — lagged far behind the reality, a gap that grew more dangerous as the skies filled.

Part of what made the gap so hard to close was the machinery, or the absence of it, for writing binding global rules at all. The United Nations body responsible for outer space operated by consensus, which in a domain of sharp great-power rivalry meant, in practice, near-paralysis: any of the major spacefaring states could block a rule it disliked, and so the strongest measures on debris and traffic remained voluntary guidelines rather than enforceable law. Into that vacuum stepped, of all things, a national telecommunications regulator. Because the United States controlled access to the world's most lucrative market and to prized slices of the radio spectrum, its Federal Communications Commission found itself setting standards that operators everywhere had to heed if they wanted to do business — a de facto global rule-maker by accident of market power rather than any grant of authority. It was an improvised and lopsided arrangement, one nation's domestic agency shaping the behavior of a planetary commons, and it satisfied no one fully: too weak to bind the operators it did not license, too unilateral to command the legitimacy of a genuine international regime. The world was governing its most crowded new frontier, in effect, through a patchwork of national permits and the leverage of the largest customer, while the binding, universal framework that the situation cried out for remained unwritten. That such a framework was needed almost everyone agreed; that anyone could assemble it, across the widening chasm of great-power rivalry and commercial competition, no one could confidently promise.

The deeper problem was that space had become a commons — a shared resource, belonging to all humanity and to future generations, on which many actors now operated — without the governance that a commons requires to be managed sustainably. The old law, written for a sparse and government-dominated space age, could not manage a crowded, commercial, competitive one, and the institutions and agreements needed to fill the gap — to govern resources, traffic, debris, security, and the fair sharing of the domain — did not exist or were weak, contested, and lagging. This governance gap was one of the genuine dangers of the new space age, less visible than the rockets and the constellations but potentially as consequential, for without adequate governance, the shared domain of space risked being crowded, contested, and despoiled, its benefits captured by the powerful and its costs borne by all, its long-term usability sacrificed to short-term competition. SpaceX, as the dominant actor, sat at the center of these questions — its constellations crowding the skies, its ambitions reaching toward lunar resources, its dominance raising the stakes of governance — and how the world chose to govern the space that SpaceX had done so much to transform would shape whether the new space age fulfilled its promise or squandered its commons. There was, too, a smaller and more immediate way that ordinary people were beginning to meet the consequences of the crowding: astronomers, who found the long exposures of their telescopes streaked by the bright trails of passing satellites, and stargazers, who noticed trains of Starlink spacecraft marching across the dusk where once there had been only stars — a quiet, visible sign that the commons overhead now belonged, in practice, to whoever filled it first.

The challenge of space governance is, in the end, a challenge of collective action, of the kind that has always been hardest for humanity to meet — the management of a shared resource for the common and long-term good against the pull of individual and short-term interest. The new space age, by making space so much more accessible and so much more crowded, made this challenge urgent, and it found humanity's institutions unprepared, its laws outdated, its agreements weak. Whether the world would rise to the challenge — would build the governance needed to manage the orbital commons, to share the resources of the Moon peacefully, to keep space secure and sustainable — remained an open and pressing question, one of the great unresolved problems of the transformed space age, and one on which the long-term fate of humanity's presence beyond the Earth would, in no small part, depend.

Chapter 64: The Founder Problem

No account of SpaceX can avoid the problem of its founder, and by the 2020s that problem had grown acute. Elon Musk was, without question, the indispensable figure in the company's history — the source of its vision, the provider of its founding capital, the driver of its most audacious bets, the demanding presence who pushed his engineers to accomplish what they had thought impossible. Without Musk, there would have been no SpaceX, no reusable rockets, no Starlink, no Starship; the entire transformation this book has described flowed, in large part, from his singular determination and his willingness to risk his fortune and stake his life's energy on a goal that nearly everyone thought foolish. To that extent, the concentration of the company around one visionary individual had been its greatest strength, the thing that allowed it to move fast, think from first principles, and pursue a coherent long-term mission free of the timidity and short-termism that afflicted its competitors. But the same concentration that had been SpaceX's strength became, as Musk's public conduct grew more erratic and more polarizing, one of its most serious liabilities and one of the new space age's most genuine risks.

Over the 2010s and into the 2020s, Musk transformed from an admired if eccentric entrepreneur into one of the most controversial and divisive public figures in the world. He accumulated an enormous following and an enormous fortune, becoming at times the richest person on Earth, his wealth at moments exceeding the annual output of entire nations — a scale of personal fortune with few precedents in history, and one that freed him from every ordinary check that markets and creditors impose on lesser figures. He used his platform — especially after his purchase, in October 2022, of the social-media company Twitter, for the staggering sum of forty-four billion dollars, which he renamed X — to broadcast his views on an ever-widening range of subjects, many of them far removed from rockets or cars. There was a symbolic weight in the acquisition itself: a man who had built his reputation making physical things of enormous difficulty — rockets, cars, engines — had now spent a fortune to own the megaphone, and the megaphone increasingly owned him. His public statements grew more provocative, more political, and more erratic; he embraced controversial political causes and figures, feuded publicly with critics, spread claims that many found false or inflammatory, and behaved in ways that struck a growing share of observers as reckless, cruel, or unstable. He became deeply enmeshed in politics, and in early 2025 took on a formal role in the American government, leading a newly created effort — the Department of Government Efficiency, established by executive order in January 2025 — to slash federal spending and reshape the government bureaucracy. It was an extraordinary conflation of the interests of a private businessman, whose companies held billions in government contracts, with the exercise of public power, and it raised profound questions about conflicts of interest and the concentration of influence. The venture also proved a cautionary tale in its own right: Musk had promised to cut two trillion dollars from federal spending, departed after roughly four months in the spring of 2025 with a fraction of that claimed and much of it disputed, and afterward conceded that the effort had been only "somewhat successful" and that he would not do it again — a rare admission from a man not given to them, and a telling measure of the gap between the sweep of his pronouncements and the stubbornness of reality. His conduct alienated many, energized others, and made him a lightning rod, his name and his views inescapably entangled with the companies he led.

The pattern of impulsiveness was long-established and well documented, and a few episodes stand in for many. In 2018, in a fit of pique during the rescue of a boys' soccer team trapped in a flooded Thai cave, Musk publicly and baselessly branded one of the British divers who had led the rescue a "pedo guy," a slander that led to a defamation suit and revealed a startling willingness to lash out at anyone who crossed him. That same summer he fired off a tweet claiming he had "funding secured" to take Tesla private at four hundred and twenty dollars a share — a number widely understood as a marijuana joke — when in fact no such funding existed; the Securities and Exchange Commission charged him with fraud, and the settlement forced him to pay twenty million dollars, surrender the chairmanship of Tesla's board, and submit certain of his tweets to oversight by company lawyers, a supervision the press promptly dubbed the "Twitter sitter." These were not the missteps of a man under some singular late-life strain; they were, over more than a decade, a consistent signature — the same impulsiveness, the same disregard for consequence, the same conviction that the ordinary rules constraining public statements did not apply to him. What changed by the 2020s was not the temperament but the stakes, as the companies bound to that temperament grew essential to the security and communications of nations, so that conduct which had once threatened only a share price now brushed against the operations of governments and armies.

This mattered for SpaceX in concrete ways, not merely as a matter of reputation. Musk's control over SpaceX was, by design, nearly absolute; through his ownership and his supervoting shares, he wielded decisive authority over a company that had become critical to national security, to the economy of space, and to the communications of millions, in a way that no ordinary public-company chief executive, answerable to a board and to shareholders, could match. This meant that his judgment, his priorities, and his temperament were not private matters but factors of public and strategic consequence, because a single individual's decisions could shape the behavior of an entity on which so much depended. The Starlink-in-Ukraine episodes had already demonstrated the danger — a private individual, on his own judgment, shaping what a nation's military could and could not do in the middle of a war. But the concern extended further. What if Musk's political entanglements compromised the company's relationships or its reliability? What if his erratic conduct damaged the company, drove away talent, or led to decisions driven by ego or ideology rather than sound judgment? What if the concentration of so much critical capability in a company controlled by so volatile a figure proved, in some crisis, to be a catastrophic vulnerability? These were not idle worries. They were the natural consequence of a situation without precedent: a single, individually controlled private company had become essential to the space capabilities of the world's leading power, and that individual had become one of the most unpredictable public figures alive.

There was a further, subtler twist to the problem, and it was the reverse of the usual relationship between a government and its contractors. Ordinarily a company that depended on federal business for its survival could be disciplined by that dependence — a regulator's frown, the threat of a lost contract, the pressure of public accountability all serving to keep it in line. But SpaceX had become so indispensable that the leverage ran partly the other way. The government could not easily punish, ground, or alienate a company that provided the only Western ride to orbit for its astronauts, launched its spy satellites, and connected its forces in the field, because there was no ready substitute to turn to; and so officials found themselves in the uncomfortable position of needing to accommodate a contractor whose founder publicly insulted the very agencies he served. The normal mechanisms of accountability strained against the reality of dependence, and the result was a peculiar forbearance — a government tolerating from Musk what it would tolerate from no ordinary supplier, precisely because he had made himself too essential to discipline. That dynamic, more than any single outburst, was what made the founder problem structural rather than merely personal.

It is tempting, and common, to treat the founder problem as a matter of Musk's politics — to suppose that those troubled by it are simply troubled by the particular views he came to champion, and that a change of his opinions would dissolve the concern. This is a misreading, and an important one to correct. The problem is structural, not partisan. It would exist in the same form, and pose the same danger, if Musk's politics ran in precisely the opposite direction, because the hazard lies not in the content of one man's opinions but in the fact that so much consequential capability had come to rest on the judgment of one man at all, unchecked by the accountability that governs public institutions and answerable, ultimately, to no one. A commentator who defended Musk's every utterance and a commentator who deplored them ought, if they were thinking clearly, to arrive at the same unease about the concentration itself — for the next founder, or the same one in a different mood, might hold views either admired, and the structure that made his whim consequential would be unchanged.

There was a genuine tension here that honest observers had to hold in mind, and it resists easy resolution. On one hand, Musk's singular vision and drive had produced achievements of enormous value — the reusable rockets, the restoration of American human spaceflight, the connectivity that Starlink brought to millions, the reopening of the frontier. It was precisely his willingness to think differently, to reject conventional wisdom, to pursue a grand mission with fanatical determination, that had made these things possible, and a more conventional, more cautious, more agreeable leader might well have achieved far less. The qualities that made Musk difficult and dangerous were inseparable from the qualities that made him effective; one did not get the reusable rocket without the reckless, driven, uncompromising personality that built it. On the other hand, those same qualities, unchecked by the accountability that governs public institutions and amplified by unprecedented wealth and power, posed real risks — to the company, to the interests that depended on it, and to the public good. The concentration of so much capability under the control of a single, erratic individual was a genuine problem, not a partisan talking point, and it would have been a problem regardless of the particular content of Musk's politics. It was a structural danger, inherent in the situation, and it would persist as long as so much depended on one person's judgment.

There was, to be fair, one significant mitigation, and it deserves acknowledgment. SpaceX was not, in its daily operations, a one-man band; it was run, day to day, by a deep and formidable professional organization, and above all by its long-serving president, Gwynne Shotwell, who had for years managed the company's operations, contracts, and customer relationships with a steadiness that stood in marked contrast to her founder's turbulence. Much of what made SpaceX reliable — its cadence, its execution, its relationships with the government customers on whom it depended — rested on that professional machinery, which continued to function through the storms of Musk's public life. This tempered the founder problem without dissolving it, for the ultimate authority, the final decisions, and the supervoting control still rested with one man, and no lieutenant, however capable, could override him. The mitigation was real; the concentration remained.

It is worth noting how widely the concern was shared, because that breadth is itself evidence that it was something more than partisanship. Officials of rival political persuasions, allied governments, defense strategists, and even some of Musk's own admirers arrived independently at the same disquiet — not because they agreed about his politics, on which they differed sharply, but because they could all see the same structural fact: that too much now hinged on the moods and choices of one unaccountable man. When observers who agree about almost nothing else converge on a single worry, the worry is usually rooted in the situation rather than in the observers, and so it was here. The founder problem was one of the few things about Musk on which his critics and his champions could nearly agree.

How this tension would resolve — whether Musk's leadership would continue to drive SpaceX to greater achievements, or whether his conduct would damage the company or endanger the interests that relied on it, whether the concentration of power in his hands would prove a manageable feature or a catastrophic bug — remained one of the great uncertainties hanging over the new space age. It was possible that Musk would continue to lead SpaceX to triumph after triumph, that his flaws, however glaring, would not undermine the company's mission, and that history would remember him, on balance, as the flawed genius who reopened the space frontier. It was equally possible that his erratic conduct, his political entanglements, or his sheer unpredictability would, at some point, do serious damage — to the company, to its relationships, to the interests that depended on it — and that the concentration of so much power in so volatile a figure would come to be seen as a dangerous mistake. The honest chronicler, writing in mid-2026, could not know which. What could be said with confidence was that the founder problem was real, that it grew more acute as the company grew more essential and the man grew more erratic, and that it represented, alongside the crowded skies and the concentrated power, one of the genuine shadows over the achievements this book has celebrated. And it was against this complicated backdrop — of triumph and dominance, of dependence and danger, of a visionary founder become a lightning rod — that SpaceX, in June 2026, took the step that would open its finances to the world and set the seal on its transformation from a company that could not afford a fourth launch into the most valuable rocket company in the history of the world.

Chapter 65: Tesla, X, and the Musk Empire

SpaceX did not exist in isolation. It was one node in a sprawling constellation of enterprises controlled by a single individual, and to understand SpaceX fully — its resources, its risks, its entanglements — one must understand the empire of which it was part, and the ways in which the fortunes and controversies of Elon Musk's other ventures reached into the rocket company and shaped its story. By the 2020s, Musk controlled or dominated a remarkable array of companies: SpaceX, the rocket and satellite giant; Tesla, the electric-car and energy company that had made him, at times, the richest person on Earth; the social-media platform X, formerly Twitter, which he had bought in 2022 for forty-four billion dollars; Neuralink, a brain-computer-interface venture; the Boring Company, which dug tunnels; and xAI, an artificial-intelligence startup founded in 2023 to compete at the frontier of that most consequential technology. This concentration of enterprises under one person was without real precedent, and the interconnections among them — of money, of attention, of reputation, of risk — were a defining feature of the whole.

The reach of the empire is easier to grasp through its particulars. Neuralink was not a thought experiment: in January 2024 it implanted its first device in a human being, a young man paralyzed from the shoulders down after a diving accident, who learned within weeks to move a computer cursor and play online chess by thought alone — a genuine and startling milestone in the effort to wire the brain directly to a machine, and a demonstration that Musk's ambitions extended past the outward frontiers of space into the inward frontier of the mind. The Boring Company had, with characteristic literalism, built a system of passenger tunnels beneath Las Vegas in which cars ferried convention-goers from hall to hall, a modest realization of a grander promise to dig humanity out of traffic. And xAI, founded in 2023, thrust Musk into the most consequential technological contest of the era, building large artificial-intelligence models and a chatbot, Grok, wired directly into the X platform — so that the same man who launched more than half the world's rockets was also placing a serious bet on the technology that many believed would reshape civilization itself. Woven among these were older entanglements, including Tesla's controversial 2016 absorption of SolarCity, a solar-energy company founded by Musk's cousins and rescued, critics charged, at Tesla shareholders' expense — one more thread binding the ventures together through their common owner.

The scale of the wealth that underwrote all this defied ordinary comparison. At his peak Musk was not merely the richest person on Earth but the possessor of a fortune that, at moments, approached and then surpassed sums no private individual had ever before commanded, larger in nominal terms than the annual economic output of many nations — a concentration of personal resources that freed him, uniquely, from the disciplines that markets and lenders impose on everyone else. A man who can, if he chooses, absorb losses that would bankrupt a large corporation, and who answers to no board with the power to remove him, operates under a different set of constraints than any ordinary executive, and that freedom was the source of both his boldness and the unease he provoked. It allowed him to fund a rocket company through the years when it earned nothing, to buy a social network on a whim, to gamble across half a dozen industries at once; and it meant that the checks which normally temper great power — the fear of ruin, the discipline of investors, the accountability of office — pressed upon him more lightly than upon almost anyone alive.

The most fundamental connection was financial and historical, and it traced back to the origins of Musk's fortune. He had made his first real money in the early internet, selling a software company and then co-founding the venture that became PayPal, from whose sale to eBay in 2002 he emerged with something on the order of a hundred and eighty million dollars — a fortune that he then, with a recklessness that still astonishes, poured almost entirely into two ventures the conventional wisdom deemed near-certain to fail: a rocket company and an electric-car company. SpaceX and Tesla had thus been born together, funded from the same PayPal fortune, and they had very nearly died together in the crisis of 2008. That year is the hinge of the whole story, and it is worth dwelling on, because it shows how close the entire edifice came to never being built. Through 2008 both companies were burning through cash and running out of it; Falcon 1 had failed on its first three launches, each failure a public humiliation and a drain on dwindling reserves; Tesla could not raise the money it needed as the global financial system seized up; and Musk, in the middle of it all, was going through a divorce and was, by his own account, borrowing money from friends to pay his rent. He later called 2008 the worst year of his life. Then, in the closing months, the fortunes turned with almost theatrical timing: on the twenty-eighth of September, Falcon 1 reached orbit on its fourth attempt, becoming the first privately built liquid-fueled rocket ever to do so; Musk split his last thirty million dollars between the two companies rather than let either die, choosing to imperil both rather than sacrifice one; and two days before Christmas, NASA awarded SpaceX a resupply contract worth some one-and-a-half billion dollars, a lifeline that arrived, quite literally, in the final hours. Both companies were saved, within days of each other, at the close of that terrible year — and everything this book has described followed from that narrow escape.

The two companies had grown up entangled, their fates linked through their common owner, and their successes had reinforced each other, building Musk's fortune, his reputation, and his ability to fund ambitious ventures. Here, though, a common misconception is worth correcting. It is sometimes assumed that Tesla's profits bankroll SpaceX, that the car company's soaring value flows directly into the rocket company's coffers — but the two are, in law and in fact, entirely separate enterprises, with separate shareholders, separate finances, and no direct channel of funding between them. Tesla had become the primary source of Musk's personal wealth, its stock value making him extraordinarily rich, and that wealth indirectly underwrote his capacity to raise money, to reassure investors, and to weather crises across his ventures; but the link ran through the man, not through the balance sheets. The health of Tesla mattered to SpaceX not through any direct financial transfer but through their common owner, whose fortune and whose capacity to fund and support his ventures depended significantly on the electric-car company's fortunes. The empire was interconnected, its parts sharing an owner whose wealth and attention and reputation flowed among them, even where their books did not.

The purchase of Twitter in 2022, and its transformation into X, marked a turning point that reached into SpaceX in troubling ways. The acquisition was enormously expensive and enormously controversial, and it consumed a vast amount of Musk's attention and capital and reputation. Under his ownership, the platform became a vehicle for his increasingly provocative and political pronouncements, a megaphone for views that alienated many and energized others, and a source of constant controversy that attached itself to Musk's name and, through him, to all his companies. The reputational entanglement was real: SpaceX, a company dependent on government contracts and public trust, found its reputation increasingly bound up with Musk's conduct on X and in the political arena, its standing affected by the controversies its owner generated far from the domain of rockets. The distraction was real too — the concern that Musk's attention, spread across so many ventures and consumed by the drama of X and of politics, might be diverted from the demanding work of running a rocket company, though SpaceX's strong operational leadership under its president, Gwynne Shotwell, mitigated this risk, the professional organization beneath the founder absorbing much of the shock of his turbulence. And the political entanglement deepened dangerously when Musk, in early 2025, took on a formal role in the American government, leading an effort to cut federal spending — an extraordinary conflation of a private businessman, whose companies held billions in government contracts including with the very agencies he was reshaping, with the exercise of public power, raising acute questions of conflict of interest that touched SpaceX directly, given its deep dependence on government business.

The concentration of so many critical enterprises under one increasingly controversial individual created risks that extended beyond any single company, and it is worth naming just how varied and consequential the holdings were. SpaceX had become essential to national security, to the space economy, to the communications of millions; Tesla was a major force in the automotive and energy industries; X was a significant platform for public discourse; Neuralink reached toward the human brain itself; and xAI aimed squarely at artificial intelligence, perhaps the most transformative and most feared technology of the age. That all of these were controlled by a single person, whose conduct had grown erratic and whose politics had grown divisive, concentrated an extraordinary amount of power and influence in one pair of hands, and it meant that the fortunes and the failures of the whole empire were linked through their common owner. A crisis in one company could reach the others through the shared owner's finances, attention, or reputation; a scandal or a misjudgment by Musk could damage all his ventures at once; the concentration of so much in one person created a systemic fragility, a dependence of many critical enterprises on the judgment and stability of a single individual. This was a novel and unsettling situation, and it made the concerns about the founder, examined elsewhere in this book, not merely a matter of SpaceX's governance but of a whole empire of consequential companies bound together under one controversial and powerful man.

History offers analogies, though none quite fits. The industrial titans of an earlier age — a Rockefeller whose Standard Oil controlled the flow of a nation's fuel, a Carnegie who commanded its steel — concentrated extraordinary economic power, and the republic eventually answered them with antitrust law and the breaking up of their trusts. The aviation pioneer Howard Hughes ran an eccentric empire spanning aircraft, film, and defense contracts, and his later instability became a cautionary tale of what happens when a single erratic man controls enterprises of national importance. One might reach even further back, to the chartered companies like the East India Company that once wielded quasi-sovereign power beyond the reach of ordinary government. But Musk's position had a novelty that none of these precedents captured: he combined the economic concentration of a Rockefeller, the personal volatility of a Hughes, and a direct entanglement in the national security of the world's leading power, all amplified by a personal megaphone that reached hundreds of millions instantly, and all under a control so nearly absolute that no board could check him. The old remedies — antitrust, regulation, the discipline of markets — had been designed for concentrations of a single kind, economic power in a single industry. They were not obviously equal to a concentration that spanned rockets, cars, communications, artificial intelligence, and the machinery of government at once.

There was, to be fair, a case that the empire's interconnections were a source of strength as well as risk, and that the ventures genuinely fed one another. The common ownership allowed for the sharing of talent, technology, and resources among the companies — engineers moving between them, manufacturing lessons learned building cars applied to building rockets and satellites, the artificial-intelligence work feeding the self-driving car and the humanoid robot alike. The ventures reinforced each other's capabilities and Musk's ability to pursue ambitious, long-horizon goals free of the short-term pressures that constrained ordinary companies; and the concentration of vision and control enabled a coherence and a boldness across the empire that more diffuse ownership might not have allowed. Musk's defenders argued that his ability to build and run multiple transformative companies at once was a mark of extraordinary capability, that the empire's achievements — the rockets, the electric cars, the rest — spoke for themselves, and that the concentration that critics feared was the very thing that made such achievements possible. There was force in this, as there was force in the concerns it answered. The empire was, like so much about Musk and his companies, genuinely double-edged: a source of remarkable capability and coherence, and a source of concentration, entanglement, and systemic risk, the two inseparable, the same concentration producing both the strength and the danger.

What is certain is that SpaceX cannot be understood in isolation from the empire of which it is part. Its resources, its risks, its reputation, and its fate are bound up with those of Tesla, X, and Musk's other ventures, linked through their common owner, whose fortune funds them, whose attention divides among them, whose reputation attaches to all of them, and whose conduct shapes their standing. The rocket company that reopened the space frontier was one part of a sprawling and unprecedented concentration of enterprises under a single, powerful, controversial individual, and the concerns and the strengths of that concentration reached into SpaceX as into all the rest. To reckon with SpaceX is thus to reckon with the Musk empire, and with the novel and unsettling reality of so much power and consequence gathered in one person's hands — a reality that the world, in the middle of 2026, had not yet learned how to govern, and that stood as one of the defining and most troubling features of the age that SpaceX had done so much to create.

Chapter 66: Going Public

In June 2026, SpaceX did something it had avoided for the whole of its existence: it went public. For nearly a quarter of a century, SpaceX had remained a privately held company, its shares closely controlled, its finances guarded, funding its enormous ambitions through private investment and its own revenues rather than the public markets. Musk had long resisted taking the company public, wary of the short-term pressures and the disclosure requirements that public markets impose, and famously concerned that public shareholders, focused on quarterly results, would balk at the long-horizon, high-risk investments — above all the staggering cost of the Mars mission — that the company's founding purpose required. He had said as much for years, warning employees that a publicly traded SpaceX would be pressured to abandon the very missions that justified its existence, and pointing to the quarterly torments of Tesla as a cautionary example of what public ownership could do to a founder's freedom. For years, the question of when, or whether, SpaceX would have an initial public offering had been a subject of endless speculation, and the answer had always been: not yet. It had even been widely assumed that if any part of the empire were ever floated, it would be Starlink, spun off on its own as a tidy, cash-generating business, while the riskier rocket company stayed private; in the event, it was the whole of SpaceX that came to market. In June 2026, the answer changed, and SpaceX made its debut on the public markets in the largest initial public offering in history.

The numbers were staggering, and they set the seal on the company's astonishing trajectory. SpaceX priced its shares at a level that valued the company at close to one and three-quarter trillion dollars — approaching two trillion — making it, at a stroke, one of the most valuable companies in the world, and by far the most valuable enterprise ever built in the space or aerospace industry. The offering raised an immense sum, reported on the order of seventy-five billion dollars, the largest amount ever raised in a public offering, and it gave ordinary investors, for the first time, the chance to buy a piece of the company that had transformed spaceflight. The valuation reflected the market's assessment of what SpaceX had become and what it might yet be: a company that dominated global launch, that operated the largest and most valuable satellite constellation in history with Starlink and its ten million subscribers, that held critical contracts with governments and militaries, and that was building, in Starship, a vehicle that might — if it succeeded — transform the economics of space yet again and open the Moon and Mars to human settlement. Investors were betting not just on the company's present dominance but on its future potential, on the possibility that SpaceX would continue to grow, that Starlink would keep expanding, that Starship would eventually work, and that the company would extend its lead into the coming decades.

The public offering also, for the first time, forced a flood of real financial information into the open, replacing years of estimate and speculation with actual figures — and the figures confirmed what shrewd observers had suspected: that the business case now rested less on rockets than on the constellation the rockets had made possible. Starlink, dismissed by many at its inception as a distraction or a boondoggle, had grown into the engine of the whole enterprise, generating well over half of SpaceX's revenue and having tipped from years of heavy losses into genuine profitability as its subscriber base swelled past ten million across more than a hundred and fifty countries. This was the quiet financial truth beneath the spectacle: the reusable rockets had been the means, but the business the world could finally see and value was, above all, a global internet utility beamed from low orbit — a company that launched astronauts and landed boosters, yes, but that paid its way by selling connectivity to farmhouses, ships, aircraft, and armies. The disclosures confirmed that SpaceX had grown into an enormously large and, increasingly, profitable enterprise, and that the wager on Starlink, so widely doubted, had proved the shrewdest bet of all.

Behind the revenue figures lay a human transformation that the accountants could not fully capture. Starlink had brought fast internet to places the world's telecommunications had always skipped — remote villages beyond the reach of any cable, ranches and islands and mountain valleys, ships far out at sea and airliners crossing oceans, scientific stations at the ends of the Earth. In disaster zones where storms had flattened the towers and cut the lines, emergency crews unpacked flat rectangular dishes and were online within minutes. And in Ukraine, under bombardment, the constellation had kept a nation's soldiers and civilians connected when its ground infrastructure lay in ruins, a demonstration of the technology's strategic weight that no marketing could have bought. Ten million subscribers was a business statistic; it was also ten million households and vessels and outposts that had crossed, often for the first time, from disconnection into the networked world. The company that the public could finally buy a share of was, in this human sense, not merely a rocket-maker or a defense contractor but a utility that had rewired the map of who on Earth could reach everyone else — and the IPO, by opening its ownership to ordinary investors, extended a small share of that enterprise to the public whose lives it had begun to change.

The symbolism of the IPO was almost too neat, and it invites a moment's reflection on the arc it completed. A company that, in December 2008, had been days from bankruptcy — its founder splitting his last reserves between SpaceX and Tesla, its survival hanging on a single successful launch and a rescue contract from NASA that arrived two days before Christmas — had, in June 2026, become the most valuable rocket company in the history of the world, worth close to two trillion dollars, its public offering the largest ever. The distance traveled in those eighteen years was almost impossible to comprehend: from a small startup that could not afford a fourth Falcon 1 launch to a colossus that launched more than half the world's payloads and operated the majority of its satellites; from a company the industry had dismissed as a rich man's folly to the dominant force in global spaceflight; from a firm whose rockets kept exploding to one that had landed and reflown its boosters hundreds of times — a single booster reaching a twenty-ninth flight — carried astronauts to orbit, connected millions to the internet from space, and built the largest rocket ever flown. The reusable rocket that everyone had said was impractical had become the foundation of a nearly two-trillion-dollar enterprise and a new space age.

The milestones strung along that arc are worth recalling in sequence, because each had seemed, in its moment, to be the improbable summit, and each turned out to be merely a foothold for the next. The little Falcon 1, launched from a coral speck in the Kwajalein Atoll, failed three times before its fourth flight reached orbit in 2008. In 2012 a SpaceX Dragon became the first commercial spacecraft ever to reach the International Space Station, delivering cargo where only governments had gone before. In December 2015 a Falcon 9 first stage, having lofted its payload, turned around and settled back onto a landing pad in Florida, upright and intact — the maneuver the entire industry had called a stunt and a waste — and in March 2017 SpaceX flew a used booster for the first time, proving that the landing had been not a trick but the opening of a new economics. In May 2020, with the

Demo-2 mission, SpaceX carried two NASA astronauts to orbit, restoring to the United States a capability it had lost when the Space Shuttle retired in 2011 and had been forced, humiliatingly, to rent from Russia in the years since. Each of these had been dismissed, in advance, as beyond a private company's reach; each was accomplished; and the cumulative effect, by the middle of the 2020s, was a company that had normalized one impossibility after another until the impossible had become its ordinary business.

It is worth remembering where the wager began, because the origin is as improbable as the outcome. In 2001, a newly rich Musk, casting about for a way to reignite public enthusiasm for Mars, had traveled to Russia with the half-formed idea of buying refurbished intercontinental ballistic missiles to use as launch vehicles for a philanthropic Mars demonstration. The Russians, by the accounts that later circulated, treated the young American with something between amusement and contempt, and quoted him prices that made no sense. On the long flight home from Moscow, rebuffed and irritated, Musk opened a spreadsheet on his laptop and began working through the actual physics and economics of rocketry from first principles — the cost of the raw materials, the propellant, the fabrication — and reached a conclusion that would reorganize an entire industry: that the materials in a rocket were only a small fraction of its price, that the rest was the accumulated inefficiency of an industry that threw its vehicles away after a single use, and that a company built to attack that inefficiency could cut the cost of reaching space by an order of magnitude. He founded SpaceX the following year. The wager Musk had made on that flight home — that the cost of space could be transformed if only someone were bold enough to try — had paid off beyond any reasonable expectation, and the IPO of June 2026 was, in a sense, the market's final ratification of a calculation first scratched out at thirty thousand feet somewhere over the Atlantic.

The IPO carried a significance beyond SpaceX itself, for it set a valuation on the entire idea of commercial space. A private rocket company worth close to two trillion dollars was a fact that reorganized how investors, governments, and rival entrepreneurs thought about the domain: no longer a money-losing province of national prestige but, demonstrably, one of the great industries of the age, capable of generating fortunes on the scale of the largest technology firms. That validation would ripple outward, drawing capital toward competitors and startups, encouraging other nations to nurture their own commercial sectors, and confirming that the transformation SpaceX had begun was not a single company's fluke but the opening of a genuine new economy. The largest public offering in history was, in this sense, a milestone not only for one firm but for the proposition — long doubted, now priced — that space was a place to build lasting and enormous businesses.

And yet, characteristically, the story was not one of completion but of a threshold, and the IPO marked not an ending but a passage into a new and uncertain phase. For all its dominance and all its wealth, SpaceX had not yet achieved the goal for which it had been founded. Mars remained a promise, not an achievement; no human had set foot there, and the vision of a self-sustaining Martian city remained a dream whose realization, if it ever came, lay decades in the future and depended on solving problems not yet solved. Starship, the vehicle meant to carry humanity to Mars, was still a developmental rocket in 2026, losing boosters and getting grounded by regulators, its critical orbital-refueling capability still unproven, its timelines still uncertain. The return to the Moon, in which Starship was meant to play a central role, had slipped repeatedly and remained years away, with a determined Chinese program racing toward its own lunar deadline. And the shadows over the company's achievements — the crowded and contested skies, the concentration of critical infrastructure and power in a single private firm, the unpredictable and increasingly controversial founder — remained real and, if anything, grew more prominent as the company grew more essential. There was, too, a fresh tension latent in the very act of going public: a company that had justified itself by the willingness to spend fortunes on a Martian dream that might never earn a dividend had now invited millions of shareholders whose patience for such spending was, by the nature of public markets, finite. SpaceX had transformed spaceflight, but it had not finished the work it had set out to do, and the largest questions about its future, and about the future of the new space age it had created, remained open.

That, perhaps, is the truest note on which to leave the story, because it is where the story actually stands. SpaceX had accomplished something genuinely historic — it had rewritten the economics of reaching space, ended a half-century of stagnation, restored American leadership, connected millions, and reopened a race for the future of humanity beyond Earth that most had thought long over. These were real achievements, of enormous consequence, and they had been won against odds that had seemed, at the outset, impossible. But the enterprise remained unfinished, its grandest promises unfulfilled, its future uncertain, its shadows unresolved. The company that had learned to land rockets and fly them again, that had filled the sky with satellites and carried astronauts to orbit, that had become the most valuable rocket company in history, still stood on the threshold of its ultimate ambitions, the Moon and Mars still ahead, still promises rather than accomplishments. Whether it would fulfill them — whether Starship would work, whether humans would return to the Moon and reach Mars, whether the new space age would deliver on its promise or falter — no one, in the middle of 2026, could yet say. The story of SpaceX and the new space race, so far, is a story of the impossible made real and the unfinished still pending, of triumph shadowed by risk, of a frontier reopened and not yet crossed. It is, in the deepest sense, a story still being written, and the most important chapters may be the ones not yet begun.

Epilogue: The View from Here

Stand, for a moment, at the edge of the present and look at what has changed. A little over two decades ago, in 2002, reaching space meant what it had meant for half a century: an enormously expensive, government-run enterprise conducted a few times a year, in which each rocket was built at ruinous cost and thrown away after a single flight, and in which the price of putting a kilogram into orbit had barely moved since the 1960s. The Moon had not been visited by human beings since 1972 and showed no sign of being visited again. The United States, the world's leading space power, was a decade away from losing even the ability to launch its own astronauts. The sky held a few thousand satellites, and the idea that a private company might land rockets and fly them again, or fill the heavens with ten thousand spacecraft, or carry ordinary citizens to orbit, belonged to science fiction. That was the world, and it had been the world, in its essentials, for a very long time.

Now look again. Rockets land themselves, standing upright on their tails, and fly again and again — one booster has flown twenty-nine times. A single private company launches more than half of all the mass humanity puts into orbit, more than the entire rest of the world combined. The sky holds more than ten thousand satellites belonging to that one company alone, beaming internet to ten million subscribers on every continent and to a war in Ukraine. Astronauts fly to orbit again from American soil, on a private spacecraft, and private citizens follow them — to the space station, to record altitudes, into the vacuum on the first commercial spacewalk. The largest rocket ever built rises on a column of fire from the Texas coast, and its booster is caught out of the sky by the arms of a tower. Two great powers race once more for the Moon, and the language of the space race, silent for fifty years, is spoken again. And the company that began all of it, once days from bankruptcy, is worth close to two trillion dollars. The transformation is real, and it happened faster than almost anyone imagined possible.

It is worth being precise about what caused it, because the temptation is to attribute so sweeping a change to a single genius or a single stroke of insight, and the truth is both simpler and more instructive. The hinge on which the new space age turned was reusability — the decision to stop throwing rockets away, to recover and refly the expensive hardware, and thereby to break the cost barrier that had gated everything above the atmosphere. Everything else followed from that: the cheap, frequent launch that made mega-constellations economical; the Starlink business that funded the larger ambitions; the human spaceflight, the Moon contracts, the Mars vision. And reusability, in turn, succeeded not because of a single invention but because of a way of working — iterative, vertically integrated, tolerant of failure, willing to build and test and blow up and build again, in public, until the impossible became routine. That method, as much as any rocket, is SpaceX's real legacy, and it has begun to spread, reshaping how an entire industry approaches the building of hard things.

But the honest observer, surveying all this achievement, must also hold the shadows in view, because they are as real as the triumphs and they will shape what comes next. The same reusability that opened space crowded the sky, streaking the astronomers' images and raising the specter of an orbital environment choked with debris. The same dominance that restored American leadership concentrated critical infrastructure and power in a single private company, controlled by a single volatile individual, creating dependence and vulnerability where there had been plurality. The same drive that built the rockets imposed real costs — on the workers who burned out, on the fragile coast at Boca Chica, on the communities beneath the falling debris. And the grandest promises remained unkept: Mars a dream, the Moon still ahead, Starship still failing, the future still pending. The new space age is not a finished triumph but an unfinished experiment, its benefits and its costs still accumulating, its ultimate meaning not yet determined.

What, then, is the view from here, from the vantage of the middle of 2026? It is a view of a frontier reopened but not yet crossed, of a transformation real but incomplete, of a story whose most important chapters may be the ones not yet written. Whether humanity will return to the Moon and reach Mars, whether the cost revolution will carry space travel from the wealthy few to the many, whether the concentration of orbital power will be governed wisely or abused, whether the crowded skies will be managed or despoiled, whether the founder's flaws will prove manageable or catastrophic — all of this remains open, to be decided by choices not yet made and events not yet unfolded. The one thing that can be said with confidence is that the world has changed, that reaching space is no longer what it was, and that the change was wrought, against enormous odds, by a company that learned to do the one thing the whole industry had ruled out: it landed the rocket, and flew it again. The rest of the story — the Moon, Mars, the fate of the new space age — belongs to the future, and to those who will write the chapters we cannot yet read.

Timeline of SpaceX and the New Space Race

Dates for developmental milestones are approximate where the record is ambiguous, and a few early founding dates are disputed by weeks. Figures for 2025–2026 are the most recent and fastest-moving, and are given as of mid-2026.

The first space age - **1926** — Robert Goddard launches the first liquid-fuel rocket, a spindly device that rises a few dozen feet over a Massachusetts field. - **1942** — Germany's V-2 becomes the first human-made object to cross the boundary of space. - **1957** — The Soviet Union launches *Sputnik* (October 4), the first artificial satellite; the original space race begins. - **1958** — The United States orbits *Explorer 1* (January) and creates NASA (October) in response. - **1961** — Yuri Gagarin becomes the first human in orbit (April 12); weeks later President Kennedy commits America to landing a man on the Moon before the decade is out. - **1962** — At Rice University, Kennedy tells the nation "we choose to go to the Moon," framing the effort as a test of national will. - **1965** — Alexei Leonov makes the first spacewalk, tethered outside his Voskhod capsule. - **1967** — The Outer Space Treaty enters into force, declaring space the province of all humankind and barring national appropriation of the Moon and planets. - **1969** — Apollo 11 lands the first humans on the Moon (July 20), launching from Kennedy Space Center's Pad 39A. - **1971** — Elon Musk is born in Pretoria, South Africa (June 28). - **1972** — Apollo 17 leaves the Moon (December); no human returns to the lunar surface for over half a century. - **1981** — The first Space Shuttle, *Columbia,* flies (April); its promised low-cost "reusability" is undermined from the start by costly refurbishment. - **1986** — *Challenger* is lost with its seven crew (January), a wound to national confidence in the Shuttle. - **1993** — The Delta Clipper (DC-X), an early reusable prototype, makes its first vertical take-off and landing. - **1998** — Assembly of the International Space Station begins in orbit. - **2003** — *Columbia* breaks apart on reentry with its seven crew (February 1), grounding the Shuttle fleet and hastening its retirement.

From code to rockets - **1989** — At seventeen, Musk leaves South Africa for Canada, later transferring to the University of Pennsylvania to study physics and economics. - **1995** — Musk drops out of a Stanford graduate program and, with his brother Kimbal, founds the internet firm Zip2. - **1999** — Compaq buys Zip2 for roughly $300 million; Musk founds the online-banking venture X.com. - **2000** — X.com merges with Confinity to form what becomes PayPal; Jeff Bezos founds Blue Origin. - **2001** — Musk, active in the Mars Society, conceives "Mars Oasis," a plan to land a small greenhouse on Mars; he travels to Moscow to buy surplus missiles. - **2002 (early)** — A second Moscow trip fails; on the flight home Musk calculates that he could build rockets himself for a fraction of the price. - **2002 (May)** — SpaceX is founded; eBay buys PayPal for $1.5 billion (October), giving Musk the fortune he stakes on the company. Early hires include Tom Mueller and, later that year, Gwynne Shotwell. - **2003** — SpaceX unveils the Falcon 1 and begins engine testing at its McGregor, Texas, site. - **2004** — Musk testifies to the U.S. Senate in favor of commercial spaceflight; SpaceX wins its first military launch contract. - **2005** — SpaceX establishes a launch site on Omelek Island at the Kwajalein Atoll in the Pacific.

Falcon 1 and the near-death years - **2006 (March 24)** — Falcon 1 Flight 1 fails seconds after liftoff, felled by a corroded fuel-line nut. - **2006 (August)** — NASA's Commercial Orbital Transportation Services (COTS) program awards SpaceX a development partnership, a crucial early lifeline. - **2007 (March 21)** — Falcon 1 Flight 2 reaches space but is lost to propellant slosh before orbit. - **2008 (August 3)** — Flight 3 fails when the stages collide at separation; the loss nearly ends the company. - **2008 (September 28)** — **Flight 4 reaches orbit**, the first privately developed liquid-fuel rocket to do so. - **2008 (December 23)** — NASA awards SpaceX a ~$1.6 billion Commercial Resupply Services (CRS) cargo contract, the deal that saves the company. - **2009 (July 14)** — Falcon 1's fifth and final flight launches Malaysia's RazakSAT; the little rocket is retired in favor of the Falcon 9.

The workhorse and the first landings - **2010 (April)** — President Obama tours Cape Canaveral and throws his support behind commercial spaceflight, over the objections of parts of Congress. - **2010 (June 4)** — Falcon 9 debuts successfully from Cape Canaveral. - **2011 (July)** — The Space Shuttle flies for the last time; the United States loses the ability to launch its own astronauts, beginning a nine-year dependence on Russian Soyuz rockets. - **2010 (December 8)** — A Dragon capsule orbits Earth and is recovered, a first for a commercial spacecraft. - **2012 (May 25)** — Dragon berths with the International Space Station, the first commercial craft to do so. - **2012 (October)** — CRS-1, the first operational cargo run, begins routine resupply of the station. - **2012 (September)** — The Grasshopper test vehicle makes its first short hop at McGregor, the earliest step toward reusable rockets. - **2012–2013** — Grasshopper demonstrates progressively higher vertical takeoffs and landings. - **2013** — Falcon 9 v1.1 debuts, longer and more powerful, and reaches geostationary transfer orbit (SES-8); Grasshopper makes its highest hop, rising some 744 meters before setting back down. - **2014** — A Falcon 9 first stage performs a controlled soft touchdown on the ocean surface, proving the descent profile; NASA awards Commercial Crew contracts to SpaceX and Boeing (September). - **2015 (January)** — Google and Fidelity invest roughly $1 billion in SpaceX, in part to fund Starlink; a drone-ship landing attempt ends in a crash. - **2015 (April)** — A second drone-ship attempt tips over and explodes on the deck. - **2015 (June 28)** — The CRS-7 Falcon 9 breaks apart in flight, destroyed by a failed strut in the upper stage. - **2015 (December 21)** — **The first booster landing**, upright at Landing Zone 1, after launching Orbcomm satellites — the night the rocket came home. - **2016 (April 8)** — **First drone-ship landing** at sea, on *Of Course I Still Love You*, during the CRS-8 mission. - **2016 (September 1)** — The AMOS-6 Falcon 9 explodes on the pad during fueling, traced to a burst helium vessel in super-chilled oxygen. - **2016 (September)** — At the International Astronautical Congress, Musk unveils the Interplanetary Transport System, the grandiose first sketch of what becomes Starship. - **2017 (March 30)** — **First reflight of a used booster**, launching SES-10 and landing again — reuse proven. - **2017 (June)** — SpaceX reflies a Dragon cargo capsule for the first time, extending reuse to spacecraft as well as boosters. - **2017 (September)** — Musk presents a scaled-down "BFR," the design that will carry the program to Boca Chica. - **2017** — SpaceX flies a record eighteen times in the year and begins routine booster recoveries.

Falcon Heavy, Starlink, and human spaceflight - **2017 (January)** — SpaceX returns to flight after the AMOS-6 explosion, launching the first Iridium NEXT batch and landing the booster at sea. - **2018 (February 6)** — **Falcon Heavy** flies for the first time, sending Musk's Tesla Roadster and its mannequin "Starman" toward Mars orbit; the two side boosters land in synchrony. - **2018 (May 11)** — The Block 5 Falcon 9, engineered for rapid reuse, debuts with Bangabandhu-1. - **2018 (November)** — The BFR is renamed Starship and Super Heavy; late in the year Musk decides to build it of stainless steel rather than carbon fiber. - **2018 (December)** — SpaceX launches its first national-security payload, a GPS III satellite for the U.S. military, breaking into a market long closed to it. - **2019 (March)** — Crew Dragon's uncrewed Demo-1 docks with the ISS and returns safely. - **2019 (April)** — A Crew Dragon capsule is destroyed during a ground test of its abort thrusters, delaying the first crewed flight. - **2019 (May 23)** — The first sixty operational Starlink satellites launch on a single Falcon 9. - **2019 (August)** — The squat Starhopper prototype makes a tethered and then untethered hop at Starbase. - **2019 (September)** — Musk unveils the first full-scale Starship prototype, Mk1, before a crowd at Boca Chica. - **2019** — First deep-cryo Raptor engine tests; the Starship program moves to Boca Chica, Texas. - **2020 (January)** — Crew Dragon's in-flight abort test demonstrates the launch-escape system. - **2020 (May 30)** — **Demo-2** carries astronauts Bob Behnken and Doug Hurley to orbit, ending the nine-year U.S. crew gap and marking the first crewed orbital launch by a private company. - **2020 (August 2)** — Behnken and Hurley splash down in the Atlantic, completing Demo-2 and the first crewed water landing by an American spacecraft since Apollo. - **2020 (October)** — Starlink opens a public beta, wryly branded "Better Than Nothing," to its first users. - **2020 (November 15)** — Crew-1 begins operational astronaut rotations aboard Crew Dragon. - **2020 (August–September)** — The tank-like SN5 and SN6 prototypes make 150-meter hops on a single Raptor, the first flights of Starship-class hardware. - **2020 (December 9)** — Starship prototype SN8 flies to altitude, performs its "belly-flop" descent, and explodes on landing. - **2021 (April 16)** — NASA selects Starship as the Human Landing System for Artemis, choosing SpaceX to return astronauts to the lunar surface. - **2021 (May 5)** — **SN15 lands intact**, the first Starship prototype to survive. - **2021 (May)** — China lands its Zhurong rover on Mars, becoming the second nation to operate a rover there. - **2021 (July)** — Richard Branson (July 11) and Jeff Bezos (July 20) reach the edge of space on their own suborbital craft, opening a summer of billionaire spaceflight. - **2021 (September 15–18)** — **Inspiration4**, the first all-civilian orbital mission, commanded and funded by Jared Isaacman. - **2021 (April–November)** — Crew Dragon settles into routine six-month crew rotations with Crew-2 and Crew-3. - **2021** — Starlink passes 100,000 subscribers; SpaceX's valuation climbs past $100 billion. - **2022 (February 24)** — Russia invades Ukraine; within days Starlink terminals become vital to Ukraine's defense and communications. - **2022 (April)** — Axiom's Ax-1 flies the first fully private crew to the ISS aboard Crew Dragon. - **2022 (May)** — Rocket Lab attempts to catch an Electron booster in mid-air with a helicopter, probing its own route to reuse. - **2022 (October)** — Musk completes his roughly $44 billion purchase of Twitter, later renamed X, entangling SpaceX's founder in a fresh and lasting controversy. - **2022 (November 16)** — NASA's Artemis I sends an uncrewed Orion around the Moon. - **2022 (late)** — China completes assembly of its Tiangong space station, giving it a permanent crewed outpost in orbit. - **2022 (December)** — Starlink passes roughly one million subscribers.

Starship and the new race - **2023 (February)** — A full Super Heavy booster fires thirty-one of its engines in a static test, the most powerful rocket ever ignited to that point. - **2023 (April 20)** — **Starship Flight 1** clears the pad, loses engines, tumbles, and is destroyed, cratering the launch mount and scattering debris. - **2023 (August)** — India's Chandrayaan-3 makes the first soft landing near the lunar south pole, widening the field of spacefaring nations. - **2023 (November 18)** — Flight 2 achieves hot-stage separation before both stages are lost. - **2023** — SpaceX flies ninety-plus orbital missions in the year, more than the rest of the world combined. - **2024 (January)** — Axiom's Ax-3 flies the first all-European commercial crew to the ISS aboard Crew Dragon. - **2024 (February 22)** — Intuitive Machines' Odysseus makes the first commercial soft landing on the Moon, launched by a Falcon 9. - **2024 (March 14)** — Flight 3 reaches near-orbital velocity and coasts halfway around the world before breaking up on reentry. - **2024 (June 5)** — Boeing's Starliner Crew Flight Test launches Butch Wilmore and Suni Williams to the ISS; thruster failures soon leave the capsule unable to safely bring them home. - **2024 (June 6)** — Flight 4's booster makes a controlled splashdown and the upper stage survives a searing reentry to a soft Indian Ocean landing. - **2024 (June)** — China's Chang'e 6 returns the first samples from the Moon's far side. - **2024 (September 10–15)** — **Polaris Dawn** reaches the highest Earth orbit since Apollo and stages the first commercial spacewalk, led by Isaacman with SpaceX engineer Sarah Gillis. - **2024 (October 13)** — **First "chopsticks" catch**: the launch tower's arms seize the returning Super Heavy booster from the air on Flight 5. - **2024 (November 19)** — Flight 6 flies with President-elect Trump watching; the ship makes a daylight splashdown. - **2024** — Starlink surpasses four million subscribers, begins Direct-to-Cell service to ordinary phones, and launches more satellites in the year than all other operators combined. - **2025 (January 16)** — **Blue Origin's New Glenn** reaches orbit on its first flight, though its booster is lost on the return; the same month, Starship Flight 7 (the troubled second-generation V2 vehicle) is lost. - **2025 (March 6)** — Starship Flight 8 fails again, showering debris over the Caribbean and disrupting air traffic. - **2025 (March 18)** — Boeing's Starliner crew, Butch Wilmore and Suni Williams, finally return aboard a Crew Dragon after roughly nine months, their eight-day test flight having stretched into a saga. - **2025 (May 3)** — Boca Chica votes to incorporate as **Starbase, Texas**, a company town at the end of the Texas coast. - **2025 (May 27)** — Flight 9 reflies a booster but loses both stages, closing a punishing run of failures. - **2025 (late)** — Flights 10 and 11 finally succeed, restoring the program's momentum. - **2025 (October)** — A single Falcon 9 booster flies for the thirty-first time, extending a reuse record no other rocket approaches. - **2025 (November)** — New Glenn's second flight lands its booster and launches NASA's ESCAPADE Mars probes. - **2025** — SpaceX flies **165 orbital launches**, roughly 51 percent of the world's total; consolidated revenue reaches about $18 billion, most of it from Starlink.

The present - **2026 (February)** — Starlink passes **10 million subscribers.** - **2026 (March)** — Starlink passes **10,000 active satellites**, more than half of all working spacecraft in orbit. - **2026 (April)** — New Glenn suffers a partial failure, deepening Blue Origin's difficult year. - **2026 (May 22)** — **Starship Flight 12** — the first V3 vehicle, the first with Raptor 3 engines, and the first to fly from Pad 2 — succeeds with its upper stage but loses its booster, and is grounded by the FAA pending investigation. - **2026 (May 28)** — A New Glenn static-fire test explodes and destroys its launch pad, a severe blow to Blue Origin. - **2026 (June)** — A single Falcon 9 booster flies for the thirty-fifth time as cumulative Falcon 9 flights pass six hundred and fifty, records no other rocket approaches. - **2026 (June)** — SpaceX targets the first orbital-propellant-transfer demonstration between two Starships. - **2026 (June)** — **SpaceX's initial public offering** — around $135 per share, roughly $75 billion raised, at a valuation near $1.8 trillion: the largest IPO in history. - **2026 (mid-year)** — NASA's Artemis III is reframed as a 2027 Earth-orbit demonstration, with the first crewed lunar landing slipping to a later mission; China holds to its goal of landing astronauts on the Moon before 2030.

Who's Who

The principal figures in the story of SpaceX and the new space race, grouped by the part they played. The list is selective; hundreds of engineers whose names never reached the public did the work these people are remembered for.

Forebears

Konstantin Tsiolkovsky — The self-taught Russian schoolteacher who, at the turn of the twentieth century, worked out the mathematics of rocket flight, including the equation that still governs how much a rocket must weigh in fuel to reach orbit. He grasped that a rocket could work in the vacuum of space, and dreamed of humanity spreading beyond Earth.

Robert Goddard — The American physicist who launched the first liquid-fuel rocket in 1926 and was mocked by his contemporaries for suggesting a rocket could reach the Moon. Nearly everything Falcon and Starship do rests on principles he first demonstrated.

Wernher von Braun — The German-American engineer who designed Nazi Germany's V-2 missile and, after the war, led the development of the Saturn V that carried Apollo to the Moon. His towering rocket remains the benchmark against which Starship is measured.

Sergei Korolev — The Soviet Union's secretive "Chief Designer," who built the rocket that launched Sputnik and Gagarin. His work opened the space age and set the pace of the original race.

At SpaceX

Elon Musk — Founder, chief executive, and chief engineer of SpaceX; also the founder of Tesla and, from 2022, owner of the social-media platform X. Made his fortune from Zip2 and PayPal and staked it on rockets and electric cars. The visionary, financier, and driving force behind SpaceX, and, by the 2020s, one of the world's richest and most controversial public figures. His singular control over the company is both its great strength and, as his conduct grew erratic, one of the new space age's genuine risks.

Gwynne Shotwell — President and Chief Operating Officer; an early hire (2002) who rose to run the company day to day, negotiating its most important contracts and providing the operational steadiness that complemented Musk's ambition. Widely regarded as indispensable to SpaceX's success and the person most often named as the one who keeps the enterprise on its feet.

Tom Mueller — Founding propulsion engineer and architect of the Merlin engine, recruited from the amateur rocketry world where he had built one of the largest amateur liquid-fuel engines. His engines powered every Falcon flight; he later left to found his own space company.

Jim Cantrell — Aerospace consultant and Russia specialist who advised Musk on the early Moscow trips and helped assemble the founding team as SpaceX's first head of business development. He soon parted ways with Musk and, years later, started his own launch company.

Chris Thompson — One of the very first employees, a former Boeing and McDonnell Douglas structures engineer and U.S. Marine who led early vehicle production. He built much of the Falcon 1 hardware in the company's spartan first years.

Tim Buzza — Early vice president of launch and testing, a veteran of Boeing's Delta program who ran the grueling test campaigns at McGregor and the launch operations on Kwajalein. He was central to getting Falcon 1 off the ground.

Hans Koenigsmann — Early chief of avionics and mission assurance, a steadying technical presence through the Falcon 1 years and beyond, and the man who often stood before the cameras to explain what had gone wrong.

Steve Davis — An early engineer who joined in the company's first year and worked across a remarkable range of problems; he later led Musk's tunneling venture, The Boring Company, and became one of his most trusted lieutenants.

Mark Juncosa — Vice president of vehicle engineering and one of Musk's closest collaborators, known for his frenetic energy; he drove the iterative development of Falcon reuse and then of Starship at Starbase.

Bill Riley — Senior Starship engineer who helped lead the build-test-fail-repeat campaign at Boca Chica, bringing an outsider's discipline from the world of motor racing to rocket production.

Bill Gerstenmaier — For decades NASA's most respected human-spaceflight official, who ran the agency's crewed programs before being reassigned in 2019; he then joined SpaceX, where his caution and experience shaped its approach to flight safety and crew missions.

Kiko Dontchev — Vice president of launch, who rose from an internship to oversee the relentless Falcon 9 cadence that came to define SpaceX's operations.

Jonathan Hofeller — Longtime commercial executive who led Starlink and launch sales, turning the constellation into the revenue engine that funds the company's deeper ambitions.

Sarah Gillis — SpaceX lead space-operations engineer who trained the company's astronauts and then flew on the Polaris Dawn mission (2024), performing one of the first commercial spacewalks and playing the violin in orbit.

Anna Menon — SpaceX lead space-operations engineer and mission director who flew as medical officer on Polaris Dawn, reading a children's book she co-wrote to viewers on Earth from orbit.

Astronauts and private flyers

Bob Behnken and Doug Hurley — NASA astronauts who flew Demo-2 (2020), the first crewed Dragon mission, ending the U.S. human-spaceflight gap. Both were veteran Shuttle fliers; their safe return validated a decade of commercial-crew effort.

Jared Isaacman — Billionaire entrepreneur, jet pilot, and private astronaut; commanded and funded Inspiration4 (2021), the first all-civilian orbital mission, and Polaris Dawn (2024), which reached record altitude and staged the first commercial spacewalk. Later nominated and appointed to lead NASA — a remarkable arc from private flyer to head of the space agency.

Scott "Kidd" Poteet — Retired U.S. Air Force fighter pilot and close associate of Isaacman, who served as mission director for Inspiration4 and piloted Polaris Dawn.

Sian Proctor, Hayley Arceneaux, and Chris Sembroski — The three citizen crewmates of Inspiration4: Proctor, a geoscientist who became the first Black woman to pilot a spacecraft; Arceneaux, a childhood-cancer survivor and physician assistant who flew as the mission's medical officer; and Sembroski, an aerospace worker who won his seat through a fundraising campaign.

Butch Wilmore and Suni Williams — NASA astronauts who flew Boeing's Starliner Crew Flight Test (2024); when Starliner proved unsafe to return them, their eight-day visit to the ISS stretched to roughly nine months, and they came home on a Crew Dragon in 2025 — the story that fixed the phrase "stranded astronauts" in the public mind, though both bristled at the word.

Yusaku Maezawa — Japanese entrepreneur and art collector who visited the ISS aboard a Soyuz and booked an early Starship flight around the Moon, the "dearMoon" project, before canceling it as the schedule slipped.

Peggy Whitson — The most experienced American astronaut, who after her NASA career commanded a private Axiom mission to the ISS aboard Crew Dragon, embodying the blurring line between government and commercial spaceflight.

NASA and government

Dan Goldin — The long-serving NASA Administrator of the 1990s whose "faster, better, cheaper" philosophy and frustration with traditional contractors helped prepare the ground for commercial spaceflight.

Mike Griffin — NASA Administrator who backed the COTS program (2006), opening the door to commercial cargo and giving the fledgling SpaceX its first great opportunity.

Lori Garver — NASA Deputy Administrator (2009–2013) and political champion of commercial crew and cargo, who fought Washington's entrenched interests to fund the approach that produced Dragon and Crew Dragon; she later chronicled the struggle in a candid memoir.

Charlie Bolden — Former astronaut and NASA Administrator (2009–2017) who oversaw the retirement of the Shuttle and the early years of Commercial Crew, defending the partnership with SpaceX against congressional skeptics.

Jim Bridenstine — NASA Administrator (2018–2021), a former congressman who set the Artemis program in motion and presided over the first crewed Dragon flight.

Bill Nelson — Former senator and Shuttle passenger who became NASA Administrator (2021–2025), once a skeptic of commercial crew who came to preside over its triumphs and over Artemis.

Kathy Lueders — The NASA engineer who managed the Commercial Crew Program and then led all of the agency's human spaceflight, steering Dragon to operational service and Artemis's landers; after leaving NASA she joined SpaceX as general manager of Starbase.

Rivals and the wider field

Jeff Bezos — Founder of Amazon and of Blue Origin (2000); Musk's great rival in private spaceflight, pursuing a vision of millions living and working in space via the Moon and orbital habitats. Blue Origin's New Shepard and New Glenn rockets and Blue Moon lunar lander are its principal vehicles. Patient where Musk is impatient, he took two decades to reach orbit.

Dave Limp — Chief executive of Blue Origin from 2023, recruited from Amazon to instill urgency; he drove New Glenn to flight and steered the company through the setbacks of 2026.

Peter Beck — Founder and chief executive of Rocket Lab, maker of the small Electron launcher and the reusable Neutron; a self-taught New Zealand engineer whose company is among the most credible of those following SpaceX's path.

Tory Bruno — Chief executive of United Launch Alliance, maker of the Vulcan Centaur rocket, the traditional American launch establishment's measured answer to SpaceX.

Richard Branson — Founder of Virgin Galactic, pioneer of suborbital space tourism via a rocket-powered spaceplane; he flew to the edge of space himself in 2021, days ahead of Bezos.

China's space leadership — The China National Space Administration and the state program racing to land astronauts on the Moon before 2030 with the Long March 10 rocket, the Mengzhou crew spacecraft, and the Lanyue lander, while operating the Tiangong space station and returning samples from the Moon and Mars.

Glossary

Ablative heat shield — A heat shield that protects a spacecraft during reentry by slowly burning and eroding away, carrying heat off with the vaporized material. Used on capsules; contrasted with the reusable tile and steel shields SpaceX sought.

AMOS-6 — The communications satellite and the Falcon 9 destroyed in a fueling explosion on the launch pad in September 2016, traced to a subtle interaction between super-chilled oxygen and a helium bottle.

Apogee and perigee — The high and low points of an elliptical orbit: apogee is the farthest distance from Earth, perigee the closest. Firing an engine at one changes the altitude of the other.

Artemis — NASA's program to return humans to the Moon and establish a sustained presence, using the Space Launch System rocket, the Orion capsule, and commercial Human Landing System landers (Starship HLS and Blue Moon). Named for the Greek goddess of the Moon, twin of Apollo.

Artemis Accords — A set of principles for cooperative and peaceful lunar exploration, led by the United States and signed by dozens of nations; China and Russia have declined to join, pursuing their own framework.

Block 5 — The final, reuse-optimized version of the Falcon

9 (from 2018), engineered so that boosters could fly many times with minimal refurbishment.

Blue Moon — Blue Origin's lunar lander, selected by NASA as a second Human Landing System provider alongside SpaceX's Starship HLS.

Boostback burn — An engine firing by a returning first stage that reverses its downrange motion and steers it back toward the launch site.

Cadence — The frequency of launches; a measure of a launch system's operational maturity. SpaceX achieved a cadence no one else could match.

Chopsticks (Mechazilla) — The pair of mechanical arms on the Starship launch tower that catch the descending Super Heavy booster in mid-air, first demonstrated in October 2024.

Commercial Crew Program — NASA's program (awards 2014) that funded SpaceX's Crew Dragon and Boeing's Starliner to carry astronauts, using fixed-price service contracts rather than government ownership.

COPV (composite overwrapped pressure vessel) — A light, strong tank of carbon fiber wrapped over a metal liner, used to store high-pressure helium inside a rocket's propellant tanks. A COPV's failure caused the CRS-7 loss in 2015 and was implicated in the AMOS-6 explosion.

COTS / CRS — Commercial Orbital Transportation Services and Commercial Resupply Services: NASA's programs that funded and then bought commercial cargo delivery to the space station, the deals that saved and sustained SpaceX.

Crew Dragon (Dragon 2) — SpaceX's crewed capsule, which carries up to four astronauts, docks autonomously, and has a built-in launch-escape system of SuperDraco engines. A cargo version resupplies the space station.

Cryogenic propellant — Propellant kept as a liquid at extremely low temperatures, such as liquid oxygen and liquid methane; efficient but difficult to store and transfer, especially in orbit.

Delta-v — Literally "change in velocity," the measure of how much a rocket can alter its motion; the currency of spaceflight. Every maneuver, from reaching orbit to landing on Mars, has a delta-v price that determines how much propellant is required.

Deorbit — To slow a spacecraft deliberately so that it falls out of orbit and reenters the atmosphere, either to land or to burn up. Low satellites like Starlink are designed to deorbit at the end of their lives.

Direct-to-Cell — Starlink satellites that communicate directly with ordinary mobile phones, acting as cell towers in the sky to eliminate coverage dead zones.

Downrange — The horizontal distance a rocket travels along its flight path away from the launch site; a booster flying too far downrange to return to land is recovered on a drone ship at sea.

Draco and SuperDraco — SpaceX's small maneuvering thrusters (Draco) and the powerful abort engines (SuperDraco) built into Crew Dragon, which could pull the capsule clear of a failing rocket in an emergency.

Drone ship (Autonomous Spaceport Drone Ship) — A sea-going platform that holds position in the ocean so that a first stage flying too far downrange to return to land can touch down on its deck. SpaceX's are named for ships in Iain M. Banks's novels.

Entry burn — A firing of a returning booster's engines high in the descent to slow it before it hits the thickening atmosphere, sparing the vehicle from destructive heating and stress.

Escape velocity — The speed an object must reach to break free of Earth's gravity entirely and travel to the Moon or beyond, about 11 kilometers per second — considerably faster than orbital velocity.

Expendable launch vehicle — A rocket used once and discarded, the industry standard before SpaceX made reuse practical.

Fairing — The protective nose cone that shields a satellite payload during the climb through the atmosphere, then splits open and is jettisoned. SpaceX recovers and reuses its fairing halves, which are surprisingly costly.

Falcon 1 / Falcon 9 / Falcon Heavy — SpaceX's small early rocket; its workhorse medium-lift rocket with nine Merlin engines and a reusable first stage; and its heavy-lift rocket made of three Falcon 9 cores.

Flight-proven — SpaceX's term for a rocket that has already flown and landed, framing reuse as a demonstrated virtue rather than a risk.

Full-flow staged combustion — A highly efficient and difficult rocket-engine cycle, used by the Raptor engine, in which both fuel and oxidizer are burned in preburners and fed to the main chamber as hot gas. Raptor is among the first such engines ever flown.

Gas-generator cycle — A simpler, less efficient engine cycle, used by the Merlin, in which a small amount of propellant is burned to drive the turbopumps and the exhaust is dumped overboard rather than fed to the main chamber.

Gateway — A small space station NASA and its partners plan to place in orbit around the Moon as a staging point for Artemis surface missions.

Geostationary orbit (GEO) — A high orbit about 36,000 kilometers up where a satellite circles the Earth once a day and so appears to hang fixed over one spot on the equator; the home of many communications and weather satellites.

Geostationary transfer orbit (GTO) — The elongated orbit into which a rocket drops a satellite bound for geostationary orbit, leaving the spacecraft to raise itself the rest of the way.

Grid fins — Lattice-like aerodynamic surfaces that deploy from a returning booster to steer it through the atmosphere toward its landing target.

Hot-staging — Igniting the upper stage's engines while it is still attached to the booster, the exhaust venting to push the stages apart; used on Starship.

Hoverslam (suicide burn) — A landing technique in which the engine lights at the last possible moment and the booster arrives at the ground exactly as its velocity reaches zero, because a Falcon 9's engines are too powerful for it to hover. There is no margin for error, hence the nickname.

Human Landing System (HLS) — NASA's program for the Artemis lunar landers; the SpaceX version is Starship HLS.

Inclination — The angle at which an orbit is tilted relative to the equator, which determines the range of latitudes a satellite passes over; reaching a high inclination costs extra energy.

In-situ resource utilization (ISRU) — Making use of local materials in space rather than carrying everything from Earth, such as manufacturing methane and oxygen propellant on Mars from its atmosphere and water ice.

Inter-satellite laser links (ISL) — Laser beams that carry data directly between satellites in orbit, letting a constellation like Starlink route traffic across the sky without touching a ground station.

ITAR — The International Traffic in Arms Regulations, U.S. rules that treat rocket and satellite technology as munitions and tightly restrict sharing it with foreign nationals; a constant constraint on the space industry.

Kessler syndrome — A hypothetical cascade in which collisions between orbital objects generate debris that causes further collisions, potentially rendering an orbit unusable; a central worry of the mega-constellation era.

Landing burn — The final engine firing of a returning booster, slowing it to a gentle touchdown on land or on a drone ship.

Latency — The delay in a communications signal; low for low-orbit satellites like Starlink, high for distant geostationary ones, which is why Starlink's low orbit matters.

Liquid oxygen (LOX) — The oxidizer burned in most large rockets, including all of SpaceX's; kept cryogenically cold, it lets the fuel combust in the vacuum of space where there is no air.

Low Earth orbit (LEO) — The region a few hundred kilometers up where the space station, Starlink, and most human spaceflight operate.

Mega-constellation — A network of hundreds or thousands of satellites, such as Starlink, made economical by cheap reusable launch.

Merlin — SpaceX's kerosene-and-oxygen engine, nine of which power the Falcon 9 first stage.

Methalox — The methane-and-liquid-oxygen propellant combination burned by the Raptor engine; methane can, in principle, be manufactured on Mars.

New Glenn / New Shepard — Blue Origin's orbital rocket (named for John Glenn) and its suborbital tourist rocket (named for Alan Shepard).

Orbital refueling (propellant transfer) — Transferring propellant between spacecraft in orbit; the unproven technology on which Starship's Moon and Mars missions depend.

Orbital velocity — The sideways speed an object must maintain to keep falling around the Earth without hitting it, roughly 28,000 kilometers per hour in low orbit. Reaching orbit is chiefly about going fast enough, not high enough.

Orion — NASA's crew capsule for the Artemis program, carried to lunar orbit by the Space Launch System.

Outer Space Treaty — The 1967 agreement that forms the foundation of space law, declaring space free for all to explore, barring nations from claiming the Moon or planets, and prohibiting nuclear weapons in orbit.

Overview effect — The shift in perspective many astronauts report on seeing the whole Earth from space: a sudden sense of the planet's fragility, its borderlessness, and the unity of life upon it.

Payload — The useful cargo a rocket carries to space — a satellite, a spacecraft, or supplies — as distinct from the rocket itself; a launcher's capability is measured by the payload mass it can lift to a given orbit.

Rapid reusability — The goal of turning a recovered rocket around and reflying it quickly, with minimal work, the key to truly cheap spaceflight.

Raptor — SpaceX's methane-fueled, full-flow staged-combustion engine for Starship; more than thirty power the Super Heavy booster.

Reentry — The fiery return of a spacecraft into the atmosphere, when friction with the air heats its surface to thousands of degrees and must be survived by a heat shield.

Refurbishment — The inspection and repair a used rocket requires before flying again; the Space Shuttle's was so extensive that reuse saved little, while SpaceX drove Falcon 9's down toward routine.

Reusability — The practice of recovering and reflying rocket hardware rather than discarding it after one use, the central idea on which SpaceX's economics and the new space age rest.

RP-1 (kerolox) — Refined kerosene, the fuel burned with liquid oxygen ("kerolox") in the Falcon's Merlin engines;

dense and easy to handle, but prone to leaving residue that complicates reuse.

RUD (rapid unscheduled disassembly) — SpaceX's wry engineering euphemism for an explosion, reflecting a culture that treats spectacular failures as data rather than disgrace.

Sabatier reaction — A chemical process that combines carbon dioxide and hydrogen to make methane and water; proposed as the way to manufacture Starship's propellant on Mars, using the planet's atmosphere and ice.

Specific impulse (Isp) — A measure of a rocket engine's fuel efficiency, akin to miles per gallon: how much thrust it produces per unit of propellant burned per second. Higher is better, and vacuum-optimized engines score higher than sea-level ones.

Staged combustion — An efficient engine cycle in which the propellant used to drive the turbopumps is then fed into the main chamber and burned, wasting nothing; harder to build than the gas-generator cycle, and the basis for Raptor's full-flow variant.

Staging — Discarding a spent lower stage during ascent so the rocket does not have to keep hauling empty tanks and dead engines, dramatically improving efficiency; almost every orbital rocket uses at least two stages.

Starbase — SpaceX's build-and-launch site at Boca Chica, on the South Texas coast; incorporated as a city in 2025.

Starlink — SpaceX's low-orbit satellite-internet constellation, the largest in history and the company's main revenue engine; its defense variant is **Starshield**.

Starship / Super Heavy — The fully reusable, stainless-steel super-heavy launch system: the Starship upper stage (and spacecraft) and the Super Heavy booster; the largest and most powerful rocket ever built.

Static fire — A ground test in which a rocket's engines are ignited while the vehicle is held down, to check the propulsion system before flight; a static-fire mishap destroyed AMOS-6.

Suborbital — A flight that reaches space but not orbital velocity, going up and coming straight back down; the domain of New Shepard and Virgin Galactic.

Thrust-to-weight ratio (TWR) — The ratio of an engine's thrust to the weight of the vehicle; it must exceed one for a rocket to lift off at all, and shapes how a booster accelerates and lands.

Tiangong — China's crewed space station, set to become the only permanent orbital outpost once the International Space Station is retired.

Upper stage (second stage) — The part of a rocket that fires after the booster separates, carrying the payload the rest of the way to orbital velocity; on Falcon 9 it is expended, while Starship aims to recover it too.

Vertical takeoff, vertical landing (VTVL) — The propulsive-landing technique, proven by SpaceX, in which a rocket descends and lands upright on its engines.

$/kg to orbit — Cost per kilogram to reach orbit, the headline metric that SpaceX drove down by roughly an order of magnitude, the change on which the whole new space age rests.

Appendix I: SpaceX in Numbers

Numbers do not tell this story on their own — the courage and the near-bankruptcies and the long nights are not quantities — but the numbers are astonishing enough to be worth pausing over, and some of them are so far outside the range of anything that came before that they need a little context to be believed. What follows is a tour of the figures that define SpaceX as of mid-2026. Where a number is an estimate or is disputed, it is flagged as such.

The workhorse and its cadence. The Falcon 9 first flew in June 2010. For its first several years it launched a handful of times a year, as any new rocket does. Then the curve bent upward. In 2025 SpaceX conducted about 165 orbital launches — on the order of one every fifty-three hours, sustained across the whole year — which amounted to roughly 51 percent of all the orbital launches conducted by every country and company on Earth combined. Put another way, a single private firm was flying to orbit more often than the rest of the world put together, more often in a year than the entire global industry managed in most years of the twentieth century. By 2026 the cumulative Falcon 9 flight count had passed six hundred, a total no other orbital rocket in history approaches.

Landing and reflight. The number that matters most is not how many times SpaceX has launched but how many times it has landed. The first booster came home upright in December 2015. Since then the company has recovered its first stages hundreds of times, running long streaks of consecutive successful landings, until the act of a seventy-meter rocket stage settling back onto a pad or a ship became so routine that broadcasters stopped treating it as news. The fleet leader — a single booster designated B1067 — has flown more than thirty times, each flight followed by a landing and a refurbishment measured in days or weeks rather than the months the Space Shuttle required. When SpaceX first began reusing boosters it hoped for perhaps ten flights each; it has now flown individual stages three times that, and the practical ceiling is still unknown.

The cost that changed everything. The single figure at the heart of the whole story is the cost of putting a kilogram into orbit. In the Space Shuttle era, that cost worked out to roughly $54,000 per kilogram to low Earth orbit — a number so large that it quietly foreclosed most of what people imagined doing in space. A reusable Falcon 9 brought the published price to somewhere between about $1,500 and $3,000 per kilogram, depending on how one counts, roughly a ninety-fold reduction. The list price of a Falcon 9 launch settled near $67 million; SpaceX has suggested that the internal cost of reflying a booster it already owns is far lower, on the order of $15 million a flight, though that figure is a company estimate and not independently audited. Whatever the precise number, the order of magnitude is not in dispute, and it is the hinge on which everything else — Starlink, mega-constellations, the return to the Moon — turns.

Falcon by the pound. The Falcon 9 stands about seventy meters tall and lifts off on nine Merlin engines producing around 1.7 million pounds of thrust. It can place roughly 22.8 tonnes into low orbit when flown expendably, or about 17 tonnes while reserving enough propellant to fly its booster home. The Falcon Heavy, three of those cores strapped together, ignites twenty-seven engines and some five million pounds of thrust at liftoff, enough to loft on the order of 64 tonnes to low orbit — making it, since the retirement of the Saturn V, the most powerful rocket in operational service, though not the most powerful ever built.

The scale of Starship. Starship is a different order of thing. Stacked, the vehicle stands about 121 metres tall — taller than the Saturn V — and at liftoff its Super Heavy booster fires thirty-three Raptor engines producing roughly 74 meganewtons, about 16.7 million pounds of thrust: close to twice the thrust of the Saturn V and the most powerful rocket ever flown. The booster alone holds about 3,400 tonnes of super-cooled methane and liquid oxygen; the full stack carries well over four thousand tonnes of propellant, most of the vehicle's mass being fuel. The upper stage adds six more Raptors. The engine itself is a marvel of the figures: the Raptor 2 produced about 230 tonnes of force, and the Raptor 3 that debuted on Flight 12 targets roughly 280 tonnes, at chamber pressures higher than any operational engine before it. Starship's design goal is to carry a hundred tonnes or more to orbit and, uniquely, to bring both stages home — the number that, if achieved at scale, could drive cost per kilogram down by another order of magnitude toward figures that today sound like fantasy.

Starlink. The constellation is where the rockets turn into revenue. As of early 2026 Starlink served more than ten million subscribers across scores of countries, up from a single million just over three years earlier, and by March 2026 the network had passed ten thousand satellites in orbit — around nine thousand of them active, which by itself accounted for more than half of all the working spacecraft circling the Earth. The system generated on the order of $11 billion in revenue in 2025, a figure that represented the majority of SpaceX's roughly $18 billion in consolidated revenue and, for the first time, gave the company a large and growing profit center to set against the enormous cost of developing Starship. It is a striking inversion: the rockets that were the point of the company have become, in a sense, the delivery trucks for the satellites that pay for them.

The company itself. SpaceX employs on the order of thirteen to fourteen thousand people, a workforce that would have been unthinkably large for a launch startup and is still lean by the standards of the aerospace primes it has displaced. Its valuation traces one of the steepest curves in the history of private enterprise: a company worth a few hundred million dollars in its precarious early years, about $100 billion by 2021, several hundred billion by 2024, and, on the eve of its public offering, a figure on a wholly different scale.

The offering. In June 2026 SpaceX went public in the largest initial public offering in history. The shares priced at around $135, the company raised on the order of $75 billion, and the market set a valuation near $1.8 trillion — a number that placed a rocket-and-internet company, founded to make humanity a multi-planet species, among the most valuable enterprises on Earth. It is worth holding two of these figures side by side to feel the arc of the story: a company that in 2008 had one working rocket and enough money for a single further launch, and that in 2026 was valued at nearly two trillion dollars. Between those two numbers lies everything this book has tried to describe.

A closing caution about all of this. The launch and hardware figures are firm; the financial figures, particularly the internal per-launch costs and the exact valuation, come substantially from the company and from a fast-moving market, and should be read as the best available estimates rather than audited certainties. The habit of treating SpaceX's own numbers as gospel is one this book has tried to avoid. But even discounted, even hedged, the quantities remain the most dramatic in the modern history of flight.

Quick reference (approximate, mid-2026):

- **Falcon 9:** ~70 m tall; 9 Merlin engines; ~1.7 million lbf at liftoff; ~22.8 t to LEO expendable, ~17 t reusable.

- **Falcon Heavy:** 27 engines; ~5 million lbf; ~64 t to LEO; most powerful rocket in operational service.
- **Starship (stacked):** ~121 m tall; 33 + 6 Raptor engines; ~74 MN (16.7 million lbf) at liftoff; 100+ t target payload.
- **Cost per kilogram to LEO:** Shuttle era ~$54,000; reusable Falcon 9 ~$1,500–3,000 (roughly a 90-fold fall).
- **2025 launches:** ~165 orbital flights, ~51% of the world's total.
- **Starlink (early 2026):** 10 million-plus subscribers; 10,000-plus satellites; ~$11 billion in 2025 revenue.
- **Workforce:** ~13,000–14,000 employees.
- **June 2026 IPO:** ~$135 per share; ~$75 billion raised; valuation ~$1.8 trillion — the largest IPO in history.

Appendix II: How the Rockets Work

Almost everything most of us think we know about rockets is a little wrong, because the intuitive picture — a rocket goes up, and space is up, so a rocket goes to space by going up — misses the single most important fact about spaceflight. This appendix is an attempt to explain, without equations, how the machines in this book actually do what they do: how a rocket reaches orbit, why landing one again is so hard, what is happening inside the engines, and why methane, of all things, might be the fuel that gets us to Mars.

Orbit is about going sideways. Start with the fact that undoes the intuition. Getting to space is easy; staying there is the hard part. Space, in the sense of the near-vacuum where satellites fly, begins only about a hundred kilometers up — a distance you could drive in an hour if the road went straight up. A powerful enough firework can reach it. But reaching that altitude and *staying* there are entirely different problems, because an object that simply goes straight up and runs out of push falls straight back down. To stay in space, you have to enter orbit, and orbit is not about height. It is about speed — sideways speed.

The old thought experiment, due to Isaac Newton, is still the clearest. Imagine a cannon on a very tall mountain, firing a ball horizontally. Fire it gently and the ball arcs down and lands a mile away. Fire it harder and it lands further off, the curve of its fall stretched out. Now fire it so hard that the ball's downward curve exactly matches the curve of the Earth falling away beneath it. The ball keeps falling — gravity never switches off — but the ground keeps curving out from under it at the same rate, so it never gets any closer, and it circles the planet forever. That is an orbit: a state of perpetual falling, missing the ground because you are moving sideways fast enough that the ground keeps sliding away. The speed required is enormous, about 28,000 kilometers per hour, roughly twenty-five times the speed of sound. A rocket's real job is not to lift its payload high but to hurl it sideways that fast. This is why a rocket rises vertically for only a short time before pitching over and spending most of its flight tilting toward the horizontal, chasing speed rather than altitude.

Why rockets come in stages. Reaching that speed runs into a cruel piece of arithmetic. A rocket has to carry the fuel it will burn, and that fuel is heavy, and carrying it requires burning more fuel to accelerate its weight, which requires still more fuel — a spiral that means a rocket on the pad is almost entirely propellant, with the payload a tiny fraction at the top. One trick tames the spiral: throwing away the parts you have finished with. A first stage, or booster, does the heavy lifting off the pad, burns until its tanks are nearly empty, and then separates and drops away, so the second stage does not have to drag the dead weight of empty tanks and spent engines the rest of the way to orbital speed. This is *staging*, and nearly every orbital rocket uses it. It works, but it is also the origin of the industry's central wastefulness: for sixty years, those enormous, expensive boosters were simply discarded into the ocean after a few minutes of use, as if an airline threw away the airplane after every flight.

Why reusing them is so hard. If the answer is obviously to bring the booster back and fly it again, one might wonder why it took until 2015 for anyone to do it. The difficulty is that a returning booster is not a gently gliding airplane. It is an empty metal tube the size of a fifteen-storey building, falling out of the sky, having just been through the violence of launch, and it must be slowed from several times the speed of sound and set down softly on a precise spot — sometimes a floating platform in the ocean — using nothing but its own engines, with no second chance and no room to hover. Everything about it fights the attempt. The booster is falling fast enough to burn from air friction; it is aerodynamically unstable, wanting to tumble; it has very little fuel left, so every gram of margin matters; and its rocket engines, built to heave a fully fueled stack off the ground, are far too powerful to let a nearly empty booster simply hover down. Solving all of this at once, reliably, is the achievement on which SpaceX's whole story rests.

How a booster lands. The sequence, once you know it, is a kind of choreography. Minutes after liftoff, the booster shuts down and separates from the upper stage, which carries on to orbit. The booster, now coasting high above the atmosphere, flips itself around using small thrusters so its engines point in the direction of travel. If it is returning to the launch site, it performs a *boostback burn,* firing its engines to reverse its downrange motion and aim itself back toward home; if it is landing on a drone ship far downrange, it skips this and simply falls toward the waiting platform. As it plunges back into the thickening air, it lights some of its engines again in an *entry burn,* slowing down so the atmosphere does not tear it apart or cook it. Then, in the last stretch, grid fins — lattice-like panels near the top — unfold and tilt to steer it like a falling dart toward the target. Finally comes the *landing burn,* and here is the hardest trick of all. Because the engines are too strong to hover, the booster cannot ease down; instead it lights its engine at the last possible second and decelerates so sharply that it arrives at the pad exactly as its downward speed reaches zero, legs deploying just in time. Engineers call it the *hoverslam,* or, less reassuringly, the *suicide burn.* Light the engine a moment too early and the booster stops in mid-air and climbs again; a moment too late and it hits the ground still moving. There is no margin. That a computer does this, repeatedly, on a barge in the ocean, is one of the quiet marvels of the age.

What is happening inside the engine. All of this depends on engines, and engines differ chiefly in how they feed themselves. A rocket engine must pump propellant into its combustion chamber at tremendous pressure, and it uses spinning turbopumps to do so, and those pumps must themselves be driven by burning some propellant. The question of what to do with the exhaust from that pump-driving burn defines the engine's "cycle." In the simplest arrangement, the *gas-generator cycle* used by SpaceX's Merlin, a little propellant is burned in a small side-chamber to spin the pumps, and that exhaust is then simply dumped overboard — you can see it as a darker trail beside the main flame. It is reliable and relatively easy, but it throws away energy. A more efficient approach, *staged combustion*, routes that pump-driving exhaust into the main chamber to be burned again, wasting nothing, at the price of far greater complexity and heat. The Raptor engine that powers Starship uses the most demanding version of all, *full-flow staged combustion*, in which both the fuel and the oxidizer are separately pre-burned to drive their own pumps and then fed to the main chamber as hot gas. It is the most efficient practical cycle known, notoriously hard to build, and had scarcely ever flown before Raptor. Think of it as the difference between an engine that lets some of its energy escape and one that captures nearly all of it — at the cost of being a fiendishly difficult machine to keep from destroying itself.

Why methane, and why it matters for Mars. SpaceX's earlier engines burn refined kerosene, which is dense and convenient. Raptor burns methane, and the choice is deliberate and points all the way to Mars. Methane and oxygen burn cleanly, without the sooty residue that fouls kerosene engines and makes rapid reuse harder. More important, methane is a fuel you could, in principle, *make on Mars*. Mars has a carbon-dioxide atmosphere and buried water ice, and a well-known chemical process — the Sabatier reaction — combines carbon dioxide and hydrogen to yield methane and water. A ship that runs on methane could, one day, land on Mars, manufacture its own return propellant from the planet itself, and fly home, rather than hauling every drop of return fuel across the solar system. The fuel choice is not an engineering footnote; it is the plan.

How Starlink beams the internet down. The satellites in this story do something conceptually simpler but no less clever. Traditional communications satellites sit in a very high orbit, so high that they hang fixed over one spot on Earth — convenient, but so distant that the signal takes a noticeable fraction of a second to make the round trip, the lag that makes satellite internet feel sluggish. Starlink instead uses thousands of small satellites flying low, only a few hundred kilometers up, where the signal has a much shorter distance to travel and the lag nearly vanishes. The catch is that a low satellite races across the sky in minutes, so no single one can serve you for long; you need a whole swarm, thousands strong, so that as one drops below the horizon another is already rising to take over, and a dish on the ground can hand off seamlessly between them. The satellites also talk to one another directly through laser links, passing your data across the sky from satellite to satellite before dropping it to a ground station near its destination — a network in the sky. It only becomes affordable because launching thousands of satellites is only affordable if launch is cheap, which loops the whole enterprise back to the reusable rocket.

Refueling in orbit. One last piece, still unproven as this book closes, holds the deep-space future together. A Starship can reach orbit, but once there it has burned nearly all its propellant and cannot go on to the Moon or Mars. The plan is to refill it in space: launch a Starship to orbit, then send up a series of tanker Starships that rendezvous with it and transfer propellant, topping off its tanks so the now-full ship can depart for its destination with a fresh load. Transferring thousands of tonnes of cryogenic, super-cold liquid between two spacecraft in weightlessness — where fuel does not obligingly settle to the bottom of a tank — is something no one has done at scale, and the first demonstrations were only being attempted as this story ended. If it works, it turns Starship from a rocket that can reach orbit into a ship that can go anywhere. If it does not, the grandest parts of the plan wait. The whole architecture of the coming decade rests on this one difficult trick, which is a fitting note on which to end: the rockets we can already watch land are the easy part, and the future depends on the maneuvers we have not yet seen.

Appendix III: Myths and Misconceptions Corrected

Few subjects generate as much confident misinformation as spaceflight, and SpaceX, sitting at the intersection of rockets, celebrity, and the internet, attracts more than its share. Some of the beliefs below are honest confusions about difficult physics; others are half-remembered headlines that hardened into folklore. Each is stated as it is commonly held, and then corrected against the record.

"SpaceX was the first to land a rocket." Not quite. Blue Origin's New Shepard booster made a controlled vertical landing on 23 November 2015, several weeks before a Falcon 9 first stage landed on 21 December. But the two feats are not the same feat, and the difference is the whole point. New Shepard is suborbital: it goes up to the edge of space and falls straight back, reaching a speed of perhaps three times the speed of sound. The Falcon 9 booster is part of a machine flinging a payload sideways fast enough to orbit, and it separates travelling far faster and higher, then has to fly back across a much greater distance and shed vastly more energy. The energy involved in reaching orbit is on the order of a hundred times that of a suborbital hop. Blue Origin landed a rocket first; SpaceX landed an orbital-class rocket first, which is the harder and more consequential thing.

"The Space Shuttle proved that reusable rockets don't work." The Shuttle proved something narrower: that *that particular* reusable design did not save money. It was partly reusable, but refurbishing it between flights — above all inspecting and replacing thousands of fragile heat-shield tiles and overhauling the main engines — was so labor-intensive that each flight cost around a billion and a half dollars, more than many expendable rockets. The lesson was not that reuse is impossible but that reuse only pays if the turnaround is quick and cheap. SpaceX took exactly that lesson to heart, and its whole approach is a rebuke to the idea that the Shuttle settled the question.

"Falcon Heavy is the most powerful rocket ever built." It is not, and never was. NASA's Saturn V, which launched Apollo to the Moon, produced roughly a third more thrust than Falcon Heavy and could lift far more. Starship, in turn, produces more than twice the Saturn V's thrust and now holds the record for the most powerful rocket ever flown. The accurate claim, true for several years, is that Falcon Heavy was the most powerful rocket in *operational service* — a real distinction, but a narrower one than the myth.

"Elon Musk invented the reusable rocket." He did not invent the idea, which is decades older than SpaceX. NASA and McDonnell Douglas flew the DC-X, a vertical-takeoff-and-landing prototype, in the early 1990s; engineers had sketched reusable boosters since the dawn of the space age. What SpaceX did was make it *work at orbital scale and economically* — carrying it from Grasshopper's tethered hops through years of crashed boosters to routine landings. The achievement is enormous, but it is one of execution and persistence, not of original invention.

"Rockets are expensive because they're made of exotic, costly materials." Overwhelmingly, no. The aluminum, steel, and propellant in a rocket are a small fraction of its price; a Falcon 9's propellant load costs only a few hundred thousand dollars. Rockets have been expensive chiefly because the entire vehicle was thrown away after a single use — the equivalent of scrapping a jetliner after one flight and charging every passenger accordingly. SpaceX's insight was that the cost lay in the discarding, not the materials, and that recovering the hardware could collapse the price. Starship's use of ordinary stainless steel, cheaper than aerospace-grade carbon fiber, underlines the point.

"Starlink is permanently filling space with junk." The satellites are deliberately designed to do the opposite. They fly low enough that atmospheric drag pulls them down within a few years if they fail or are retired, and they burn up completely on reentry, leaving nothing behind. This is a genuine contrast with debris in higher orbits, which can linger for centuries. That said, the concern is not baseless: the sheer number of satellites raises real worries about collision risk and about the streaks they leave across astronomers' images, and those are legitimate criticisms. But the specific claim that Starlink is creating permanent orbital junk gets the physics backward.

"Musk switched off Starlink to sabotage a Ukrainian attack on Crimea." The widely repeated version has Musk turning off active service in the middle of an operation. The documented account is different: Starlink coverage was never enabled over Crimea in the first place, and when asked to extend it to allow a Ukrainian drone-boat strike on the Russian fleet, Musk declined to switch that coverage on, fearing entanglement in a major escalation. The distinction between refusing to activate a capability and shutting down a live one matters, both factually and morally, even as the episode raises entirely fair questions about a private citizen holding such power.

"Getting to space is mostly about going high enough." This is the most common misconception of all, and the most important to correct. Space begins only about a hundred kilometers up, a distance a car could cover in an hour. The hard part is not the altitude but the *speed*: to stay in orbit rather than fall back, a spacecraft must travel sideways at some 28,000 kilometers per hour. Orbit is a matter of velocity, not height, which is why rockets spend most of their flight tilting toward the horizontal rather than climbing.

"The rocket that lands has been to space and back." For a Falcon 9, the stage you see land is only the first stage, and it does not reach orbit. It falls away a couple of minutes into flight, having done the initial heavy lifting, and returns from the upper atmosphere. The second stage, which actually carries the payload all the way to orbital velocity, keeps going and is not recovered. Starship aims to change this by bringing both stages home, but for the Falcon that famous landing is of the booster alone.

"The Starliner astronauts were stranded in space." Butch Wilmore and Suni Williams were never stranded in the sense of having no way home. Once Boeing's Starliner was judged unsafe to carry them back, they became part of the regular station crew, with a Crew Dragon docked and available to return them in an emergency at any time. Their eight-day test flight did stretch to some nine months, an awkward and unwanted delay, but "stranded" implies a danger and a helplessness that did not exist — a point both astronauts made repeatedly themselves.

"NASA owns SpaceX," or "SpaceX is just a government contractor spending taxpayer money." SpaceX is a private company. NASA has been a crucial early customer and funder, but through fixed-price contracts to *buy services* — cargo and crew delivered to the station — rather than by owning the vehicles or the company, a deliberate break from the old model in which the government paid for and owned everything. Much of SpaceX's later growth, above all Starlink, is funded by private capital and commercial revenue, not federal appropriations.

"Reused rockets must be more dangerous than new ones." The record suggests the opposite can be true. A booster that has already flown and been inspected is, in a sense, a known quantity, its weaknesses revealed rather than hidden; airlines have long relied on the same logic. Flight-proven Falcon 9 boosters have carried astronauts and compiled a strong reliability record, and SpaceX argues, plausibly, that flying hardware repeatedly *improves* its understanding of how that hardware fails.

"Elon Musk's Tesla Roadster is orbiting the Earth" (or **"heading for the asteroid belt"**). The Roadster launched on the first Falcon Heavy in 2018 is not orbiting Earth at all; it is in an elliptical orbit around the Sun, looping out past the orbit of Mars and back in toward Earth's. It does not reach the asteroid belt, contrary to a persistent claim, and it poses no danger to anyone. It is, most likely, still out there, slowly circling the Sun.

"Astronauts float because there is no gravity in space." There is nearly as much gravity a few hundred kilometers up as there is on the ground — that gravity is precisely what holds the station in orbit. Astronauts float because they, and their spacecraft, are in continuous free fall, endlessly falling around the Earth together, so nothing pushes them against the floor. Weightlessness is the sensation of falling, not the absence of gravity.

"All those Starship explosions mean the program is failing." By the standards of traditional rocket development, a string of vehicles lost in flight would indeed spell disaster. SpaceX works differently, building cheap prototypes and flying them deliberately to the edge of failure to learn what breaks, accepting spectacular losses as the price of fast progress. The dramatic fireballs are, by design, part of the method; the meaningful question is not whether early Starships blow up but whether each failure teaches something the next one fixes.

"China and the other spacefaring nations are hopelessly behind and don't really matter." This is complacent and increasingly wrong. China operates its own crewed space station, has returned samples from both the Moon and Mars, and holds to a credible goal of landing astronauts on the Moon before 2030 with hardware it is actively flying. Others — India, Europe, Japan, and a crowd of new companies — are advancing quickly. SpaceX's dominance is real, but it is not the whole of spaceflight, and treating the rest of the world as a footnote misreads the race this book describes.

Appendix IV: Questions and Answers

Over the years, certain questions come up again and again — from readers, at dinner tables, in the comments beneath every launch video. Here are the ones asked most often, answered as plainly as the facts allow.

Why does SpaceX land some rockets on ships at sea instead of bringing them back to land? It comes down to fuel and physics. Flying a booster all the way back to its launch site costs propellant — the stage has to cancel its downrange momentum and fly home — and that propellant is subtracted from the payload the rocket could otherwise carry. For heavier payloads, there simply is not enough margin to return to land, so SpaceX instead stations a drone ship far downrange, directly beneath where the booster is already heading, and lets it land there with much less fuel spent fighting its way back. The choice between a land landing and a sea landing is essentially a trade between recovery convenience and how much the rocket can lift.

How much does a Falcon 9 launch actually cost? The advertised price to a commercial customer has hovered around $67 million for a standard mission, though the figure varies with the payload and the orbit. That is the price SpaceX charges; what it costs the company internally to fly a booster it already owns is lower — Musk has suggested something on the order of $15 million per flight, though that is a company estimate rather than an audited number. Either way, it is a fraction of what comparable rockets cost, and the gap is the reason SpaceX came to dominate the launch market.

Is Starlink actually profitable? By the middle of the 2020s, yes. After years of heavy investment, Starlink crossed into profitability and became the largest single source of SpaceX's revenue — on the order of $11 billion in 2025 — with a growing operating profit. That matters enormously to the wider story, because Starlink's cash is what allows SpaceX to pour money into Starship without depending on outside investors or government contracts for its most ambitious work.

When will humans actually walk on Mars? No one honestly knows, and anyone who gives you a confident date should be treated with caution — including, at times, Musk himself, whose predictions have consistently proved optimistic. The honest answer as this book closes is that the enabling pieces are still unproven: Starship must fly reliably, orbital refueling must be demonstrated, life support for a months-long voyage must be built, and much else besides. A crewed Mars landing in the 2030s is possible if everything goes well; delays into the 2040s or beyond are entirely plausible. The direction is real; the timeline is not yet trustworthy.

Is space tourism safe? It is far safer than the early days of spaceflight, but it is not safe in the way flying on an airliner is safe, and it would be misleading to pretend otherwise. Suborbital hops of the kind Blue Origin and Virgin Galactic sell are the gentlest form; orbital flights like those SpaceX flies are more demanding and carry the real, if small, risks inherent in riding a rocket. Crew Dragon has an excellent record and a launch-escape system designed to save its passengers if the rocket fails, but the honest framing is that human spaceflight remains an inherently hazardous endeavor that engineering has made acceptably, not perfectly, safe.

Who owns the Moon? No one, by law. The Outer Space Treaty of 1967, which nearly every spacefaring nation has signed, forbids any country from claiming the Moon or any celestial body as its territory. What the treaty does *not* clearly settle is whether a company or country may extract and use resources it finds there — mining ice or minerals — and that ambiguity is the subject of active dispute, with the U.S.-led Artemis Accords and a rival Chinese-Russian framework offering competing visions. You cannot buy a plot of the Moon; whether you can one day sell what you dig out of it is genuinely unresolved.

What happens to the International Space Station? After more than a quarter-century in orbit, the ISS is nearing the end of its life and is planned to be retired around the end of the decade, deliberately steered down to burn up over a remote stretch of the Pacific — a task for which SpaceX was contracted to build a special deorbit vehicle. The hope is that commercial space stations will replace it, keeping a continuous human presence in low orbit, though those successors are still being developed. China's Tiangong station, meanwhile, would become the only permanent crewed outpost in the interim.

How is SpaceX different from NASA? They are different kinds of thing that increasingly work as partners. NASA is a government agency that sets goals, funds research, and — in the modern model — buys services; SpaceX is a private company that builds and operates rockets and spacecraft, sometimes for NASA, sometimes for commercial customers, sometimes for itself. The crucial shift this book describes is that NASA stopped owning and operating its own rockets for tasks like cargo and crew delivery, and instead became a customer buying rides from companies like SpaceX, freeing itself to focus on exploration further afield.

Why is Starship built of stainless steel, of all things? Because, counterintuitively, steel turned out to be the better material for a rapidly reusable Mars ship. It is far cheaper than the aerospace-grade carbon fiber SpaceX first planned to use; it holds its strength at the cryogenic temperatures of the propellant and, crucially, at the searing temperatures of reentry, where carbon fiber weakens; and it is easy to weld and repair in an open field rather than a pristine factory. The shiny, almost retro-futuristic look of Starship is a direct consequence of a hard-nosed engineering decision.

What is the "overview effect"? It is the name given to a psychological shift many astronauts report on seeing the entire Earth from space: the planet appears as a small, fragile, borderless sphere hanging in blackness, and the sight tends to produce a powerful sense of its preciousness and of the artificiality of the divisions upon it. It has been described by fliers across nationalities and eras, and it is one of the reasons some believe that sending more people to space could subtly change how humanity sees itself.

Why does SpaceX deliberately fly Starships until they blow up? Because it is faster and, oddly, cheaper to learn by breaking things than by trying to perfect them on paper. Rather than spend years analyzing a design to near-certainty before a single flight, SpaceX builds relatively inexpensive prototypes and flies them hard, treating each failure as a source of data about what actually breaks under real conditions. The fireballs look like disasters but are, within this philosophy, expected steps, and the approach has repeatedly turned early chaos into rapid progress — though it demands a tolerance for public failure that traditional programs never had.

Can I invest in SpaceX now that it has gone public? With the June 2026 offering, shares in SpaceX became available to public investors for the first time, so in principle yes. It is worth understanding, though, that buying the stock means buying the whole enterprise — launch, Starlink, and the long, uncertain Mars ambition together — and that the valuation the market has placed on it is extraordinary and rests heavily on expectations about the future. This book offers no investment advice beyond a general caution about any asset priced for perfection.

How do astronauts get to the space station now? For American and partner-nation crews, chiefly aboard SpaceX's Crew Dragon, which since 2020 has been the workhorse of Western human spaceflight, ferrying astronauts up in crews of four for months-long stays. Russia continues to fly its Soyuz, and the two systems have at times carried each other's crews for redundancy. Boeing's Starliner was meant to be a second American option but has been beset by problems.

Will humans really return to the Moon soon, and who will get there first? NASA's Artemis program aims to land astronauts near the lunar south pole using a version of Starship as the lander, but the schedule has slipped repeatedly, and as this book closes the first crewed landing had been pushed past its original targets. China, meanwhile, holds firmly to a goal of putting its own astronauts on the Moon before 2030. Who steps onto the surface first is, genuinely, an open question — the first real contest of a second Moon race.

Is all of this ruining the night sky? It is a real and growing concern rather than a settled catastrophe. Large constellations like Starlink add thousands of moving points of light that can streak across telescope images and, briefly after launch, are visible to the naked eye as trains of dots. SpaceX has taken steps to dim its satellites, and astronomers continue to press for more, but the tension between cheap global internet and an unspoiled sky is one of the genuine costs of the new space age, not easily dismissed by either side.

Could a piece of a satellite fall on me? The risk to any individual is exceedingly small. Most satellites, Starlink's included, are designed to burn up entirely on reentry, and the Earth's surface is overwhelmingly ocean and empty land. Larger objects occasionally survive to reach the ground, and as the number of satellites grows the issue deserves attention, but the odds of debris harming any particular person remain vanishingly low.

What does Elon Musk actually do at SpaceX day to day? He is not a figurehead; by most accounts he functions as the company's chief engineer as much as its chief executive, deeply involved in technical decisions about the rockets, especially Starship, while others — above all Gwynne Shotwell — handle much of the running of the business. His attention is also famously divided among Tesla, X, and his other ventures, and the question of how much any one company depends on a single overextended person is one this book treats as a real risk rather than a strength.

Appendix V: Watching a Launch and Visiting the New Space Age

There is a moment, the first time you see a rocket launch in person, when the abstraction of everything in this book becomes suddenly, viscerally real. The light reaches you first, silent and impossibly bright; then, seconds later, the sound arrives across the miles as a physical thing, a crackling roar you feel in your chest as much as hear. No video conveys it. This appendix is a practical guide, for the general reader, to seeing that for yourself — and, short of travel, to following the ongoing story as it unfolds. It is written to stay useful over time, so it avoids specific dates and schedules, which change constantly; the tools it points to are where you will find those.

Where launches happen. Almost all of the launches in this book fly from one of a small number of places, and each welcomes spectators differently. The busiest by far is Florida's "Space Coast," the stretch of Atlantic shore around Cape Canaveral and NASA's adjacent Kennedy Space Center, from which most Falcon 9 and Falcon Heavy missions rise. On the other side of the country, Vandenberg Space Force Base on the central California coast handles flights that need to head south into polar orbits, over the ocean rather than over land. And in the far south of Texas, at Boca Chica near Brownsville, sits Starbase, the build-and-launch site for Starship — a place unlike any other, where the largest rocket ever made is assembled and flown at the end of a lonely two-lane road.

Watching on the Space Coast. Florida is the easiest place to begin, because it is set up for visitors and because launches are frequent. The Kennedy Space Center Visitor Complex, near Titusville, is both a first-rate museum and, on launch days, an official viewing venue that sells tickets to see rockets rise from a few miles off — about as close as the public can safely get. But you do not need a ticket to watch. The towns and beaches along the Indian River offer free public vantage points that locals have used for decades: Space View Park and the Max Brewer Bridge in Titusville look straight across the water toward the pads; Playalinda Beach and Jetty Park at Port Canaveral put you nearer still for some trajectories; and Cocoa Beach, further south, has a festive launch-watching culture of its own. The essential things to know are simple. Launches "scrub" — get postponed for weather or technical reasons — with great frequency, so build flexibility into any trip and never plan a journey around a single attempt. Arrive early, because traffic on a bridge over a narrow river does not move. And bring patience: the payoff is worth the wait, but the wait is real.

Watching at Starbase. South Texas is a rawer, more adventurous proposition, and a Starship launch is a generational spectacle. The viewing culture centers on nearby South Padre Island, across the water from the launch site, where beaches and parks such as Isla Blanca offer a distant but dramatic line of sight to the rising vehicle; the town has become an informal gathering point for the space-minded on flight days. The area immediately around Starbase itself is subject to road closures and safety exclusions when a launch or major test is imminent — State Highway 4, the only road in, is closed by the authorities, and the public cannot approach the pad — so the island is the sensible base. Because Starship flies far less often than Falcon 9 and its schedule is more uncertain, a trip built around catching one requires real luck and a tolerance for delay. But those who have seen thirty-three engines light at once describe it as unlike anything else in civilian life.

Watching at Vandenberg. California's launch site is the least tourist-oriented of the three. Vandenberg is an active military base with limited public access, and its launches head out over the Pacific, so the experience is more distant and less choreographed for spectators. Still, launches can be seen from public areas around the base and the nearby town of Lompoc, and a west-coast liftoff at dusk can produce the spectacular, spreading "twilight phenomenon" — the exhaust plume lit by a sun already set at ground level — that occasionally startles the whole of southern California. Local knowledge helps here more than anywhere; the online communities below are the best guide to where to stand.

Finding out when to look. Because schedules shift hourly, the practical skill is knowing where to check. SpaceX streams its own launches live, with commentary, on its website and social channels, and NASA carries the crewed and agency missions on its own broadcasts. For schedules, a handful of free apps and sites aggregate the upcoming manifest and send alerts — searching for "next spaceflight" or "space launch schedule" will surface the well-known ones — and independent outlets that cover the industry closely, such as Spaceflight Now and NASASpaceflight, post detailed timelines and live coverage. The single most important habit is to check again shortly before you set out, because a launch that was set for the evening may have slipped to the small hours or to next week entirely.

Seeing Starlink from your own backyard. You do not have to travel to witness the new space age; some of it passes over your head most nights. In the day or two after a Starlink launch, the freshly deployed satellites fly in a tight line before they spread out and raise their orbits, and this "train" — a string of evenly spaced points of light sliding silently across the sky — is visible to the naked eye under a dark sky, and genuinely uncanny the first time you catch it. Free trackers, most famously a site commonly found by searching "find Starlink," will tell you when a train should pass over your location and where to look. The best times are the hour or so after dusk and before dawn, when the sky above you is dark but satellites high overhead are still catching sunlight. The same principle lets you spot the International Space Station, which is brighter than any star and crosses the sky in a few minutes; NASA's "Spot the Station" service and sites such as Heavens-Above will predict its passes for wherever you live. Watching the ISS sail over — knowing there are people aboard it — is, for many, the gateway to caring about all of this.

Museums and preserved hardware. When there is no launch to catch, the history is on display across the country, and some of it is staggering. Three genuine Saturn V rockets survive, laid on their sides in vast halls: at the Kennedy Space Center Visitor Complex in Florida, at Space Center Houston beside NASA's Johnson Space Center in Texas, and at the U.S. Space and Rocket Center in Huntsville, Alabama, where much of the rocket was designed. To stand beneath one — longer than a football field, taller than a thirty-storey building when it stood upright — is to understand the Apollo program in a way no photograph allows. The retired Space Shuttles are similarly scattered and worth a pilgrimage: *Atlantis* at the Kennedy visitor complex, dramatically displayed tilted as if in flight; *Discovery* at the Smithsonian's Udvar-Hazy Center outside Washington; *Endeavour* at the California Science Center in Los Angeles; and the test vehicle *Enterprise* aboard the Intrepid museum in New York.

For the newer chapter of the story, SpaceX has displayed a recovered, flown Falcon 9 booster standing upright outside its headquarters in Hawthorne, California — a chance to see the very thing this book is about, a rocket that went to the edge of space and came home, close enough to touch the soot on its legs. Flown boosters and Dragon capsules have appeared at various museums and events as the company's history lengthens, and the Smithsonian's National Air and Space Museum on the National Mall in Washington remains the essential collection for the whole sweep of flight, from the Wright Flyer to the space age. None of these requires waiting for good weather or a successful countdown; the hardware is simply there, patient, waiting to be understood.

Following the ongoing story. This book had to end somewhere, but the story did not stop when it did, and part of the pleasure of the subject is that it keeps moving. Following it well means choosing sources that value accuracy over excitement, of which there is no shortage in this field. A number of dedicated outlets cover spaceflight seriously and continuously — the same publications that carry live launch coverage also report the industry's business and engineering in depth — and several careful journalists have made the beat their life's work, offering the kind of skeptical, informed reporting that cuts through both boosterism and doom. The official channels of SpaceX, NASA, and the other agencies and companies provide the primary material, best consumed with the healthy caution this book has tried to model, since every organization narrates its own triumphs generously. Read widely, weigh the claims, and remember that in a field moving this fast the confident prediction is usually the one to distrust.

Mostly, though, go outside and look up. The most extraordinary thing about this moment is how much of it is visible to anyone willing to learn where and when to point their eyes: a chain of new satellites drifting through the dusk, the station arcing over in the dark, and, if you are lucky and patient, a rocket climbing off the coast on a pillar of fire and returning, minutes later, to land. People will tell their grandchildren they saw it happen. You still can.

A Note on Sources

This book is a work of popular history and synthesis, written for the general reader, and it rests on the public record rather than on original archival research or exclusive interviews. Its purpose is to gather the sprawling, fast-moving story of SpaceX and the new space race into a single, current, balanced narrative, and to be honest about what is known, what is estimated, and what remains uncertain. Readers who wish to go deeper will find a rich literature and an abundance of primary material, some of the most useful of which is noted here.

For the founding and early years — the near-bankruptcy, Falcon 1, and the rise of Falcon 9 — the indispensable accounts are **Eric Berger's** two books, *Liftoff* (2021), on the Falcon 1 era, and *Reentry* (2024), on the Falcon 9 and reusability years. Berger, a journalist with deep access and deep technical knowledge, has written the closest thing there is to an authoritative history of SpaceX's development, and any reader wanting more detail than this book provides should begin there. The early character and business of the company are also covered in **Ashlee Vance's** biography *Elon Musk: Tesla, SpaceX, and the Quest for a Fantastic Future* (2015) and, more recently and more expansively, in **Walter Isaacson's** *Elon Musk* (2023), an authorized biography useful for its access and best read critically. **Christian Davenport's** *The Space Barons* (2018) and **Tim Fernholz's** *Rocket Billionaires* (2018) situate SpaceX within the wider rise of commercial spaceflight and the rivalry with Blue Origin.

For the NASA side of the story — the policy revolution of commercial cargo and crew, without which SpaceX could not have survived or succeeded — **Lori Garver's** memoir *Escaping Gravity* (2022) is invaluable, an insider's account of the political fight to fund the commercial approach. The deeper background of the commercial-space movement is captured in **Julian Guthrie's** *How to Make a Spaceship* (2016), on the Ansari X Prize, and the Apollo benchmark against which the new race is measured is best conveyed by **Andrew Chaikin's** *A Man on the Moon* (1994). The intellectual case for Mars, which animates so much of Musk's thinking, is laid out in **Robert Zubrin's** *The Case for Mars* (1996 and later editions).

For the events of recent years — the Starship test campaign, Starlink's growth, Crew Dragon's flights, the Artemis delays, the rise of competitors, and the 2026 public offering — the essential sources are contemporary, and no book can stay current with a story moving this fast. The most valuable are: **SpaceX's own** launch pages, mission webcasts, and updates, and the Starlink progress reports, which provide primary mission data; **NASA's** program pages and the reports of its Office of Inspector General, which offer candid assessments of Artemis and the Human Landing System contracts; and the **Federal Aviation Administration's** statements on launch licenses and mishap investigations. Among journalists and outlets, **Eric Berger at Ars Technica** remains the essential chronicler, along with the space coverage of **Reuters**, **Scientific American**, **IEEE Spectrum**, **CNBC** (especially Michael Sheetz on the business), and specialist publications such as **SpaceNews** and **The Space Review**. For the raw numbers — launch tallies, satellite counts, orbital statistics — the work of independent trackers such as **Jonathan McDowell** and databases like **CelesTrak** is indispensable, as is the FAA's launch data.

A word of caution about figures. For most of the period this book covers, SpaceX was a privately held company that guarded its finances closely, and many of the numbers that appear in accounts of the company — its revenues, the economics of Starlink and of reuse, its internal valuations — were estimates, sometimes informed and sometimes not. This book has tried to flag such figures as estimates and to rely, where possible, on the public record. The great exception is the June 2026 initial public offering, which for the first time forced real financial data into the open; where this book gives figures for the IPO and the company's finances around it, they rest on that disclosure. As always, the reader should treat precise numbers about a fast-moving, partly private enterprise with appropriate caution, and should consult the most current sources for a story that did not stop when this book was finished.

Finally, the story told here reaches only to the middle of 2026, and several of its threads were still unspooling as it was written. The reader who comes to these pages later will know things the author could not: how Starship's grounding was resolved, whether orbital refueling succeeded, how the race to the Moon turned out, what became of the company's soaring valuation and its controversial founder. That is the nature of writing about a story still in progress. This book offers a map of the territory as it stood at a particular moment; the journey continues beyond its edges.

About the Author

E.B Cohen writes narrative non-fiction about history and technology and the forces that shaped the modern world. Drawn to the moments when a single idea or a stubborn group of people bends the course of events, the author writes for the general reader who wants a story told clearly and honestly — vivid where the evidence allows, careful where it does not, and unafraid to name what remains uncertain. *Flight Proven* joins a growing shelf of books that trace the hinges of the modern age, from the engines of the Industrial Revolution to the rockets of the new space age.

www.ingramcontent.com/pod-product-compliance
Lightning Source LLC
LaVergne TN
LVHW100132061126
851857LV00010B/59
* 9 7 9 8 2 3 5 0 2 4 6 6 3 *